RELATIVISTIC QUANTUM MECHANICS and QUANTUM FIELDS

D1681002

T-Y Wu
W-Y Pauchy Hwang

World Scientific
Singapore • New Jersey • London • Hong Kong

Published by

World Scientific Publishing Co. Pte. Ltd.
P O Box 128, Farrer Road, Singapore 9128
USA office: 687 Hartwell Street, Teaneck, NJ 07666
UK office: 73 Lynton Mead, Totteridge, London N20 8DH

Library of Congress Cataloging-in-Publication data is available.

RELATIVISTIC QUANTUM MECHANICS AND QUANTUM FIELDS

ISBN 981-02-0608-9 ✓
     981-02-0609-7 (pbk)

Printed in Singapore by Utopia Press.

RELATIVISTIC QUANTUM MECHANICS and QUANTUM FIELDS

Charles Seale-Hayne Library

University of Plymouth

(01752) 588 588

LibraryandITenquiries@plymouth.ac.uk

Foreword

For "Quantum Mechanics" (Ta-You Wu)
by Professor T. D. Lee, Columbia University

Nowdays, any serious student with an interest in science has to be acquainted with quantum mechanics. Yet quantum mechanics is much more than a useful tool; it is an intellectual achievement of the highest order. Thus the proper teaching of quantum mechanics should not consist merely of its mathematical techniques and physical facts; because the subject is so well-developed there are, however, a number of topics that must be covered. An ideal solution would be to have all of that, plus the historical perspective and at least some of the philosophical implications. Of course, this is not easy. Although there are many textbooks on quantum mechanics, none precisely fulfills such a need. That is why this new book by Professor Ta-You Wu is so welcome.

After a brief review of classical physics, Professor Wu discusses in some detail the pre-quantum mechanical period, the early development of matrix mechanics and wave mechanics, and then proceeds to the general theory of quantum mechanics. He particularly emphasizes the historical evolution of the basic ideas. His incisive description of the contributions of the giants of early times, Planck, Einstein, Bohr, Pauli, Heisenberg, Schrödinger, Dirac evokes a sense of how science is done by real people. The discovery of truth rarely follows a straight line; each new segment often contains only a partial truth which is then complemented by the next step taken by someone else. This is especially true in the case of quantum mechanics, as it is brought out clearly in Professor Wu's book.

In the general theory of quantum mechanics, Professor Wu discusses with particular care the postulates involved, with special emphasis on measurement theory and the related tools of the density matrix and Green's functions. He then starts discussions on applications to atomic structure. The chapters on multi-electron atoms, perturbative theory and time-dependent systems, would be very useful to graduate students interested in physics and chemistry. For those doing more advanced work in atomic and molecular structure, these chapters form a good introduction to Professor Wu's standard treatise on vibrational spectra and the structure of polyatomic molecules.

The remaining third of the book is mainly devoted to relativistic theories, from the Klein-Gordon equation through Dirac's theory to field quantization.* Again, the emphasis is on the mathematical formalism as well as the physical implications. A genuine regret is that the book ends at the threshold of quantum field theory, which makes the reader look forward to a sequel to the present volume.

New York   T.D.Lee

*The author had decided to remove these chapters from the volume for a possible future volume (which is the present volume).

Foreword

In my Foreword to Professor Wu's well-known volume on Quantum Mechanics, I expressed the hope that there would be a sequel which would cover, in addition to Relativistic Quantum Mechanics, also Quantum Field Theory. This wish has been amply fulfilled in the present book by Professor Wu in collaboration with Professor W-Y. Pauchy Hwang.

In reading this treatise, I was reminded of the period 1945-46 in Kunming when Professor Ta-You Wu gave us extra lessons on Quantum Mechanics and Relativistic Quantum Mechanics. It was my good fortune to be his student. In that brief time of one year, he gave me such a solid foundation in physics that it enabled me to come to the University of Chicago in the fall of 1946 and immediately start doing graduate work. Other students in that special enrichment class in Kunming were J. S. Wang, G. Y. Zhu, A. Q. Tang and B. H. Sun; all of them benefited greatly from Professor Wu's lectures and made important contributions to science later on. The publication of Professor Wu's Quantum Mechanics and the present sequel will make it possible for serious science students everywhere to share the insight and the wisdom that were instrumental in launching our careers.

This is indeed an important event for the entire physics community.

T.D. Lee
April 10, 1990
New York

PREFACE

There exists already an ample literature on relativistic quantum mechanics, quantum electrodynamics, and even on the more recent developments on the so-called "standard model" which consists of quantum chromodynamics for strong interactions and the Glashow-Weinberg-Salam $SU(2) \times U(1)$ theory for electroweak interactions. A few words may perhaps be in order for the addition of the present volume to the literature.

When in 1986 one of the present authors (Wu) published Quantum Mechanics ("Q.M."), the manuscript originally contained chapters on the theory of Klein-Gordon and of Dirac, classical fields and the quantization of free fields. It stopped just before the important development of quantum electrodynamics. This would be very unsatisfactory, and Wu decided to publish the non-relativistic part as "Quantum Mechanics" and to hold the rest of the manuscript, hoping to interest a colleague in completing it into a sequel volume that will cover relativistic quantum mechanics and an introduction to quantum electrodynamics and the standard model. In 1988, the present authors discussed and agreed upon a joint effort to prepare this sequel to "Q.M.". The first eight chapters come from the earlier manuscript as mentioned above, and follow the generally pedagogic style of "Q.M.", and the last six chapters are based on the notes of lectures given by Hwang at Indiana University and at National Taiwan University. Attempt has been made to present the latter chapters at the same level of ease (or difficulty) as the earlier ones, but an entirely smooth integration is difficult because of the nature of the subject matter itself and of the difference in style of two co-authors.

The scope of the present volume may be seen from the table of content. It ranges from relativistic quantum mechanics where the Klein-Gordon and Dirac equations are treated in detail, to an introduction to quantum field theory with quantum electrodynamics (QED) as our basic example, and to a pedagogic exposition of important issues related to the standard model. As our primary objective is to expose the students to the themes and ideas underlying the contemporary developments, we attempt to avoid using excessively mathematical tools such as the method of path integrals, which may add to mathematical elegance pertaining to the subject but without enhancing significantly the global understanding of the subject matter in terms of basic and easy-to-grasp notions. Thus, we hope that this volume will serve as a useful textbook or reference material in the area of relativistic quantum mechanics and quantum electrodynamics and an introduction to the more recent developments on gauge field theories.

The convention of choosing units such that $\hbar = c = 1$ is almost universal in the field theory literature but is less so in quantum mechanics. For the convenience of some readers, certain equations are written out also in ordinary units in the footnotes (with primed equation numbers).

In preparing the final version of the manuscript, we have generally followed closely the line of presentation given by authors in the original papers since, after all, many of these things have become so standard that they may as well appear without much distortion in a textbook on the subject. For example, the section on dimensional regularization is based on the original paper of 't Hooft and Veltman; most materials for the chapter on experimental tests of the standard model are from the 1988 publication of the Particle Data Group; and, since Hwang used in his lectures a couple of very nice books, notably the one by E.D. Commins and P.H. Bucksbaum ("Weak Interactions of Leptons and Quarks", Cambridge University Press, 1983) and the other one by F. Halzen and A. D. Martin ("Quarks and Leptons", John Wiley & Sons, 1984), it should not be a surprise that footprints of these authors may be found occasionally in the last few chapters. Wu first learned from Professor George Uhlenbeck, in 1932, about the Zitterbewegung, Klein paradox, and the "divorce" of velocity and momentum in Dirac's theory, and later studied Pauli's article. The treatments in many sections in Chapters 7 and 8 have followed the memeographed notes of lectures by Professor E. Mishkin at the Polytechnique Institute of Brooklyn in 1964. To these and many other authors from whose works we have drawn freely, we gratefully express our indebtedness.

Finally, we wish to acknowledge Ms. Bi-Lien Lin and Ms. Yu-Ru Hung for helping to prepare the entire camera-ready manuscript in TeX and Professor Wen-Kwei Cheng at State University of New York at Old Westbury and Mr. Yen-Sheng Su at National Taiwan University for offering useful remarks upon reading the manuscript.

September 1989

Ta-You Wu, Academia Sinica

W-Y. Pauchy Hwang, National Taiwan University

Table of Contents

Chapter 0. Introduction

Quantum mechanics and Einstein's theory of special relativity constitute the basis for the main body of modern physics developed during the first half of the twentieth century. In a previous volume by one of the present authors on quantum mechanics, the conceptual marriage of quantum mechanics and the principle of special relativity has been deliberately put off. We wish to take up this important task in Part A of this volume, where the Klein-Gordon equation and the Dirac equation will be introduced and elucidated.

Physical systems that can be described by relativistic wave equations include electrons, quarks, and many other elementary particles in the subatomic world. We shall start with the Klein-Gordon equation which describes spinless particles such as the pions. Then we shall present the Dirac relativistic equation in greater length. The Dirac theory for the electron has many successes: It is Lorentz covariant; it contains the electron spin with the gyromagnetic ratio $g = 2$; it predicts the anti-particle positron which is experimentally discovered; it gives the correct fine structure of the hydrogenic atom levels. Application of the Dirac equation to other leptons such as muons and neutrinos, and to quarks in the context of bag models, also leads to quantitative successes. It is clear that, unlike atomic or molecular physics where relativistic effects can often be treated as small perturbations, a suitable introduction to elementary particle physics and field theories must commence with a description of relativistic wave equations.

There are "difficulties" too with the Dirac equation. One is that of extending the theory for a single electron to a system of many electrons. Another is of an even more basic nature, namely, a theory started out to represent a single electron ends up being inseparably bound with a many-body effect on account of the infinite sea of electrons in negative energy states. In the presence of a strong electric field, an electron does not exist as a lone electron, but is a system containing infinitely many (and not constant in number) electrons and positrons. Thus the single-electron theory is, strictly speaking, valid only in the sense of a limiting case of a free electron, and must be replaced by a theory of "many-electrons".

Now the appropriate representation of a system of an infinitely large and variable number of "particles" is a "field" - just as the representation of "photons" is the electromagnetic field. Thus we may say that, starting from the particle point of view, one seeks a theory of particles that satisfies the relativity principle and finds that the appropriate theory is a many-body theory, and the appropriate representation of a many-body system is "field".

The problem is then to formulate a many-electron field. One may start from the Dirac equation for a free electron and regard it as the equation of a classical field Ψ - in this case a 4-component field. Then treat Ψ_i as the field variables (similarly to the 4-potential A_μ of the electromagnetic field), and their canonical conjugates. Then introduce the quantization condition in the form of the anticommutation relations and, as expected, one obtains the electron as the particles of the quantized field. Such a quantization procedure is called "second quantization". The quantized Ψ possesses both the wave and the particle properties - just as the quantized electromagnetic field possesses the photon and the wave property.

The quantum theory of a pure electron field presents no difficulties, but for the system of an electron coupled with the electromagnetic field, a relativistic quantum theory - known as quantum electrodynamics or QED - presents serious deep-rooted difficulties, namely the persistent presence of infinities in the results of calculations of physical quantities.

The theory of quantized electromagnetic fields begins with the work of Dirac in 1927, followed immediately by the work of Jordan and Wigner, and by Fermi in 1930. A breakthrough comes only in the mid 1940's with the work of Tomonaga in Japan and Schwinger, Feynaman, and Dyson in the United States. Although infinities still remain, the theory succeeds in "subtracting" them away in a definite, covariant way so that finite results can be obtained, which have been found to be in excellent agreement with the observed Lamb shifts and the "g anomaly". The decade from the mid 1940's to the mid 1950's is a period of fervent studies of quantum electrodynamics, both in further calculations on this "renormalization" theory and in attempts to rid the theory of the infinities (not just to isolate and bury them, so to speak). Till now, there seems not yet a satisfactory solution of this problem and in the meantime physicists have turned their interest to other areas - elementary particles and the nature of their interactions and their unification.[1]

During the decade from the mid 1950's to the mid 1960's, serious attempts were made in searching for an alternative means of describing interactions among elementary particles in terms of the so-called "S-matrix theory", in which one tries to determine the scattering amplitudes, or the S-matrix elements, using general principles such as unitarity and microscopic causality [which lead to dispersion relations] and a minimal set of dynamical assumptions. Altogether in the S-matrix approach, the question concerning the underlying dynamics must be answered, or postulated, before any quantitative predictions can be consistently made. Therefore, the notion of using "gauge field theories"[2] to describe interactions among "building blocks" of mater has scored an amazingly successful comeback since the late 1960's while progresses in the S-matrix approach, which have since been relatively limited, have become things of the past, at least for the time being.

The development of particle physics of the last two decades [since the late 1960's till the present] consists of several major breakthroughs which culminate in the general acceptance of the $SU(3)_{color} \times SU(2)_{weak} \times U(1)$ gauge field theory of strong, electromagnetic, and weak interactions as the "standard model". The Glashow-Salam-Weinberg [GSW] $SU(2)_{weak} \times U(1)$ gauge theory provides a unified description of electromagnetic and weak interactions. It predicts the existence of neutral weak interactions, the existence of the charm quark, and the existence of weak bosons $W^{\pm}$ and Z^0 [which mediate weak forces], all of

---

[1] *Dirac believes that in a completely satisfactory theory, infinities should not appear. Since the late 1940's, he set out to re-examine and reformulate classical dynamics and electromagnetic theory with a view to find a different basic theory for quantization. He believes even some new mathematics not yet known may be needed. He expressed his views only occasionally in writing, but much more freely in private discussions.*

[2] *QED is the simplest prototype of gauge field theories, in which there exist a class of so-called "gauge transformations" which do not affect physical observables. The notion of a gauge transformation will be introduced in §1.1. and §4.2. and used extensively in the second and third parts of this book.*

4

which have been substantiated experimentally.[3] In the meantime, the quantized $SU(3)_{color}$ gauge theory, or quantum chromodynamics [QCD], has been established as the candidate theory of strong interactions among quarks and gluons, which are believed to be the building blocks of all observed strongly interacting elementary particles or hadrons. QCD supports the dual picture of considering, e.g., a proton as a collection of almost noninteracting quarks, antiquarks, and gluons at high energies [because of the asymptotically free nature of QCD] and as a system of confined, dressed valence quarks [because QCD is consistent with color confinement].

In Part B of the present volume, we shall present in some detail quantum electrodynamics, the simplest prototype of all quantized gauge field theories. In Part C, we wish to describe the standard model, i.e., QCD and the GSW electroweak theory, and the experimental tests of it.

§.0.1. Building Blocks of Matter

We shall for convenient reference begin with a qualitative summary concerning the subatomic and atomic world. In the standard model, building blocks of matter are known to include (a) three generations of fermions, (b) mediators of fundamental interactions, and (c) scalar particles which are responsible for spontaneous symmetry breaking related to the physical vacuum. Specifically, fermions consist of leptons, quarks, and their antiparticles:

Leptons:

$$\begin{pmatrix} e^- \\ \nu_e \end{pmatrix}, \quad \begin{pmatrix} \mu^- \\ \nu_\mu \end{pmatrix}, \quad \begin{pmatrix} \tau^- \\ \nu_\tau \end{pmatrix}, \tag{1}$$

Quarks:

$$\begin{pmatrix} u_R \\ d_R \end{pmatrix}, \quad \begin{pmatrix} c_R \\ s_R \end{pmatrix}, \quad \begin{pmatrix} t_R \\ b_R \end{pmatrix},$$

$$\begin{pmatrix} u_Y \\ d_Y \end{pmatrix}, \quad \begin{pmatrix} c_Y \\ s_Y \end{pmatrix}, \quad \begin{pmatrix} t_Y \\ b_Y \end{pmatrix}, \tag{2}$$

$$\begin{pmatrix} u_B \\ d_B \end{pmatrix}, \quad \begin{pmatrix} c_B \\ s_B \end{pmatrix}, \quad \begin{pmatrix} t_B \\ b_B \end{pmatrix}.$$

[3] *In a short time span of ten years starting from 1976, Nobel prizes have been awarded three times to discoveries related to the GSW electroweak theory: for the discovery of the charm quark, for the successes of the GSW theory, and for the experimental observation of $W^\pm$ and Z^0.*

Note that leptons include electrons (e^-), muons (μ^-), tau-leptons (τ^-), electronlike neutrino (ν_e), and so on. Quarks come in with six possible flavors: up (u), down (d), charm (c), strange (s), bottom (b), and top (t). A quark of any given flavor is assumed to carry one of three possible colors: red (R), yellow (Y), and blue (B), or x, y, and z with

$$x = \begin{pmatrix} 1 \\ 0 \\ 0 \end{pmatrix}, \quad y = \begin{pmatrix} 0 \\ 1 \\ 0 \end{pmatrix}, \quad z = \begin{pmatrix} 0 \\ 0 \\ 1 \end{pmatrix}. \tag{3}$$

Quarks are not observed in isolation presumably because color is strictly confined, a property consistent with the conjectured two-phase picture of QCD.

Mediators of fundamental interactions include (1) the photon (γ), which mediates the well-known electromagnetic interaction, (2) three weak bosons $(W^\pm, Z^0)$, which mediate charged and neutral weak interactions, and (3) eight gluons, which mediate strong interactions among quarks and antiquarks. Gluons carry one of eight possible octet colors and cannot exist in isolation because of color confinement.

Hadrons, or strongly interacting elementary particles, are by assumption color-singlet, or colorless, composites of quarks, antiquarks, and gluons. The hadrons include mesons, baryons, glueballs, and so on. Pions $(\pi^\pm, \pi^0)$, kaons $(K^\pm, K^0, \bar{K}^0)$, etas (η, η'), rho-mesons $(\rho^\pm, \rho^0)$, ψ/J, and upsilons $(\Upsilon, \Upsilon', ...)$ all are *mesons* which, at low energies, are believed to be quark-antiquark pairs confined to within the region defined by the meson size. Nucleons (p, n), lambda (Λ), sigmas $(\Sigma^\pm, \Sigma^0)$, xi's (Ξ^-, Ξ^0), deltas $(\Delta^-, \Delta^0, \Delta^+, \Delta^{++})$, charmed lambda (Λ_c), and bottomed lambda (Λ_b) all are *baryons* which, at low energies, look like systems of three quarks confined to within the region defined by the baryon size. *Glueballs* are colorless objects consisting of gluons only.

Nucleons, i.e., protons (p) and neutrons (n), are believed to be primary building blocks of the *nuclei* of the various atoms, ranging from the proton itself [as the simplest nucleus], to the deuteron, α, ^{12}C, ^{56}Fe, ^{208}Pb, and even to neutron stars [$A = \infty$ nuclei]. Replacement of a nucleon in an ordinary nucleus by a lambda (Λ) or by a sigma-baryon $(\Sigma^\pm, \Sigma^0)$ results in a *hypernucleus*. An *atom* is a system of electrons around an ordinary nucleus, the whole system being often electrically neutral. *Molecules* are built from atoms. All matter observed terrestrially or celestially are believed to be composed of building blocks conjectured in the standard model.

The GSW $SU(2)_W \times U(1)$ electroweak theory invokes the so-called "Higgs mechanism", in which the physical vacuum, or the true ground state, differs from the trivial vacuum [where expectation values of all fields vanish identically]. This theory predicts, among other things, the existence of a Higgs particle, which has not been observed so far. However, many other important predictions of the GSW electroweak theory have been substantiated experimentally, including the existence of: (1) neutral weak interactions, (2) the charm quark, and (3) weak bosons $W^{\pm}$ and Z^0. It is very likely that any new theory which is beyond the standard model must reproduce or explain the successes of the standard model and so will contain the standard model as a limiting case.

Up to the moment of writing, building blocks of matter, as conjectured in the standard model, appear to be structureless [or, less precisely, pointlike] at the highest energy scale (or the smallest distance scale) which we are capable of probing. Qualitatively speaking, quarks and leptons can be described by Dirac equations of some sort while mediators of fundamental interactions (spin-1 particles) and the Higgs particle (spin-0 particle) are described, respectively, by gauge field theories and a generalized Klein-Gordon equation. It is clear that the presentation of relativistic quantum mechanics in Part A is most relevant in the subatomic world. Indeed, it is not clear at all whether a Dirac equation will ever be relevant in the description of a composite spin-$\frac{1}{2}$ system such as a proton or a neutron.

§.0.2. Natural Units

The brief summary in the preceding section concerning the building blocks of matter should have made it clear that a knowledge of relativistic quantum mechanics and quantum fields is most relevant in the area of elementary particle and nuclear physics. Since the kinetic energy of a particle under investigation is often more important than its rest mass in a typical particle or nuclear physics problem, it is convenient to measure a given velocity in units of the light velocity c.[4]

$$c = 2.9979 \times 10^{10} cm/sec. \tag{4}$$

---

[4]*Note that we do not quote in this book all the significant figures of many basic constants which have been measured with great precision.*

Similarly, it is convenient to express the action in units of $\hbar c$:

$$\hbar c = 197.33\,MeV - fm, \qquad (5)$$

where $1fm \equiv 10^{-13} cm$. The remaining units can be chosen as powers of MeV (or GeV) or as powers of fm or as powers of *seconds*. For instance, a delta width of $110\,MeV$ corresponds to

$$110MeV \cdot \frac{1}{\hbar c} \cdot 10^{13} fm/cm \cdot c$$
$$= 110MeV \cdot \frac{1}{197.33MeV - fm} \cdot 10^{13} fm/cm \cdot 2.9979 \times 10^{10} cm/sec$$
$$= \frac{1}{5.98 \times 10^{-24} sec},$$

or to a very short lifetime of $5.98 \times 10^{-24} sec$. The system in which quantities are measured in units of c and $\hbar c$ is referred to as "natural units". Customarily, one writes

$$\hbar = c = 1. \qquad (6)$$

For problems of present-day particle and nuclear physics, adoption of natural units leads to equations which look simpler than those obtained in ordinary units, although the physical content remains the same. Natural units are used in the present volume. The situation may be considered as different from ordinary quantum mechanics where the role played by $\hbar$ should always be emphasized.

Finally, we note that the metric used by Pauli, Feynman, and T. D. Lee will be adopted:

$$g_{\mu\nu} = \delta_{\mu\nu}; \quad \mu,\nu \in (1,2,3,4). \qquad (7)$$

For instance, the four-momentum of a particle is specified by

$$p_\mu = (\mathbf{p}, iE), \quad with \quad p_4 = ip_0 = iE. \qquad (8)$$

The inner product of two four-vectors A_μ and B_μ is given by[5]

---

[5] *As a convention, we use a bold-faced Roman letter or a Greek letter to denote a three-vector in configuration space. The arrow is reserved only for a vector in isospin space or in other internal space.*

$$A \cdot B = A_\mu B_\mu = \mathbf{A} \cdot \mathbf{B} - A_0 B_0. \qquad (9)$$

The only complication arises when one tries to take the complex conjugate of a complex four-vector. For instance, we have

$$\xi_\mu^* = (\xi^*, i\xi_0^*),$$

for a complex polarization four-vector $\xi_\mu = (\xi, i\xi_0)$, so that $\xi^* \cdot p = \xi^* \cdot \mathbf{p} - \xi_0^* p_0$. We shall write out the expressions explicitly in the case that some confusion may arise.

Finally, we note that[6] equations and results in the previous volume entitled "Quantum Mechanics"[7] will be used occasionally in this book. As such referencing occurs, we shall use notations such as "(VIII-58), Vol. I" or "p. 12, Vol. I" where "(VIII-58)" is the equation number and "p.12" indicates the page number.

---

[6] *To benefit readers in the area of atomic physics, certain equations (up to Chapter 8) are written in ordinary (m.k.s.a.) units as footnotes with primed equation numbers.*

[7] *Henceforth referred to as "Vol. I", authored by one of us, T.-Y. Wu, and published by World Scientific, 1986.*

Part A. Relativistic Quantum Mechanics

Chapter 1.
Relativistic Wave Equation: Klein-Gordon Equation

The Schrödinger equation

$$i\frac{\partial \psi}{\partial t} = H\psi, \tag{1}$$

has been the basic equation for the successful treatment of problems in many fields. But it does not conform to the special theory of relativity since, being a differential equation of first order in time and second order in the spatial coordinates, its form is obviously not invariant under Lorentz transformation.*

In quantum mechanics, the probability postulate and the consequent normalization condition are
(i)

$$w = \psi^*\psi \geq 0, \tag{2}$$

(ii)

$$\frac{d}{dt}\iiint w\,dxdydz = 0. \tag{3}$$

The condition that a dimensionless scalar quantity be an invariant under Lorentz transformation is
(iii)

$$\iiint \psi^*\psi\,dxdydz = \text{Lorentz invariant}. \tag{4}$$

Conditions (ii) and (iii) are satisfied if a 3-vector $\mathbf{I}(I_x, I_y, I_z)$ exists such that

$$(I_x, I_y, I_z, iw) \tag{5}$$

form a 4-vector (in Minkowski space), so that the 4 divergence

$$\frac{\partial w}{\partial t} + \nabla \cdot \mathbf{I} = 0 \tag{6}$$

guarantees (ii) (if $\mathbf{I}$ vanishes on the surface of a volume V). Now, from Eq. (1) and its complex conjugate, one readily obtains

$$\frac{\partial}{\partial t}(\psi^*\psi) + \nabla \cdot \{\frac{1}{2mi}(\psi^*\nabla\psi - \psi\nabla\psi^*)\} = 0, \tag{7}$$

which satisfies (i) and (ii), but the vector $\mathbf{I}$ defined by this equation and $\psi^*\psi$ do not form a 4-vector, i.e., the condition (iii) is not satisfied.

As early as 1926, Schrödinger, Klein, Gordon, de Broglie and Fock have all obtained a wave equation, known in the literature as the Klein-Gordon equation, that satisfies the Lorentz invariance. (See, e.g., references at the end of §.3.3., "Quantum Mechanics" by one of us; this volume will henceforth be referred to as "Vol. I.") It is of the second order in x, y, z, t and, as a wave equation for electrons, has difficulties. In 1928, Dirac obtains a relativistic wave equation that is of the first order in x, y, z, t and is amazingly successful in accounting for the electron spin and in leading to the prediction and discovery of the positron. In the present chapter we shall discuss the Klein-Gordon equation and, in following chapters, the Dirac theory.

§.1.1. The Klein-Gordon Equation

According to the principle of special relativity, the energy-momentum relation for a free particle of rest mass m_0 and four-momentum $(\mathbf{p}, iE)$ is given by

$$\mathbf{p}^2 - E^2 = -m_0^2 \tag{8}$$

Quantization may be accomplished by

$$\mathbf{p} \to -i\nabla, \qquad E \to i\frac{\partial}{\partial t}, \tag{9}$$

so that Eq. (8) becomes an operator equation,

$$\partial_\mu \partial_\mu \equiv \nabla^2 - \frac{\partial^2}{\partial t^2} = m_0^2. \tag{10}$$

There is a functional space on which the operator is supposed to act. In the case of Eq. (10), we have

$$(\nabla^2 - \frac{\partial^2}{\partial t^2})\psi(x) = m_0^2 \psi(x), \tag{11}$$

which is the Klein-Gordon equation for a free particle. The function $\psi(x)$ characterizes the state of the particle.

For a particle of charge e in an electromagnetic field with the four-potential $(\mathbf{A}, i\phi)$, we use the principle of "minimal substitution",

$$\mathbf{p} \to \mathbf{p} - e\mathbf{A}, \qquad E \to E - e\phi, \tag{12}$$

which becomes the gauge principle in the context of gauge field theories. [Eq. (12) amounts to the statement that, upon coupling to a gauge field, the derivative ∂_μ is to be substituted by the gauge-invariant derivative D_μ.]

Instead of Eqs. (8) and (11), we have[1]

$$(\mathbf{p} - e\mathbf{A})^2 - (E - e\phi)^2 = -m_0^2, \tag{13}$$

$$\{(-i\nabla - e\mathbf{A})^2 - (i\frac{\partial}{\partial t} - e\phi)^2\}\psi(x) = -m_0^2\psi(x). \tag{14}$$

Eq. (14) is the Klein-Gordon equation for a spinless particle of charge e moving in an external electromagnetic field $(\mathbf{A}, i\phi)$.

It is straightforward to show that under gauge transformation

$$\mathbf{A} \to \mathbf{A}' = \mathbf{A} + \nabla\chi, \tag{15a}$$

$$\phi \to \phi' = \phi - \frac{\partial\chi}{\partial t}, \tag{15b}$$

$$\psi \to \psi' = e^{ie\chi}\psi, \tag{15c}$$

the Klein-Gordon equation is invariant, i.e.,

$$\{(-i\nabla - e\mathbf{A}')^2 - (i\frac{\partial}{\partial t} - e\phi')^2\}\psi'(x) = -m_0^2\psi'(x).$$

Let us look at these equations from the criteria (i) - (iii) [Eqs. (2) - (4)]. One can form a 4-vector[2]

---

[1] *Equations labeled by primed numbers are written in ordinary units. They always appear in the footnotes. For example, we have*

$$\{(-i\hbar\nabla - \frac{e}{c}A)^2 - \frac{1}{c^2}(i\hbar\frac{\partial}{\partial t} - e\phi)^2\}\psi(x) = -m_0^2c^2\psi(x) \tag{14'}$$

[2]

$$S_1 \equiv \frac{1}{c}I_x = \frac{\hbar}{2m_0ci}(\psi^*\frac{\partial\psi}{\partial x} - \psi\frac{\partial\psi^*}{\partial x} - \frac{2i}{\hbar c}eA_x\psi^*\psi). \tag{16'}$$

$$S_1 \equiv I_x = \frac{1}{2m_0 i}(\psi^* \frac{\partial \psi}{\partial x} - \psi \frac{\partial \psi^*}{\partial x} - 2ie A_x \psi^* \psi),$$

$$x = x, y, z, \tag{16}$$

$$S_4 \equiv iw = -\frac{1}{2m_0}(\psi^* \frac{\partial \psi}{\partial t} - \psi \frac{\partial \psi^*}{\partial t} + 2ie\phi \psi^* \psi).$$

From Eq. (14), one obtains the continuity equation

$$\frac{\partial S_\alpha}{\partial x_\alpha} = 0. \quad \left(\begin{array}{c} \text{summation convention here} \\ \text{over} \quad x_\alpha = x, y, z, it \end{array} \right) \tag{17}$$

Thus, w transforms as t and the conditions (ii), (iii) are satisfied. But on account of the term $e\phi$, w is not always positive, i.e., condition (i) is not satisfied. Even for a free particle, $\phi = 0$, w is not always positive, as ψ and $\frac{\partial \psi}{\partial t}$ can be assigned arbitrary initial values, the differential equation being of the second order in t. For this reason, the Klein-Gordon equation may be treated at the classical level as a "classical wave equation" similar to the electromagnetic field equation, but must be further quantized; ψ is then regarded as an operator (like the potentials A, ϕ), instead of a probability amplitude (as in the Schrödinger equation). The problem concerning the violation of condition (i) is then resolved, so that the Klein-Gordon equation is valid as a field equation describing a relativistic spinless (scalar, or pseudoscalar) particle.

In quantum mechanics, the observables p, q, H are operators and quantization is achieved by the commutation relation

$$pq - qp = -i. \tag{18}$$

In the electromagnetic field theory, the potentials A, ϕ are treated as operators, and quantization of the electromagnetic field leads to photons. In the same way, we regard the Ψ in the Klein-Gordon equation as fields, and quantization of Ψ, called field quantization, or "second quantization", leads to pions or other scalar particles as the quantized particles. All these will be treated in the following chapters. [See, e.g., Ch. 8.]

§.1.2. Klein-Gordon equation - approximate form

From Eq. (13), one has[3]

$$E - e\phi = m_0\{1 + \frac{1}{m_0^2}(\mathbf{p} - e\mathbf{A})^2\}^{\frac{1}{2}}. \tag{19}$$

The second term of the radicand is of order $(\frac{v}{c})^2$. On expanding the expression to order $(\frac{v}{c})^2$, one has

$$E - e\phi = m_0 + \frac{1}{2m_0}(\mathbf{p} - e\mathbf{A})^2, \tag{20}$$

and using Eq. (9), the equation

$$(i\frac{\partial}{\partial t} - m_0)\Psi = (\frac{1}{2m_0}(\mathbf{p} - e\mathbf{A})^2 + e\phi)\Psi. \tag{21}$$

On making the transformation

$$\Psi = \Psi' exp(-im_0 t), \tag{22}$$

this becomes[4]

$$i\frac{\partial \Psi'}{\partial t} = (\frac{1}{2m_0}(\mathbf{p} - e\mathbf{A})^2 + e\phi)\Psi', \tag{23}$$

which is the Schrödinger equation Eq. (VIII-43) or (VII-44a), Vol. I.

The time-dependence of the "nonrelativistic" wave function $\psi'(x)$ is governed by $\exp(-i\varepsilon t)$ where ε is the eigenenergy with the rest mass m_0 subtracted. To justify the v/c expansion, ε must be much smaller than m_0 so that the oscillation frequency $\varepsilon/2\pi$ for ψ' is considerably smaller than that for ψ itself. In order to recover the nonrelativistic Schrödinger picture as a limiting case of the Klein-Gordon equation, the fast-oscillating component must be subtracted. In

---

[3]
$$E - e\phi = m_0 c^2\{1 + \frac{1}{m_0^2 c^2}(\mathbf{p} - \frac{e}{c}\mathbf{A})^2\}^{\frac{1}{2}}. \tag{19'}$$

[4]
$$i\hbar\frac{\partial \Psi'}{\partial t} = (\frac{1}{2m_0}(\mathbf{p} - \frac{e}{c}\mathbf{A})^2 + e\phi)\Psi', \tag{23'}$$

addition, we note that a choice of the sign for taking the square root in Eq. (19) has also been made. Accordingly, the physics content of Eq. (23) is already different from the Klein-Gordon equation itself.

§.1.3. Klein-Gordon equation for Coulomb field - pionic atom

In a static Coulomb field, $\mathbf{A} = 0$, and if one assumes

$$\Psi(\mathbf{r},t) = u(\mathbf{r})exp(-iEt), \tag{24}$$

Eq. (14) becomes[5]

$$\nabla^2 u = (m_0^2 - (E - e\phi)^2)u(\mathbf{r}). \tag{25}$$

Let

$$u(\mathbf{r}) = R(r)Y_{\ell m}(\theta,\varphi),$$
$$e\phi(r) = -\frac{Ze^2}{4\pi r}, \tag{26}$$
$$\alpha = \frac{e^2}{4\pi} \quad (\simeq \frac{1}{137}),$$

and express length in units of λ where[6]

$$\frac{1}{\lambda} = ZE\alpha,$$
$$\rho = \frac{r}{\lambda}, \quad \epsilon = (m_0^2 - E^2)\lambda^2 \quad (>0), \tag{27}$$

ρ and ϵ being dimensionless. Eq. (25) becomes

$$\{\frac{d^2}{d\rho^2} + \frac{2}{\rho}\frac{d}{d\rho} - \epsilon + \frac{2}{\rho} - \frac{\ell(\ell+1) - Z^2\alpha^2}{\rho^2}\}R(\rho) = 0. \tag{28}$$

Let

---

[5]
$$\nabla^2 u = ((\frac{m_0 c}{\hbar})^2 - (\frac{E - e\phi}{\hbar c})^2)u(\mathbf{r}). \tag{25'}$$

[6]
$$\rho = \frac{r}{\lambda}, \quad \epsilon = \frac{m_0^2 c^4 - E^2}{e^4}\alpha^2\lambda^2 \quad (>0). \tag{27'}$$

$$R(\rho) = \frac{1}{\rho} e^{-\sqrt{\epsilon}\rho} v(\rho). \tag{29}$$

The equation for $v(\rho)$ is

$$\frac{d^2 v}{d\rho^2} - 2\sqrt{\epsilon}\frac{dv}{d\rho} + \frac{2}{\rho}v - \frac{\ell(\ell+1) - Z^2\alpha^2}{\rho^2}v = 0. \tag{30}$$

Let

$$v(\rho) = \rho^\beta \sum_{s=0} a_s \rho^s. \tag{31}$$

The indicial equation

$$\beta(\beta - 1) - [\ell(\ell+1) - Z^2\alpha^2] = 0, \tag{32}$$

has two roots β_1, β_2 where $\beta_1 - \beta_2 \neq$ integer, so that one has two independent solutions. Take

$$\beta = \frac{1}{2} + \frac{1}{2}\sqrt{(2\ell+1)^2 - 4Z^2\alpha^2}. \tag{33}$$

The recursion relation for the a_s is

$$2\{1 - \sqrt{\epsilon}(\beta + s)\}a_s = -\{(\beta + s)(\beta + s + 1) - [\ell(\ell+1) - Z^2\alpha^2]\}a_{s+1}.$$

In order that $R(\rho)$ in Eq. (29) be quadratically integrable, the series must terminate at $s =$ an integer n_r, so that $a_{n_r+1} = 0$, i.e.,

$$\epsilon = \frac{1}{(n_r + \beta)^2}. \tag{34}$$

From this and Eq. (27), one gets

$$\frac{E}{m_0} = [1 + \frac{Z^2\alpha^2}{(n_r + \frac{1}{2} + \sqrt{(\ell+\frac{1}{2})^2 - Z^2\alpha^2})^2}]^{-\frac{1}{2}}. \tag{35}$$

For[7]

$$(Z\alpha)^2 \ll (\ell + \frac{1}{2})^2,$$

---

[7] *It is amusing to note that the energy eigenvalue becomes complex for $(Z\alpha)^2 > (\ell + \frac{1}{2})^2$. For $\ell = 0$, this means $Z > \frac{1}{2\alpha}$ or $Z \geq 69$. This result is not in*

one obtains, on expanding Eq. (35) and using $n = n_r + \ell + 1$,[8]

$$E - m_0 = -\frac{Z^2\alpha^2 m_0}{2}[\frac{1}{n^2} + \frac{Z^2\alpha^2}{n^4}(\frac{n}{\ell + \frac{1}{2}} - \frac{3}{4})...]. \tag{36}$$

The first term is the Bohr-Schrödinger formula. The second term gives the fine structure, which does not agree with the result in Eq. (VIII-53), Vol. I, nor with Dirac's theory, nor with the observed result (see Eq. (VIII-58), Vol. I).

The explanation of the above result is as follows. The operator Eq. (13) is a Lorentz invariant operator, and the Ψ of the Klein-Gordon equation Eq. (14) is a scalar (or pseudo-scalar) function which remains unchanged under (proper) Lorentz transformation, and not as a vector or tensor. It is natural to associate a vector (or tensor) Ψ with a particle (of the quantized field) with a spin angular momentum, and a scalar (or pseudo scalar) Ψ with a spinless particle. Thus a scalar Ψ of the Klein-Gordon equation does not correspond to an electron so that the energy formula Eq. (35) or (36) does not correspond to a hydrogenic atom.

On the other hand, the pions π^+, π^- are known from experiments to be spinless and pseudo-scalar. Hence Eq. (36) applies to a pionic atom, i.e., a π^- moving in the field of a nucleus Ze.

---

agreement with the known X-ray levels (K level) of heavy atoms, showing that electrons cannot be described by the Klein-Gordon equation. For pionic atoms, it is not possible in practice to satisfy the condition experimentally.

8

$$E - m_0 c^2 = -\frac{Z^2\alpha^2 m_0 c^2}{2}[\frac{1}{n^2} + \frac{Z^2\alpha^2}{n^4}(\frac{n}{\ell + \frac{1}{2}} - \frac{3}{4})...], \quad \alpha = \frac{e^2}{\hbar c}. \tag{36'}$$

§. 1.4. Relativity principle and quantum mechanics

The wave equation (14) is indeed Lorentz covariant for a scalar wave function, and so is the Dirac equation given later. But let us look more deeply into the problem of unification of the relativity principle and quantum mechanics. We shall find that a unification is not to be achieved so easily; in fact there may be a basic incompatibility between the principles of the two foundation pillars of physics, as can be seen from the following simple consideration.

In Ch. 7. Vol. I, Eq. (VII-156), it is pointed out that in classical dynamics $-E$ and t formally behave as canonical conjugate variables and in relativity theory,

$$(x, y, z, it),$$

$$(p_x, p_y, p_z, iE), \tag{37}$$

are four-vectors under Lorentz transformations. Hence Lorentz covariance suggests that if in quantum mechanics p, x are hermitian operators satisfying[9]

$$p_x x - x p_x = -i, \tag{38}$$

then one should have[10]

$$Et - tE = +i. \tag{39}$$

But in Eq. (V-108), Ch. 5, Vol. I, it is shown that no hermitian operator t exists that satisfies this last commutation relation. In quantum mechanics p, q are hermitian operators and one may have either the coordinate representation in which the operators are

$$q \quad and \quad p = -i \frac{\partial}{\partial q}, \tag{40}$$

---

[9]

$$p_x x - x p_x = -i\hbar \tag{38'}$$

[10]

$$Et - tE = i\hbar. \tag{39'}$$

or the momentum-representation in which the operators are

$$p \quad and \quad q = i\frac{\partial}{\partial p}.$$ (41)

The situation with $-H$, t is different. The Hamiltonian is a hermitian operator, but it is on plausibility ground that one takes, in Eq. (V-109), Vol. I,

$$H = i\frac{\partial}{\partial t}.$$ (42)

The variable t remains in quantum mechanics a *classical* parameter (c-number) and not an operator, and the operator

$$t = \frac{1}{i}\frac{\partial}{\partial E}$$ (43)

does *not* exist as E is not always continuous. Thus in quantum mechanics, q and t are not on the same footing: In the coordinate-time-representation the wave equation Eq. (14) is Lorentz covariant; but unlike the time-independent Schrödinger equation which can be transformed to the momentum representation, no "momentum-energy" representation exists!

That the time variable remains a classical parameter in quantum mechanics while the coordinates x, y, z are operators is an asymmetry lying at the most basic level. Until this asymmetry is removed, a full unification of the relativity principle and quantum mechanics is not at hand. But symmetrizing the basic role of the space-time coordinates in quantum mechanics will entail very basic changes in quantum mechanics.[11]

---

[11] *Of course, it is also possible that the time variable is indeed intrinsically different from the spatial coordinates, as implied by quantum mechanics. Thus, a complete equivalence between t and (x, y, z) does not exist at the most basic level, so that some minute violation of Lorentz invariance may eventually be detected. On the other hand, we note that, very recently, T. D. Lee has developed a theory of discrete classical and quantum mechanics in which the concept of discrete time is introduced. It is hoped that the theory may contribute to this problem.*

Exercises: *Chapter 1*

1. Treat the harmonic oscillator according to the Klein-Gordon equation. A Klein-Gordon harmonic oscillator may be obtained in three different ways: (i) $\mathbf{p}^2 - (E - \frac{1}{4}kr^2)^2 = -(m + \frac{1}{4}kr^2)^2$, (ii) $\mathbf{p}^2 - (E - \frac{1}{2}kr^2)^2 = -m^2$, or (iii) $\mathbf{p}^2 - E^2 = -(m + \frac{1}{2}kr^2)^2$. Use the quantization rule, Eq. (9), to obtain the Klein-Gordon equation. Solve one of the three cases and discuss the other cases.

2. (i) In the case of the Klein-Gordon equation in a Coulomb field, derive *explicitly* the equation for the radial part $R(\rho)$. That is, derive Eq. (28) from Eq. (14).

   (ii) For a given l, discuss the behavior of $R(\rho)$ near $r = 0$ and $r \to \infty$.

   (iii) In addition, discuss the case for $l = 0$ and $Z\alpha > \frac{1}{2}$ where the energy eigenvalue becomes complex.

   (iv) In the general case *except* those considered in (iii), compute the fine structure for $n = 1, 2, 3$ levels and compare the results with Eqs. (VIII-53), (VIII-54), (VIII-57), and (VIII-58) of Vol. I.

3. Let $x_\mu\{\equiv (\mathbf{x}, it)\}$ and $p_\mu\{\equiv (\mathbf{p}, iE)\}$ be four vectors. The 4-dimensional transform of $\psi(x)$ is defined by

$$\phi(p) = \int e^{-ip \cdot x} \psi(x) d^4 x,$$

or, conversely,

$$\psi(x) = \frac{1}{(2\pi)^4} \int e^{ip \cdot x} \phi(p) d^4 p.$$

(i) Show that the momentum representation of the Klein-Gordon equation, Eq. (11), is given by

$$(p^2 + m_0^2)\phi(p) = 0.$$

(ii) Show that the momentum representation of the equation

$$(p^2 + m^2)\psi(x) = F(x)\psi(x),$$

is given by

$$(p^2 + m^2)\phi(p) = \frac{1}{(2\pi)^4} \int G(p - p')\phi(p') d^4 p',$$

with

$$G(p - p') = \int e^{-i(p-p')\cdot x} F(x) d^4 x.$$

(iii) Show that $\phi(p)$ in part (ii) can be expressed as an integral equation

$$\phi(p) = \int K(p, p') \phi(p') d^4 p',$$

where the kernel $K(p, p')$ is

$$K(p, p') = \frac{1}{(2\pi)^4} \int \frac{e^{-i(p-p')\cdot x}}{p^2 + m^2} F(x) d^4 x.$$

(iv) Work out the equivalent of part (iii) in configuration space:

$$\psi(x) = \int K(x, x') \psi(x') d^4 x'.$$

Find the expression for $K(x, x')$.

4. Consider the time-independent Klein-Gordon equation with a source of unit strength at $\mathbf{r} = 0$, i.e.

$$\nabla^2 \psi(\mathbf{r}) - \frac{1}{\lambda^2} \psi(\mathbf{r}) = -\delta^3(\mathbf{r}),$$

with $\lambda = \frac{1}{m}$. Derive the Yukawa potential:

$$\psi(\mathbf{r}) = \frac{1}{4\pi r} e^{-r/\lambda}.$$

Chapter 2.
Dirac Theory for Free Electrons

§.2.1. Dirac's relativistic equations

In the preceding chapter, we have seen some of the difficulties arising from the wave equation being of the second order in time. Dirac set out in seeking for an equation of the first order in time, and hence also of the first order in x, y, and z.

To satisfy the condition (i) in Eq. (2), Ch. 1, namely, $w \geq 0$, Dirac assumes to have n components $\psi_1, \psi_2, ..., n$ being undetermined at the outset, and takes

$$
\begin{aligned}
w &= (\psi_1^*, \psi_2^*, ...) \begin{pmatrix} \psi_1 \\ \psi_2 \\ \vdots \end{pmatrix} \\
&= \sum_{k=1}^{n} \psi_k^* \psi_k.
\end{aligned} \tag{1}
$$

The most general partial differential equation of the first order in t, x, y, z may be put in the form

$$
-E\psi_k + \sum_j (\alpha_{kj} \cdot \mathbf{p})\psi_j + m_0 \sum_j \beta_{kj}\psi_j = 0,
$$

or[1]

$$
\frac{\partial \psi_k}{\partial t} + \sum_j (\alpha_{kj} \cdot \nabla)\psi_j + i m_0 \sum_j \beta_{kj}\psi_j = 0, \tag{2}
$$

where α_{kj}, β_{kj} are matrices

$$
(\alpha_x)_{kj}, \qquad (\alpha_y)_{kj}, \qquad (\alpha_z)_{kj}, \qquad \beta_{kj} \tag{3}
$$

---
[1]

$$
\frac{1}{c}\frac{\partial \psi_k}{\partial t} + \sum_j (\alpha_{kj} \cdot \nabla)\psi_j + \frac{i m_0 c}{\hbar} \sum_j \beta_{kj}\psi_j = 0. \tag{2'}
$$

which operate only on the subscript j of ψ_j, and do not operate on the momentum $\mathbf{p}$.

Let $\tilde{\alpha}$ be the transpose of α, i.e., $\tilde{\alpha}_{mn} = \alpha_{nm}$. The complex conjugate and transpose of Eq. (2) is

$$\frac{\partial \psi_k^*}{\partial t} + \sum_j (\nabla \psi_j^* \cdot \tilde{\alpha}_{jk}^*) - i m_0 \sum_j \psi_j^* \tilde{\beta}_{jk}^* = 0. \tag{4}$$

From Eqs. (2) and (4), one obtains

$$\sum_k (\psi_k^* \frac{\partial \psi_k}{\partial t} + \frac{\partial \psi_k^*}{\partial t} \psi_k) + \sum_k \sum_j \{\psi_k^* (\alpha_{kj} \cdot \nabla) \psi_j + (\tilde{\alpha}_{kj}^* \cdot \nabla \psi_j^*) \psi_k\}$$
$$+ i m_0 \sum_k \sum_j \{\psi_k^* \beta_{kj} \psi_j - \tilde{\beta}_{kj}^* \psi_j^* \psi_k\} = 0. \tag{5}$$

If α, β satisfy the conditions

$$(\alpha_x^*)_{jk} = (\alpha_x)_{kj}, \quad \text{etc.,} \qquad (\beta^*)_{jk} = \beta_{kj}, \tag{6}$$

it is possible to define w as in Eq. (1) and a 3-current

$$\mathbf{I} = \sum_{k,j} \psi_k^* \alpha_{kj} \psi_j \tag{7}$$

so that Eq. (5) becomes

$$\frac{\partial w}{\partial t} + \nabla \cdot \mathbf{I} = 0. \tag{8}$$

This satisfies the conditions (i), (ii).

The conditions Eq. (6) are also necessary conditions to obtain Eq. (8). They show that

$$\alpha_x, \quad \alpha_y, \quad \alpha_z, \quad \beta \quad \text{must be hermitian.} \tag{9}$$

Eq. (2) can be written in matrix form

$$\frac{\partial \psi}{\partial t} + (\alpha \cdot \nabla) \psi + i m_0 \beta \psi = 0, \tag{10}$$

or[2,3]

$$i\frac{\partial\psi}{\partial t} = H\psi, \tag{11}$$

$$H = (\boldsymbol{\alpha} \cdot \mathbf{p}) + \beta m_0. \tag{12}$$

To determine the number of components of ψ_k (i.e., the α and β matrices), we employ the follow of argument: The solution ψ of Eq. (10) must also satisfy the Klein-Gordon equation.[4] Operating on Eq. (10) by

$$-\frac{\partial}{\partial t} + (\boldsymbol{\alpha} \cdot \nabla) + im_0\beta, \tag{13}$$

one gets[5]

$$-\frac{\partial^2\psi}{\partial t^2} + \sum_x^z \alpha_x^2\frac{\partial^2\psi}{\partial x^2} + \sum_{x>y}(\alpha_x\alpha_y + \alpha_y\alpha_x)\frac{\partial^2\psi}{\partial x\partial y}$$
$$+ im_0 \sum_x^z (\alpha_x\beta + \beta\alpha_x)\frac{\partial\psi}{\partial x} - m_0^2\beta^2\psi = 0. \tag{14}$$

The necessary and sufficient condition for this to be identical with the Klein-Gordon equation Eq. (11), Ch. 1, is

---
2

$$\frac{1}{c}\frac{\partial\psi}{\partial t} + (\boldsymbol{\alpha} \cdot \nabla)\psi + \frac{im_0 c}{\hbar}\beta\psi = 0. \tag{10'}$$

3

$$H = c(\boldsymbol{\alpha} \cdot \mathbf{p}) + \beta m_0 c^2. \tag{12'}$$

[4] *Note that the solution of the second order Klein-Gordon equation is not necessarily a solution of the first order equation (10).*

5

$$-\frac{1}{c^2}\frac{\partial^2\psi}{\partial t^2} + \sum_x^z \alpha_x^2\frac{\partial^2\psi}{\partial x^2} + \sum_{x>y}(\alpha_x\alpha_y + \alpha_y\alpha_x)\frac{\partial^2\psi}{\partial x\partial y}$$
$$+ \frac{im_0 c}{\hbar} \sum_x^z (\alpha_x\beta + \beta\alpha_x)\frac{\partial\psi}{\partial x} - \frac{m_0^2 c^2}{\hbar^2}\beta^2\psi = 0. \tag{14'}$$

$$\alpha_x \alpha_y + \alpha_y \alpha_x = 2\delta_{xy}, \quad x, y, z \quad \text{cyclically},$$

$$\alpha_x \beta + \beta \alpha_x = 0, \quad x = x, y, z; \tag{15}$$

$$\beta^2 = 1,$$

i.e., $\alpha_x, \alpha_y, \alpha_z$, anticommute with one another. Among the 10 relations in Eq. (15), we have

$$\alpha_x^2 = \alpha_y^2 = \alpha_z^2 = 1, \tag{16}$$

so that

$$\alpha_x^{-1} = \alpha_x, \quad \beta^{-1} = \beta. \tag{17}$$

To determine the order $n \times n$ of these matrices, it can first be shown that n cannot be odd; n cannot be 2; and the lowest n is $n = 4$.[6]

---

[6] *Thus, take for example,*

$$\alpha_x \beta = -\beta \alpha_x$$
$$= -I\beta \alpha_x, \quad I = n \times n \qquad \textit{unit matrix.}$$

Take the determinant of both sides,

$$[det.\alpha_x][det.\beta] = [det.(-1)][det.\beta][det.\alpha_x]$$
$$= (-1)^n [det.\alpha_x][det.\beta].$$

Since $det.\alpha_x \neq 0$ and $det.\beta \neq 0$ [as from Eqs. (15) and (16)], we find that n must be even. Next, assume

$$\beta^{-1}\alpha_x\beta = -\alpha_x \quad (\beta^{-1} = \beta)$$

and

$$Tr \; (\beta^{-1}\alpha_x\beta) = -Tr \; \alpha_x$$

But $Tr \; (\beta^{-1}\alpha_x\beta) = Tr \; \alpha_x$, hence

$$Tr \; \alpha_x = 0. \tag{17a}$$

Taking α, β to be 4×4 matrices, one may have infinitely many choices - "representations".

(i) One representation is to use the Pauli matrices (VIII-1), Vol. I,

$$\sigma_x = \begin{pmatrix} 0 & 1 \\ 1 & 0 \end{pmatrix}, \quad \sigma_y = \begin{pmatrix} 0 & -i \\ i & 0 \end{pmatrix}, \quad \sigma_z = \begin{pmatrix} 1 & 0 \\ 0 & -1 \end{pmatrix} \tag{18}$$

and construct the 4×4 α, β matrices by

$$\alpha_1 = \alpha_x = \begin{pmatrix} 0 & \sigma_x \\ \sigma_x & 0 \end{pmatrix}, \quad \alpha_2 = \alpha_y = \begin{pmatrix} 0 & \sigma_y \\ \sigma_y & 0 \end{pmatrix},$$

$$\alpha_3 = \alpha_z = \begin{pmatrix} 0 & \sigma_z \\ \sigma_z & 0 \end{pmatrix}, \quad \alpha_4 = \beta = \begin{pmatrix} I & 0 \\ 0 & -I \end{pmatrix}. \tag{19}$$

where $I = \begin{pmatrix} 1 & 0 \\ 0 & 1 \end{pmatrix}$. Corresponding to the dimension 4×4 for α, β, there are 4 components in Ψ_k .

A related representation, as used by W. Pauli, T. D. Lee and others, are followed in the present book, is

$$\gamma_k = \begin{pmatrix} 0 & -i\sigma_k \\ i\sigma_k & 0 \end{pmatrix}, \quad k = 1, 2, 3, (x, y, z), \quad \gamma_4 = \begin{pmatrix} I & 0 \\ 0 & -I \end{pmatrix}. \tag{20a}$$

or, more explicitly,

$$\gamma_1 = \begin{pmatrix} & & & -i \\ & & -i & \\ & i & & \\ i & & & \end{pmatrix}, \quad \gamma_2 = \begin{pmatrix} & & & -1 \\ & & 1 & \\ & -1 & & \\ -1 & & & \end{pmatrix},$$

$$\gamma_3 = \begin{pmatrix} & & -i & 0 \\ & & 0 & i \\ i & 0 & & \\ 0 & -i & & \end{pmatrix}, \quad \gamma_4 = \beta. \tag{20b}$$

Similarly

$$\text{Tr } \alpha_y = \text{Tr } \alpha_z = 0, \quad \text{Tr } \beta = 0.$$

It is easy to demonstrate from Eq. (15) that the four matrices $\alpha_x, \alpha_y, \alpha_z, \beta$ are linearly independent. In the case of 2×2 matrices, there are only three linearly independent, traceless matrices, say, $\sigma_x, \sigma_y, \sigma_z$. Thus, $n = 2$ is ruled out for the choice of $\{\alpha_x, \alpha_y, \alpha_z, \beta\}$.

$$\gamma_1 = -i\beta\alpha_x, \quad \gamma_2 = -i\beta\alpha_y, \quad \gamma_3 = -i\beta\alpha_z, \quad \gamma_4 = \beta. \tag{20c}$$

(ii) Majorana representation:

The γ_μ^M in the Majorana representation are related to the Pauli γ_μ in Eq. (20) by a unitary transformation

$$\gamma_\mu^M = S\gamma_\mu S^\dagger \tag{21a}$$

where

$$S = \frac{1}{\sqrt{2}}(1 + i\gamma_2), \quad S^\dagger = S^{-1} = \frac{1}{\sqrt{2}}(1 - i\gamma_2),$$

so that, except for $\mu = 2$ (with $\gamma_2^M = \gamma_2$),

$$\gamma_\mu^M = i\gamma_2\gamma_\mu. \tag{21b}$$

Explicitly,[7]

$$\gamma_1^M = \begin{pmatrix} \sigma_z & 0 \\ 0 & \sigma_z \end{pmatrix}, \qquad \gamma_2^M = \begin{pmatrix} 0 & -i\sigma_y \\ i\sigma_y & 0 \end{pmatrix},$$

$$\gamma_3^M = \begin{pmatrix} -\sigma_x & 0 \\ 0 & -\sigma_x \end{pmatrix}, \qquad \gamma_4^M = \begin{pmatrix} 0 & -\sigma_y \\ -\sigma_y & 0 \end{pmatrix}. \tag{21c}$$

The γ_μ^M are all hermitian. $\gamma_1^M, \gamma_2^M, \gamma_3^M$ are all real and symmetric; γ_4^M is imaginary and antisymmetric.

Eq. (10) can be expressed in terms of the γ_μ's by multiplying (10), from the left, by $-i\beta$. One gets

---

[7] *If* $\gamma_1, \gamma_2, \gamma_3$ *are as in Eq. (20a) but one chooses*

$$\gamma_4 = \begin{pmatrix} 0 & -I \\ -I & 0 \end{pmatrix},$$

then $\gamma_1^M, \gamma_2^M, \gamma_3^M$ *are the same as in Eq. (21c) but*

$$\gamma_4^M = \begin{pmatrix} -\sigma_y & 0 \\ 0 & \sigma_y \end{pmatrix}. \tag{21d}$$

$$\left(\gamma_\mu \frac{1}{i}\frac{\partial}{\partial x_\mu} - im_0\right)\Psi = 0 \tag{22}$$

or[8]

$$\left(\gamma_\mu \frac{\partial}{\partial x_\mu} + m_0\right)\Psi = 0, \tag{22a}$$

where the summation convention (over repeated indices from 1 to 4) is used. The relativistic equation for an electron (charge -e) in an electromagnetic field $(A, i\phi)$ can be obtained by replacing $p_k, \frac{1}{i}\frac{\partial}{\partial x_4}$ in Eq. (22) by[9]

$$\Pi_k = \frac{1}{i}\frac{\partial}{\partial x_k} + eA_k, \quad \Pi_4 = \frac{1}{i}\frac{\partial}{\partial x_4} + ie\phi, \tag{23}$$

leading to

$$(\gamma_\mu\Pi_\mu - im_0)\Psi = 0, \tag{24a}$$

or,[10]

$$\left\{i\frac{\partial}{\partial t} + e\phi - \alpha \cdot (\mathbf{p} + e\mathbf{A}) - \beta m_0\right\}\Psi = 0. \tag{24b}$$

Note that this is just another application of the principle of minimal substitution, Eq. (12), Ch. 1, which yields a gauge-invariant interaction in the context of gauge field theories.

8

$$\left(\gamma_\mu \frac{\partial}{\partial x_\mu} + \frac{m_0 c}{\hbar}\right)\Psi = 0. \tag{22a'}$$

9

$$\Pi_k = \frac{\hbar}{i}\frac{\partial}{\partial x_k} + \frac{e}{c}A_k, \quad \Pi_4 = \frac{\hbar}{i}\frac{\partial}{\partial x_4} + i\frac{e}{c}\phi. \tag{23'}$$

10

$$\left\{\frac{\hbar}{i}\frac{\partial}{\partial t} - e\phi + c\alpha \cdot (\mathbf{p} + \frac{e}{c}\mathbf{A}) + \beta m_0 c^2\right\}\Psi = 0. \tag{24b'}$$

§.2.2. Solution of Dirac equations for a free electron

On using the α, β representation of Eq. (19), the Dirac equation (10) becomes

$$
\begin{aligned}
&(\frac{1}{i}\frac{\partial}{\partial t} + m_0)\psi_1 + \frac{1}{i}(\frac{\partial}{\partial x} - i\frac{\partial}{\partial y})\psi_4 + \frac{1}{i}\frac{\partial}{\partial z}\psi_3 = 0, \\
&(\frac{1}{i}\frac{\partial}{\partial t} + m_0)\psi_2 + \frac{1}{i}(\frac{\partial}{\partial x} + i\frac{\partial}{\partial y})\psi_3 - \frac{1}{i}\frac{\partial}{\partial z}\psi_4 = 0, \\
&(\frac{1}{i}\frac{\partial}{\partial t} - m_0)\psi_3 + \frac{1}{i}(\frac{\partial}{\partial x} - i\frac{\partial}{\partial y})\psi_2 + \frac{1}{i}\frac{\partial}{\partial z}\psi_1 = 0, \\
&(\frac{1}{i}\frac{\partial}{\partial t} - m_0)\psi_4 + \frac{1}{i}(\frac{\partial}{\partial x} + i\frac{\partial}{\partial y})\psi_1 - \frac{1}{i}\frac{\partial}{\partial z}\psi_2 = 0.
\end{aligned}
\tag{25}
$$

If one assumes plane waves for ψ_μ ,

$$
\psi_\mu = A_\mu \exp\{i(\mathbf{p}\cdot\mathbf{r} - Et)\}, \quad A_\mu = \text{const.}, \tag{26}
$$

one gets for A_μ,[11]

$$
\begin{aligned}
-(E - m_0)A_1 + &&+ p_3 A_3 &&+ (p_1 - ip_2)A_4 &= 0, \\
&-(E - m_0)A_2 + (p_1 + ip_2)A_3 &- p_3 A_4 && &= 0, \\
&p_3 A_1 \ + (p_1 - ip_2)A_2 &- (E + m_0)A_3 && &= 0, \\
&(p_1 + ip_2)A_1 - p_3 A_2 && &- (E + m_0)A_4 &= 0.
\end{aligned}
\tag{27}
$$

The condition for existence of nontrivial solution for A_μ is

$$
[E^2 - m_0^2 - (p_1^2 + p_2^2 + p_3^2)]^2 = 0, \tag{28}
$$

whose roots (each doubly degenerate) are

$$
\left.\begin{aligned} E_+ \\ E_- \end{aligned}\right\} = \pm\sqrt{\mathbf{p}^2 + m_0^2}. \tag{28a}
$$

Eq. (28) is the relativistic energy-momentum relation, and the positive and negative energy are also there in the classical theory. In classical theory, E_- are independent and one may ignore E_- as physically meaningless. But in the theory of Dirac, there are transitions between E_+ and E_- states and one may

---

[11]

$$
-(E - m_0 c^2)A_1 + cp_3 A_3 + c(p_1 - ip_2)A_4 = 0, \text{ etc.} \tag{27'}
$$

no longer ignore the negative energy E_- states. We shall come back to this in the following section.

On substituting E_+ into Eq. (27), one obtains[12]

$$A_1^{(+)} = \frac{1}{E_+ - m_0}\{p_3 A_3^{(+)} + (p_1 - ip_2)A_4^{(+)}\},$$
$$A_2^{(+)} = \frac{1}{E_+ - m_0}\{(p_1 + ip_2)A_3^{(+)} - p_3 A_4^{(+)}\},$$

(29a)

or,

$$A_3^{(+)} = \frac{1}{E_+ + m_0}\{p_3 A_1^{(+)} + (p_1 - ip_2)A_2^{(+)}\},$$
$$A_4^{(+)} = \frac{1}{E_+ + m_0}\{(p_1 + ip_2)A_1^{(+)} - p_3 A_2^{(+)}\}.$$

(29b)

If $E_+ - m_0 \ll m_0$, then

$$E_+ + m_0 \gg E_+ - m_0,$$

(30)

and

$$A_3^{(+)}, \quad A_4^{(+)} \quad \ll \quad A_1^{(+)}, \quad A_2^{(+)}.$$

(31)

$A_1^{(+)}, A_2^{(+)}$ are called the big components; $A_3^{(+)}, A_4^{(+)}$ the small components.

On substituting E_- into Eq. (27), one has only to replace E_+ by E_- in Eqs. (29a)-(29b).[13]

---

[12]

$$A_2^{(+)} = -\frac{c}{E_+ - m_0 c^2}\{(p_1 + ip_2)A_3^{(+)} - p_3 A_4^{(+)}\}, etc.$$

(29a')

[13] *The results (29)-(32) can be expressed in a compact form. Let the column matrix ψ_μ in Eq. (26) be put in the form*

$$\psi_\mu(\mathbf{r}, t) = \begin{pmatrix} \psi_1(\mathbf{r}) \\ \psi_2(\mathbf{r}) \\ \psi_3(\mathbf{r}) \\ \psi_4(\mathbf{r}) \end{pmatrix} e^{-iEt} \equiv \begin{pmatrix} \Phi \\ \varphi \end{pmatrix} e^{-iEt},$$

(33)

where

$$\Phi = \begin{pmatrix} \Phi_1 \\ \Phi_2 \end{pmatrix}, \quad \varphi = \begin{pmatrix} \varphi_1 \\ \varphi_2 \end{pmatrix}.$$

(33a)

$$A_1^{(-)}, \ A_2^{(-)} \ll A_3^{(-)}, \ A_4^{(-)}. \tag{32}$$

Let $\chi_\pm$ be the eigenfunction of σ_z ,

$$\sigma_z \chi_\pm = \begin{pmatrix} 1 & 0 \\ 0 & -1 \end{pmatrix} \chi_\pm = \pm \chi_\pm. \tag{37}$$

The eigenvalues of σ_z are ± 1 .

The normalized $\psi(x)$, including the spin function, are, from Eq. (33) and Eq. (36),

$$\psi = N \left(\begin{array}{c} \chi_\pm \\ \frac{(\sigma \cdot p)}{\lambda |E| + m_0} \chi_\pm \end{array} \right) \frac{1}{(2\pi)^{\frac{3}{2}}} \exp\{i(\mathbf{p} \cdot \mathbf{r} - Et)\}. \tag{38}$$

Using the α, β of Eq. (19) in Eq. (10), one obtains

$$(E - m_0)\Phi - (\sigma \cdot \mathbf{p})\varphi = 0,$$

$$-(\sigma \cdot \mathbf{p})\Phi + (E + m_0)\varphi = 0, \tag{34}$$

where σ are the Pauli matrices Eq. (18). To solve for Φ and φ, one may make use of the identity

$$(\sigma \cdot \mathbf{A})(\sigma \cdot \mathbf{B}) = (\mathbf{A} \cdot \mathbf{B}) + i(\sigma \cdot [\mathbf{A} \times \mathbf{B}]) \tag{35}$$

and obtain the two roots E_+, E_- in Eq. (28a) again, and[13]

$$\varphi = \frac{(\sigma \cdot \mathbf{p})}{\lambda |E| + m_0} \Phi, \quad \lambda = \pm \ \text{for} \ \left\{ \begin{array}{c} E_+ \\ E_- \end{array} \right. , \tag{36}$$

This is the same as Eqs. (29)-(32) above.

§.2.3. Properties of negative energy states

(1) *Divorce of momentum from velocity*

From Eqs. (11) and (12)[15]

$$i\frac{\partial \Psi}{\partial t} = H\Psi$$
$$= \{(\alpha \cdot p) + \beta m_0\}\Psi, \tag{39}$$

and the equations of motion in Heisenberg's picture, one gets[16]

$$\dot{x}_k = i(Hx_k - x_k H) = \alpha_k, \tag{40}$$

$$\dot{p}_k = i(Hp_k - p_k H) = 0. \tag{41}$$

The expectation value v_x of the velocity $\dot{x}$ is hence

$$v_x = \int \tilde{\Psi}^* \alpha_x \Psi d\tau. \tag{42}$$

If

$$\Psi(\mathbf{r}, t) = \psi(\mathbf{r})e^{-iEt}, \tag{43}$$

then Eq. (39) gives

$$E\psi - (\alpha \cdot \mathbf{p})\psi - \beta m_0 \psi = 0. \tag{44}$$

The complex conjugate and transpose of this is ($\tilde{\alpha}^* = \alpha$, etc)

$$E\tilde{\psi}^* - \mathbf{p}\tilde{\psi}^* \cdot \alpha - m_0\tilde{\psi}^*\beta = 0. \tag{45}$$

---

[15]
$$i\hbar\frac{\partial \Psi}{\partial t} = H\Psi$$
$$= \{c(\alpha \cdot \mathbf{p}) + \beta m_0 c^2\}\Psi. \tag{39'}$$

[16]
$$\dot{x}_k = \frac{i}{\hbar}(Hx_k - x_k H) = c\alpha_k. \tag{40'}$$

Multiplying Eq. (44) from the left by $\tilde{\psi}^*\alpha_k$, and Eq. (45) from the right by $\alpha_k\psi$, integrating, and using Eq. (15), one gets

$$E \int \tilde{\psi}^*\alpha_k\psi d\tau = \int \tilde{\psi}^* p_k\psi d\tau. \qquad (46)$$

Hence, from Eq. (42), one obtains[17]

$$v_k = \frac{1}{E} < p_k > . \qquad (47)$$

For E_- , the average $\mathbf{v}_-$ and the average $\mathbf{p}$ are in opposite directions! In classical dynamics one would associate a *negative* mass with an electron in a negative energy state! Such a particle has the strange property that when a force is applied in one direction, the acceleration is in the opposite direction.

This divorce between the "velocity" $\dot{x}_k$ and the momentum p_k can already be seen from Eqs. (40) and (41). The α_k's do not commute among themselves, nor with H. Hence the $\dot{x}_k$ are unlike any physical velocity in the classical sense. The p_k are constants of motion, as seen from Eq. (41).

(2) The Zitterbewegung of Schrödinger

The Heisenberg equations of motion are

$$\dot{\alpha}_k = i(H\alpha_k - \alpha_k H),$$
$$\dot{\beta} = i(H\beta - \beta H). \qquad (48)$$

From Eq. (12) and Eq. (15), one obtains

$$H\alpha_k + \alpha_k H = 2p_k,$$
$$H\beta + \beta H = 2m_0. \qquad (49)$$

Hence[18]

$$\dot{\alpha}_k = 2i(p_k - \alpha_k H) = 2i(H\alpha_k - p_k),$$
$$\dot{\beta} = 2i(m_0 - \beta H) = 2i(H\beta - m_0). \qquad (50)$$

---

[17]
$$v_k = \frac{c^2}{E} < p_k > . \qquad (47')$$

[18]
$$\dot{\alpha}_k = \frac{2i}{\hbar}(cp_k - \alpha_k H) = \frac{2i}{\hbar}(H\alpha_k - cp_k). \qquad (50')$$

From Eq. (41), $\dot{p}_k = 0$, and $\dot{H} = 0$, hence, on differentiating Eq. (50) ,

$$\ddot{\alpha}_k = -2i\dot{\alpha}_k H, \tag{51}$$

and on integration,

$$\dot{\alpha}_k = (\dot{\alpha}_k)_{t=0} e^{-2iHt}. \tag{52}$$

Substituting this into Eq. (50), one gets

$$\alpha_k = \frac{1}{2iH}(\dot{\alpha}_k)_{t=0} e^{-2iHt} + \frac{1}{H} p_k. \tag{53}$$

From Eq. (40), one gets[19]

$$\dot{x}_k = \frac{1}{2iH}(\dot{\alpha}_k)_{t=0} e^{-2iHt} + \frac{1}{H} p_k. \tag{54}$$

The velocity $\dot{x}_k$ has a classical part $\frac{1}{H} p_k = \frac{\partial E}{\partial p_k} = v_k$, and a high frequency $2H \geq 2m_0$ part found by Schrödinger. This Zitterbewegung (trembling motion) arises from the interference effect between the positive and the negative energy states, as we shall see below.

Let $\Psi(\mathbf{r}, t)$ and its Fourier transform $\Phi(p, t)$ be

$$\Psi(\mathbf{r}, t) = (2\pi)^{-3} \int \Phi(\mathbf{p}, t) e^{-i\mathbf{p}\cdot\mathbf{r}} d\mathbf{p}, \tag{55a}$$

$$\Phi(\mathbf{p}, t) = \int \Psi(\mathbf{r}, t) e^{i\mathbf{p}\cdot\mathbf{r}} d\mathbf{r}. \tag{55b}$$

The equation (39) in p-representation is then

$$i\frac{\partial \Phi}{\partial t} - (\alpha \cdot \mathbf{p})\Phi - m_0\beta\Phi = 0. \tag{56}$$

Let

$$\Phi = B_+ e^{-i|E|t} + B_- e^{i|E|t}. \tag{57}$$

19

$$\dot{x}_k = \frac{\hbar c}{2iH}(\dot{\alpha}_k)_{t=0} e^{-2iHt/\hbar} + \frac{c^2}{H} p_k. \tag{54'}$$

Then

$$(E_+ - (\alpha \cdot \mathbf{p}) - m_0\beta)B_+ = 0,$$
$$(E_- - (\alpha \cdot \mathbf{p}) - m_0\beta)B_- = 0.$$

(58a)

Taking the complex conjugate and transpose, one gets

$$E_+\tilde{B}_+^* - \tilde{B}_+^*((\alpha \cdot \mathbf{p}) + m_0\beta) = 0,$$
$$E_-\tilde{B}_-^* - \tilde{B}_-^*((\alpha \cdot \mathbf{p}) + m_0\beta) = 0.$$

(58b)

From Eqs. (58a)-(58b),[20]

$$E_+\tilde{B}_+^* B_- E_- = \tilde{B}_+^*\{(\alpha \cdot \mathbf{p}) + m_0\beta\}^2 B_-$$

Using Eqs. (28a) and (15), one gets

$$-(p^2 + m_0^2)\tilde{B}_+^* B_- = (p^2 + m_0^2)\tilde{B}_+^* B_-$$

so that

$$\tilde{B}_+^* B_- = \sum (\tilde{B}_+^*)_\mu (B_-)_\mu = 0,$$

(59)

i.e., the positive and negative energy wave functions are orthogonal.

Let us calculate the expectation value v_x of the velocity $\dot{x}$ of an electron in the momentum representation. From Eqs. (40) and (42), one gets[21]

$$\begin{aligned}
v_k &= \int \tilde{\Phi}^* \alpha_k \Phi dp \\
&= \int \tilde{B}_+^* \alpha_k B_+ dp + \int \tilde{B}_-^* \alpha_k B_- dp \\
&\quad + e^{2i|E|t} \int \tilde{B}_+^* \alpha_k B_- dp + e^{-2i|E|t} \int \tilde{B}_-^* \alpha_k B_+ dp \\
&= \frac{1}{E_+} < p_k > + \frac{1}{E_-} < p_k > + e^{2i|E|t} \int \tilde{B}_+^* \alpha_k B_- dp \\
&\quad + e^{-2i|E|t} \int \tilde{B}_-^* \alpha_k B_+ dp.
\end{aligned}$$

(60)

---

[20] B_+, B_- are column, $\tilde{B}_+^*, \tilde{B}_-^*$ are row matrices, and

$$\tilde{B}_+^* B_- = \sum_\mu (B_+^*)_\mu (B_-)_\mu.$$

[21] For the first two terms following the last equal sign, see Eq. (47).

This shows that the Zitterbewegung of Eq. (54) has its origin in the interference of E_+ and E_- states.

Before the discovery of the positron in 1932, the presence of the negative-energy states, the "divorce" between velocity and momentum, and the Zitterbewegung, were considered as troublesome features of the Dirac equation. However, quantization of the Dirac field resolves conceptual difficulties associated with these problems.

(3) The Klein paradox

Consider the motion of an electron travelling from left to right in one dimension in a potential $V(x)$ given by

$$V(x) = \begin{cases} 0, & x < 0; \\ V = \text{constant}, & 0 < x. \end{cases} \tag{61}$$

For $x < 0$, the wave function is

$$\psi_i = A \ \exp\{i(p_0 x - Et)\}. \tag{62a}$$

At $x = 0$, there are a reflected and a transmited wave

$$\psi_r = B \ \exp\{i(-p_0 x - Et)\}, \tag{62b}$$

$$\psi_t = C \ \exp\{i(px - Et)\} \tag{62c}$$

ψ_i, ψ_r, ψ_t and A , B , C are all column matrices. Dirac's equations are[22]

$$(E - \alpha p_0 - \beta m_0)A = 0, \tag{63a}$$

$$(E + \alpha p_0 - \beta m_0)B = 0, \tag{63b}$$

$$(E - V - \alpha p - \beta m_0)C = 0. \tag{63c}$$

The energy-momentum relations are

$$E^2 = p_0^2 + m_0^2, \quad \text{for } x < 0,$$

---

[22]

$$(E - c\alpha p_0 - \beta m_0 c^2)A = 0, \text{ etc.} \tag{63a'}$$

$$(E - V)^2 = p^2 + m_0^2, \quad \text{for } 0 < x. \tag{64}$$

At $x = 0$, the continuity of ψ requires that

$$A + B = C. \tag{65}$$

From Eq. (63) and this, one gets

$$(E - \beta m_0)(A + B) = \alpha p_0 (A - B),$$

$$(E - \beta m_0)(A + B) = (V + \alpha p)(A + B).$$

Hence

$$(V + \alpha(p_0 + p))B = -(V - \alpha(p_0 - p))A.$$

Multiplying both sides by $V - \alpha(p_0 + p)$ and using Eq. (64), one gets

$$B = -\frac{2V(E - \alpha p_0)}{V^2 - (p_0 + p)^2} A. \tag{66}$$

Taking the adjoints of Eqs. (62a)-(62c), one obtains the complex conjugate and transpose of Eqs. (63a)-(63c),

$$\tilde{A}^*(E - \alpha p_0 - \beta m_0) = 0, \tag{67a}$$

$$\tilde{B}^*(E + \alpha p_0 - \beta m_0) = 0, \tag{67b}$$

$$\tilde{C}^*(E - V - \alpha p - \beta m_0) = 0. \tag{67c}$$

Multiplying Eq. (63a) by $\tilde{A}^*\alpha$ from the left, Eq. (67a) by αA from the right, one gets, on adding and using Eqs. (15) and (19),

$$\tilde{A}^*\alpha A = \frac{1}{E} p_0 \tilde{A}^* A, \tag{68}$$

which is once more Eq. (47).

In a procedure similar to that in obtaining Eq. (66), one gets

$$\tilde{B}^* = -\tilde{A}^* \frac{2V(E - \alpha p_0)}{V^2 - (p_0 + p)^2}. \tag{69}$$

From Eqs. (66) and (69), one obtains

$$\tilde{B}^* B = (\frac{2V}{V^2 - (p_0 + p)^2})^2 \tilde{A}^* (E - \alpha p_0)^2 A$$

$$= (\frac{2V}{V^2 - (p_0 + p)^2})^2 \{(E^2 + p_0^2)\tilde{A}^* A - 2Ep_0 \tilde{A}^* \alpha A\}$$

$$= (\frac{2V}{V^2 - (p_0 + p)^2})^2 (E^2 - p_0^2)\tilde{A}^* A, \quad \text{by Eq. (68)}$$

The reflection coefficient R is

$$R = \frac{\tilde{B}^* B}{\tilde{A}^* A} = (\frac{2V m_0}{V^2 - (p_0 + p)^2})^2. \tag{70}$$

For a fixed E, when V increases from 0 to $E - m_0$, R increases from 0 to the value 1 , i.e., the transmitted electron has $p = 0$ and the electron is totally reflected.

When V continues to increase from $E - m_0$ to the value $E + m_0$, it is seen from Eq. (64) that p is pure imaginary. Thus,

$$p = iq, \quad q = \text{real}, \quad \text{for } E - m_0 < V < E + m_0. \tag{71}$$

Eq. (62c) for the transmitted wave is now

$$\psi_t = D \quad \exp\{-(qx + iEt)\},$$
$$\psi_t^* = D^* \quad \exp\{-(qx - iEt)\}. \tag{72}$$

In Eq. (66), p is replaced by iq , and in Eq. (69) , p is replaced by $-iq$, so that the reflection coefficient R in Eq. (70) is now[23]

$$R = \frac{(2V)^2(E^2 - p_0^2)}{[(V + p_0)^2 + q^2][(V - p_0)^2 + q^2]} \tag{73}$$
$$= 1, \quad \text{for} \quad E - m_0 \leq V \leq E + m_0,$$

since it can be verified from Eq. (64) that

---

[23]

$$R = \frac{(2V)^2(E^2 - c^2 p_0^2)}{[(V + cp_0)^2 + c^2 q^2][(V - cp_0)^2 + c^2 q^2]}. \tag{73'}$$

$$(V \pm p_0)^2 + q^2 = 2V(E \pm p_0).$$

For $V > E + m_0$, Eq. (64) shows that $p^2 > 0$ so that R is again given by Eq. (70). But from Eq. (64) ,

$$E - V = \pm\sqrt{p^2 + m_0^2},$$

and for $V > E + m_0$, $E - V$ is negative. Hence we must take

$$E - V = -\sqrt{p^2 + m_0^2} < 0. \tag{74}$$

From the relation Eq. (47), one now has[24]

$$v = \frac{\partial E}{\partial p} = \frac{1}{E - V}p, \quad E - V < 0, \tag{75}$$

so that v and p are in *opposite* directions. But this is the property of an electron in a negative energy state found in Eq. (47) .

Intuitively, we may visualize the situation as follows: For $x \leq 0$, we have $V = 0$ so that the positive-energy spectrum starts at $E = m_0$ while the negative-energy solutions appear for $E \leq -m_0$. For $x > 0$, the entire energy spectrum is lifted upward by an amount equal to the potential $V(> 0)$. A given incident electron wave of energy $E(> 0)$ is reflected totally by a potential barrier V with $E - m_0 \leq V \leq E + m_0$ since, for $x > 0$, E lies between (and outside) the positive and negative spectra. For $V \leq E - m_0$, transmission arises from transition into a positive energy solution in the $x > 0$ region . For $V > E + m_0$, transmission occurs because transition into a negative-energy solution in the $x > 0$ region becomes possible.

Thus when an electron of positive energy E impinges on a potential $V > E + m_0$, it penetrates into the barrier but makes a transition to a negative energy state.[25]

---

[24]

$$V = \frac{\partial E}{\partial p} = \frac{c^2}{E - V}p. \tag{75'}$$

[25] *It was shown by F. Sauter, Zeits. f. Phys. 69, 742; 73, 547 (1931), that if the gradient of V at $x = 0$ in Eq. (61) is not infinite but is smaller than*

When V becomes infinitely high, one sees from Eq. (70) that

$$\lim_{V \to \infty} R = \frac{E - p_0}{E + p_0}. \tag{76}$$

(4) The "hole theory" of positron

The possibility of transitions of an electron from E_+ to E_- energy states, as shown by Klein's study, shows that the negative energy states cannot be regarded as completely independent of the positive (ordinary) energy states. The question why all the electrons do not fall into the negative energy states is answered by the theory of Dirac (1930) that all these states are filled up in the sense of the Pauli principle so that a completely filled sea of E_- states constitutes the norm-the vacuum. When an electron in this sea is excited into an ordinary state (by γ -rays, say, of energy $> 2m_0c^2$), a hole is left in the *sea,* and the absence of a negative charge from the norm reveals itself as a positive charge $+e$. Dirac at first suggested that this positive charge is our proton; but it was soon pointed out by R. Oppenheimer (1930) and H. Weyl (1931) that the mass of this positive charge must be the same as that of our ordinary electron.

In 1932, C. D. Anderson discovered in a Wilson cloud chamber of cosmic rays a track which, from its direction of motion and its curvature in a magnetic field, can be definitely identified as that due to a positive charge $+e$ with the mass of an electron. This was immediately identified as the "hole" in the Dirac theory, and is named "positron".

Historically, before the discovery by Anderson, there already existed some experimental evidence for the production of positrons. It was found by C. Y. Chao at California Institute of Technology in 1930 in experiments on the absorption of hard γ -rays that there is some excess absorption not accounted for by the then known processes such as photoelectric and Compton effect. This excess absorption is later readily accounted for by the process.

$$h\nu \;\to\; e^+ + e^-, \quad h\nu \;>\; 2m_0c^2 \tag{77a}$$

$\frac{2m_0c^2}{\lambda}, \lambda = \frac{h}{m_0 c}$, *the compton wavelength, the Klein transition to negative energy states will not occur. But the existence of this transition in principle is relevant here, since one can no longer simply ignore the E_- states as in classical theories.*

in which γ-rays, of energy greater than the sum of the rest energy of the electron and the positron, excite, in the intense field in the immediate neighborhood of an atomic nucleus, say, an electron from the sea of electrons in negative energy states into an ordinary (E_+) states, thereby producing a pair of $e^- + e^+$. This process is known as "pair production".

The reverse of the pair production process of Eq. (77a) also takes place

$$e^+ + e^- \rightarrow h\nu + h\nu \tag{77b}$$

in which e^+ and e^- annihilate each other, giving rise to two (occasionally three) quanta of γ-rays (for energy and momentum conservation.) Accurate measurements, including coincidence counter devices, definitely establish Eq. (77b).

These experimental verifications of Dirac's theory of positrons (or, anti-electron) are of course most satisfying; but at the same time they also call emphasis on the view that, since an infinite sea of electrons in negative energy states are always present, and especially in the presence of strong electric fields (such as near an atomic nucleus) the positive and negative energy states are not distinctly separated as in the case of free electrons, the Dirac theory cannot, from the rigorous point of view, be regarded as an exact theory of a single electron. In other words, a theory of the electron is intimately bound to a theory of fields.

§.2.4. Electron spin

We shall show that the Dirac relativistic equation already contains the electron spin.

In Eq. (19), we have defined the Dirac $\alpha_1, \alpha_2, \alpha_3, \alpha_4 = \beta$ matrices in terms of the Pauli matrices $\sigma_x, \sigma_y, \sigma_z$.

$$\alpha_1 = \begin{pmatrix} 0 & \sigma_x \\ \sigma_x & 0 \end{pmatrix}, \quad \alpha_2 = \begin{pmatrix} 0 & \sigma_y \\ \sigma_y & 0 \end{pmatrix}, \quad \alpha_3 = \begin{pmatrix} 0 & \sigma_z \\ \sigma_z & 0 \end{pmatrix}, \quad \alpha_4 = \beta = \begin{pmatrix} I & 0 \\ 0 & -I \end{pmatrix}.$$
(78)

Now to follow the literature (Dirac), we define the 4×4 ρ_1, ρ_2, ρ_3 matrices

$$\rho_1 = \begin{pmatrix} 0 & I \\ I & 0 \end{pmatrix}, \quad \rho_2 = i \begin{pmatrix} 0 & -I \\ I & 0 \end{pmatrix}, \quad \rho_3 = \begin{pmatrix} I & 0 \\ 0 & -I \end{pmatrix} = \beta,$$

$$\rho_1^2 = \rho_2^2 = \rho_3^2 = \begin{pmatrix} I & 0 \\ 0 & I \end{pmatrix}.$$
(79)

and the 4×4 $\sigma_1, \sigma_2, \sigma_3$ matrices (not to be confused with the Pauli 2×2 matrices $\sigma_x, \sigma_y, \sigma_z$ in Eq. (18))

$$\sigma_k = \rho_1 \alpha_k, \quad k = 1, 2, 3,$$
(80a)

or, explicitly in terms of Pauli matrices[26]

---

[26] *The* $\sigma_1, \sigma_2, \sigma_3$ *matrices satisfy the following relations:*

$$\sigma_1 = -i\sigma_2\sigma_3 = -i\alpha_2\alpha_3 = -i\gamma_2\gamma_3,$$
(80c)

$$\sigma_2 = -i\sigma_3\sigma_1 = -i\alpha_3\alpha_1 = -i\gamma_3\gamma_1,$$
(80d)

$$\sigma_3 = -i\sigma_1\sigma_2 = -i\alpha_1\alpha_2 = -i\gamma_1\gamma_2,$$
(80e)

$\sigma_1, \sigma_2, \sigma_3$ *are hermitian,*

$$\alpha_k = \rho_1 \sigma_k, \quad k = 1, 2, 3,$$
(80f)

$$\alpha_4 = \rho_3 = \beta;$$

$$\sigma_i\sigma_j + \sigma_j\sigma_i = 2\delta_{ij}, \quad i, j = 1, 2, 3,$$
(81)

$$\sigma_i\sigma_j = i\sigma_k, \quad \text{cyclic permutation of } i, j, k = 1, 2, 3.$$

$$\sigma_1 = \begin{pmatrix} \sigma_x & 0 \\ 0 & \sigma_x \end{pmatrix}, \quad \sigma_2 = \begin{pmatrix} \sigma_y & 0 \\ 0 & \sigma_y \end{pmatrix}, \quad \sigma_3 = \begin{pmatrix} \sigma_z & 0 \\ 0 & \sigma_z \end{pmatrix}. \tag{80b}$$

The Hamiltonian

$$H = (\boldsymbol{\alpha} \cdot \mathbf{p}) + \beta m_0$$

in terms of the σ_k matrices Eq. (80b) is[27]

$$H = \rho_1(\boldsymbol{\sigma} \cdot \mathbf{p}) + \alpha_4 m_0. \tag{84}$$

The angular momentum $\mathbf{L}(L_x, L_y, L_z)$

$$\mathbf{L} = \mathbf{r} \times \mathbf{p} \tag{85}$$

satisfy the relations

$$L_x p_x - p_x L_x = 0, \tag{86a}$$

$$L_x p_y - p_y L_x = i p_z, \tag{86b}$$

$$L_x p_z - p_z L_x = -i p_y, \tag{86c}$$

and relations obtained by cyclic permutations of x, y, z.

$$\rho_j \sigma_k - \sigma_k \rho_j = 0, \quad k, j = 1, 2, 3 \tag{82}$$

$$\rho_1 \alpha_k - \alpha_k \rho_1 = 0, \tag{83a}$$

$$\rho_2 \alpha_k + \alpha_k \rho_2 = 0, \tag{83b}$$

$$\rho_3 \alpha_k + \alpha_k \rho_3 = 0. \tag{83c}$$

[27]

$$H = c\rho_1(\boldsymbol{\sigma} \cdot \mathbf{p}) + \alpha_4 m_0 c^2. \tag{84'}$$

44

From Eqs. (84) and (86b)-(86c) ,

$$L_x H - H L_x = \rho_1 [L_x(\sigma \cdot \mathbf{p}) - (\sigma \cdot \mathbf{p})L_x]$$
$$= \rho_1 [\sigma \cdot (L_x \mathbf{p} - \mathbf{p}L_x)]$$
$$= i\rho_1(\sigma_2 p_3 - \sigma_3 p_2)$$
$$L_y H - H L_y = i\rho_1(\sigma_3 p_1 - \sigma_1 p_3)$$
$$L_z H - H L_z = i\rho_1(\sigma_1 p_2 - \sigma_2 p_1)$$

or,[28]

$$\mathbf{L}H - H\mathbf{L} = i\rho_1 \; [\sigma \times \mathbf{p}]. \tag{87}$$

Similarly,

$$\sigma_1 H - H\sigma_1 = \sigma_1 \rho_1(\sigma \cdot \mathbf{p}) - \rho_1(\sigma \cdot \mathbf{p})\sigma_1 + (\sigma_1 \alpha_4 - \alpha_4 \sigma_1)m_0$$
$$= -2i\rho_1(\sigma_2 p_3 - \sigma_3 p_2),$$
$$\sigma_2 H - H\sigma_2 = -2i\rho_1(\sigma_3 p_1 - \sigma_1 p_3),$$
$$\sigma_3 H - H\sigma_3 = -2i\rho_1(\sigma_1 p_2 - \sigma_2 p_1),$$

or[29]

$$\sigma H - H\sigma = -2i\rho_1 \; [\sigma \times \mathbf{p}]. \tag{88}$$

From Eqs. (87) and (88), one obtains

$$(\mathbf{L} + \frac{1}{2}\sigma)H - H(\mathbf{L} + \frac{1}{2}\sigma) = 0, \tag{89}$$

showing that

$$\mathbf{J} \equiv \mathbf{L} + \frac{1}{2}\sigma \tag{90}$$

---

[28]

$$\mathbf{L}H - H\mathbf{L} = i\hbar c \rho_1 \; [\sigma \times \mathbf{p}]. \tag{87'}$$

[29]

$$\sigma H - H\sigma = -2ic\rho_1 \; [\sigma \times \mathbf{p}]. \tag{88'}$$

is a constant of motion, while the orbital angular momentum **L** is not. The natural interpretation of **J** is that it is the total angular momentum of the electron, and this suggest that $\frac{1}{2}\sigma$ is an intrinsic angular momentum - the spin. If we introduce the components

$$S_1 = \frac{1}{2}\sigma_1, \quad S_2 = \frac{1}{2}\sigma_2, \quad S_3 = \frac{1}{2}\sigma_3, \tag{91}$$

then

$$\mathbf{J} = \mathbf{L} + \mathbf{S}. \tag{92}$$

From Eqs. (84) and (88), one can show that[30]

$$(L_z + \frac{1}{2}\sigma_3)H - H(L_z + \frac{1}{2}\sigma_3) = 0, \tag{93}$$

$$J^2(L_z + \frac{1}{2}\sigma_3) - (L_z + \frac{1}{2}\sigma_3)J^2 = 0, \tag{94}$$

and $H, J^2, (L_3 + \frac{1}{2}\sigma_3)$ have simultaneous eigenvectors.

In spherical polar coordinates (r, ϑ, φ)

$$J_3 = L_3 + S_3 = \frac{1}{i}\frac{\partial}{\partial\phi} + \frac{1}{2}\sigma_3.$$

$$\sigma_3\Psi = \begin{pmatrix} 1 & 0 & 0 & 0 \\ 0 & -1 & 0 & 0 \\ 0 & 0 & 1 & 0 \\ 0 & 0 & 0 & -1 \end{pmatrix} \begin{pmatrix} \psi_1 \\ \psi_2 \\ \psi_3 \\ \psi_4 \end{pmatrix} = \begin{pmatrix} \psi_1 \\ -\psi_2 \\ \psi_3 \\ -\psi_4 \end{pmatrix}. \tag{95}$$

The equation

$$(L_3 + \frac{1}{2}\sigma_3)\Psi = m\Psi$$

---

[30]

$$(L_z + \frac{1}{2}\hbar\sigma_3)H - H(L_z + \frac{1}{2}\hbar\sigma_3) = 0, \text{ etc.} \tag{93'}$$

becomes[31]

$$(\frac{1}{i}\frac{\partial}{\partial\varphi} + \frac{1}{2} - m_0)\begin{pmatrix} \psi_1 \\ \psi_3 \end{pmatrix} = 0,$$

$$(\frac{1}{i}\frac{\partial}{\partial\varphi} - \frac{1}{2} - m_0)\begin{pmatrix} \psi_2 \\ \psi_4 \end{pmatrix} = 0,$$

(96)

so that eigenvalues m are given by

$$m \pm \frac{1}{2} = \pm\text{integer}.$$

(97)

On defining

$$L_\pm = L_1 \pm iL_2, \quad S_\pm = S_1 \pm iS_2,$$

(98)

then

$$\mathbf{J}^2 = \mathbf{L}^2 + \mathbf{S}^2 + 2(\mathbf{L} \cdot \mathbf{S})$$
$$= \mathbf{L}^2 + \mathbf{S}^2 + L_+S_- + L_-S_+ + 2L_zS_z.$$

(99)

From Eqs. (80b) and (91), one finds[32]

$$S_+ = \begin{pmatrix} 0 & 1 & 0 & 0 \\ 0 & 0 & 0 & 0 \\ 0 & 0 & 0 & 1 \\ 0 & 0 & 0 & 0 \end{pmatrix}, \quad S_- = \begin{pmatrix} 0 & 0 & 0 & 0 \\ 1 & 0 & 0 & 0 \\ 0 & 0 & 0 & 0 \\ 0 & 0 & 1 & 0 \end{pmatrix},$$

(100)

$$S^2 = \frac{1}{4}(\sigma_1^2 + \sigma_2^2 + \sigma_3^2) = \frac{3}{4}I,$$

(101)

I being a 4×4 unit matrix.

To obtain the eigenvalues ξ of J^2, from Eqs. (99), (95), and (100), one obtains the 4 equations

$$(\mathbf{L}^2 + \frac{3}{4} + L_3 - \xi)\begin{pmatrix} \psi_1 \\ \psi_3 \end{pmatrix} + L_-\begin{pmatrix} \psi_2 \\ \psi_4 \end{pmatrix} = 0,$$

---

[31]

$$\hbar(\frac{1}{i}\frac{\partial}{\partial\varphi} + \frac{1}{2} - m_0)\begin{pmatrix} \psi_1 \\ \psi_3 \end{pmatrix} = 0, \text{ etc.}$$

(96')

[32]

$$S^2 = \frac{1}{4}(\sigma_1^2 + \sigma_2^2 + \sigma_3^2)\hbar^2 = \frac{3}{4}\hbar^2 I.$$

(101')

$$(\mathbf{L}^2 + \frac{3}{4} - L_3 - \xi)\begin{pmatrix} \psi_2 \\ \psi_4 \end{pmatrix} + L_+\begin{pmatrix} \psi_1 \\ \psi_3 \end{pmatrix} = 0, \tag{102}$$

which can be written[33]

$$(\mathbf{L}^2 + \frac{3}{4} + L_3 - \xi)\psi_1 + L_-\psi_2 = 0,$$

$$(\mathbf{L}^2 + \frac{3}{4} - L_3 - \xi)\psi_2 + L_+\psi_1 = 0, \tag{103}$$

and two other equations in which ψ_3, ψ_4 replace the ψ_1, ψ_2 above respectively.

In spherical polar coordinates (r, ϑ, φ) , we have

$$L_\pm = \pm e^{\pm i\varphi}\{\frac{\partial}{\partial \theta} \pm i\cot\vartheta\frac{\partial}{\partial \varphi}\},$$

$$\mathbf{L}^2 = -\{\frac{1}{\sin\vartheta}\frac{\partial}{\partial \vartheta}(\sin\vartheta\frac{\partial}{\partial \vartheta}) + \frac{1}{\sin^2\vartheta}\frac{\partial^2}{\partial \varphi^2}\}, \tag{104}$$

and Eq. (103) are

$$\{-[\frac{1}{\sin\vartheta}\frac{\partial}{\partial \vartheta}(\sin\vartheta\frac{\partial}{\partial \vartheta}) - \frac{1}{\sin^2\vartheta}(m - \frac{1}{2})^2] + \frac{3}{4} + (m - \frac{1}{2}) - \xi\}\psi_1$$

$$- e^{-i\varphi}[\frac{\partial}{\partial \vartheta} + (m + \frac{1}{2})\cot\vartheta]\psi_2 = 0,$$

$$\{-[\frac{1}{\sin\vartheta}\frac{\partial}{\partial \vartheta}(\sin\vartheta\frac{\partial}{\partial \vartheta}) - \frac{1}{\sin^2\vartheta}(m + \frac{1}{2})^2] + \frac{3}{4} - (m + \frac{1}{2}) - \xi\}\psi_2 \tag{105}$$

$$+ e^{i\varphi}[\frac{\partial}{\partial \vartheta} - (m - \frac{1}{2})\cot\vartheta]\psi_1 = 0.$$

The method of solving this pair of equations is similar to that of solving the Schrödinger equation. The results are:[34]

$$\xi = j(j + 1), \tag{106a}$$

---

[33]

$$(\mathbf{L}^2 + \frac{3}{4}\hbar^2 + L_3\hbar - \xi)\psi_1 + L_-\hbar\psi_2 = 0. \tag{103'}$$

[34]

$$\xi = j(j + 1)\hbar^2. \tag{106a'}$$

$$-j \leq m \leq j, \tag{106b}$$

$$m = \pm\frac{1}{2}, \pm\frac{3}{2}, \pm\frac{5}{2}, ..., \tag{106c}$$

$$j = \frac{1}{2}, \frac{3}{2}, \frac{5}{2}, ..., \tag{106d}$$

$$\psi_1 = \sqrt{\frac{j+1-m}{2j+2}}\, Y_{j+\frac{1}{2},m-\frac{1}{2}}(\vartheta,\varphi)f(r) + \sqrt{\frac{j+m}{2j}}\, Y_{j-\frac{1}{2},m-\frac{1}{2}}(\vartheta,\varphi)g(r), \tag{107a}$$

$$\psi_2 = -\sqrt{\frac{j+1+m}{2j+2}}\, Y_{j+\frac{1}{2},m+\frac{1}{2}}(\vartheta,\varphi)f(r) + \sqrt{\frac{j-m}{2j}}\, Y_{j-\frac{1}{2},m+\frac{1}{2}}(\vartheta,\varphi)g(r), \tag{107b}$$

$$Y_{\ell,m}(\vartheta,\varphi) = (-1)^m \sqrt{\frac{(2\ell+1)!(\ell-m)!}{4\pi(\ell+m)!}}\, P_\ell^m(\cos\vartheta)e^{im\varphi}. \tag{107c}$$

The functions $f(r)$, $g(r)$ must be obtained from the Dirac equation (24b).

The functions ψ_3, ψ_4 of Eq. (102) are given by a pair of equations similar to Eq. (105) and have the same angular parts as Eqs. (107a)-(107b), but the radial parts are $F(r)$, $G(r)$ in place of the $f(r)$ and $g(r)$. The relationship between $f(r)$, $g(r)$ and $F(r)$, $G(r)$ is similar to that between the ψ_1, ψ_2 and ψ_3, ψ_4 in the case of the free electrons. If the state is a positive energy state, $F(r)$, $G(r)$ are the "small", and $f(r)$, $g(r)$ the "big" components, and vice versa for negative energy states. We shall see this when we treat the hydrogen atom problem in Chapter 5.

§.2.5. Foldy-Wouthuysen representation

In Sect. 2, Eqs. (33)-(38), we have expressed the 4-component Dirac wave function $\Psi_\mu, \mu = 1,2,3,4$, in two 2-component functions $\Phi_k, \varphi_k, k = 1,2$.

Now, we shall introduce a unitary transformation operator[35]

---
[35]

$$U = \frac{\beta H + |E|}{\sqrt{2|E|(|E|+m_0c^2)}}. \tag{108'}$$

$$U = \frac{\beta H + |E|}{\sqrt{2|E|(|E|+m_0)}}, \tag{108}$$

and the transformation of

$$F = U\Psi. \tag{109}$$

Consequent upon this transformation, any operator Q is transformed into Q_F

$$Q_F = UQU^\dagger, \tag{110}$$

$$\Lambda_F = U\Lambda U^\dagger, \tag{111}$$

$$H_F = UHU^\dagger. \tag{112}$$

Here Λ (or Λ_F) is the energy projection operator. Calculations give the following results[36]

$$
\begin{aligned}
\mathbf{p}U - U\mathbf{p} = 0, \quad & \beta H\beta = 2m_0\beta - H, \\
H_F = |E|\beta, \quad & H_F^2 = E^2, \\
\Lambda_F = \beta. &
\end{aligned}
\tag{113}
$$

If one takes $E = E_+$,

$$m_s = \frac{1}{2}, \quad F = \begin{pmatrix} 1 \\ 0 \\ 0 \\ 0 \end{pmatrix} \frac{1}{(2\pi)^{\frac{3}{2}}} e^{ip_z z}, \tag{114a}$$

$$m_s = -\frac{1}{2}, \quad F = \begin{pmatrix} 0 \\ 1 \\ 0 \\ 0 \end{pmatrix} \frac{1}{(2\pi)^{\frac{3}{2}}} e^{ip_z z}, \tag{114b}$$

If one takes $E = E_-$,

36

$$\mathbf{p}U - U\mathbf{p} = 0, \quad \beta H\beta = 2m_0 c^2 \beta - H. \tag{113'}$$

$$m_s = \frac{1}{2}, \quad F = \begin{pmatrix} 0 \\ 0 \\ 0 \\ 1 \end{pmatrix} \frac{1}{(2\pi)^{\frac{3}{2}}} e^{ip_z z}, \tag{114c}$$

$$m_s = -\frac{1}{2}, \quad F = \begin{pmatrix} 0 \\ 0 \\ 1 \\ 0 \end{pmatrix} \frac{1}{(2\pi)^{\frac{3}{2}}} e^{ip_z z}. \tag{114d}$$

On introducing the operators

$$\frac{1}{2}(1 + \beta), \quad \frac{1}{2}(1 - \beta), \quad \text{and} \quad F = \begin{pmatrix} f_1 \\ f_2 \\ f_3 \\ f_4 \end{pmatrix} \tag{115}$$

then

$$\frac{1}{2}(1 + \beta)F = \begin{pmatrix} f_1 \\ f_2 \\ 0 \\ 0 \end{pmatrix} \equiv F_+ \tag{116a}$$

$$\frac{1}{2}(1 - \beta)F = \begin{pmatrix} 0 \\ 0 \\ f_3 \\ f_4 \end{pmatrix} \equiv F_- \tag{116b}$$

In the Foldy-Wouthuysen ($F_\pm$) representation, the positive energy states $\binom{f_1}{f_2}$ and the negative energy states $\binom{f_3}{f_4}$ do not mix. The 4-component Ψ_μ is now made up of two 2-component *spinors*

$$u = \begin{pmatrix} f_1 \\ f_2 \end{pmatrix}, \quad v = \begin{pmatrix} f_3 \\ f_4 \end{pmatrix} \tag{117}$$

The two components of u (and v) come from the spin angular momentum $m_s = \pm\frac{1}{2}$ values.

The separation of Ψ by U into unmixed positive and negative energy states, however, is possible only in the case of free electrons, for only in this case, the H and Λ do not mix the E_+ and E_- states. In the presence of external fields (as in the hydrogen atom, or in the Klein problem of Sect. 3. (3) above), the E_+, E_- states are always mixed up and their separation cannot be achieved by the transformation U . In the jargon of the subject, one says that in an external field, there is a constant production of *virtual* electron-positron pairs.

Strictly speaking, in the pressence of external fields, it is not possible to regard the Dirac equation as a single-electron theory; the many-body feature is always present through the mixing with the negative energy states. Only in the non-relativistic limit is the one-particle theory valid; a fully relativistic treatment calls for the theory of quantized interacting electron-electromagnetic fields - namely, quantum electrodynamics.

References

Dirac, P. A. M., *Proc. Poy. Soc. London* **A 117**, 610; **118**, 351 (1928), gives the relativistic (linear) wave equations of four-component wave function.

Dirac, P. A. M., *ibid.* **A 126**, 360 (1930), hole theory of positron.

Oppenheimer, J. R., *Phys. Rev.* **35**, 939 (1930), hole theory of positron.

Pauli, W. Article in *Handb. der Physik*, Bd 24, 2nd ed. (1933), or Encycl. of Physics, Vol. 5/1, (1958).

Breit, G., *Proc. Nat Acad. Sci.*, (1928). The divorce between velocity and momentum in negative energy state.

Schrödinger, E., *Berlin Berichte*, 418 (1930); 63 (1931); Zeits. f. Physik **70**, 808 (1931). The Zitterbewegung, the even and odd operators. Also see Fock, V., *Zeits. f. Physik* **55**, 127 (1929); **68**, 527; **70**, 811 (1931).

Klein, O., *Zeits. f. Physik* **53**, 157 (1929), the "Klein paradox". See also Sauter, F., ibid. **69**, 742; **73**, 547 (1931).

Foldy, L. L. and Wouthuysen, S. A., *Phys. Rev.* **78**, 29 (1950); **87**, 688 (1952); see also B. Kursunoglu, *ibid.* **101**, 1419 (1956).

Exercises: *Chapter 2*

1. Clarification of the following aspects is useful for the materials covered in the present chapter:

   (i) show that the four matrices $\{\alpha, \beta\}$ are linearly independent.

   (ii) The u-spinor is specified by

   $$u(\mathbf{p}, s) = N \begin{pmatrix} 1 \\ \frac{\sigma \cdot \mathbf{p}}{E + m_0} \end{pmatrix} \chi_s,$$

   with N the normalization factor and χ_s the two-compenent Pauli spinor. Rewrite Eq. (38) in terms of the u-spinor.

   On the other hand, the v-spinor is specified by

   $$v(\mathbf{p}, s) = N \begin{pmatrix} -\frac{\sigma \cdot \mathbf{p}}{|E| + m_0} \\ 1 \end{pmatrix} \chi_{-s}.$$

   Use the v-spinor to write out solutions to the free Dirac equation. (in a form analogous to Eq. (38).)

   Obtain the condition which ensures the orthogonality between the u-spinors and v-spinors.

   Find the expression for N if these spinors are normalized to unity.

   (iii) On the problem of Klein paradox, prove Eq. (73). That is, the reflection coefficient R is unity for $E - m_0 \le V \le E + m_0$.

2. Two operators Λ^+ and Λ^- are defined by

   $$\Lambda^+ = \frac{\gamma_\mu p_\mu + i m_0}{2 i m_0},$$

   and

   $$\Lambda^- = \frac{-\gamma_\mu p_\mu + i m_0}{2 i m_0}.$$

   (i) Show that Λ^+ and Λ^- are projection operators, i.e.,

   $$(\Lambda^\pm)^2 = \Lambda^\pm,$$

   $$\Lambda^+ \Lambda^- = \Lambda^- \Lambda^+ = 0.$$

(ii) Show that

$$\sum_s u(\mathbf{p}, s)\bar{u}(\mathbf{p}, s) \propto \Lambda^+,$$

and

$$\sum_s v(\mathbf{p}, s)\bar{v}(\mathbf{p}, s) \propto \Lambda^-,$$

with $\bar{u} \equiv u^\dagger \gamma_4$ and $\bar{v} \equiv v^\dagger \gamma_4$.

3. Decide the condition under which one may obtain the simultaneous eigenvalues of the Hamiltonian and σ_3 of a free Dirac particle. Write down the solutions and their eigenvalues explicitly.

4. Show that the Dirac equation, Eq. (24b), is invariant under gauge transformations.

Chapter 3.
γ_μ matrices; Helicity; Charge Conjugation

§.3.1. Properties of the γ_μ matrices

In Eqs. (20a) and (20b), Ch. 2, we have defined the γ_μ, $\mu = 1, ..., 4$, matrices which have the property

$$\gamma_\mu \gamma_\nu + \gamma_\nu \gamma_\mu = 2\delta_{\mu\nu}, \quad \mu, \nu = 1, ..., 4, \tag{1a}$$

$$\gamma_\mu^\dagger = \gamma_\mu. \tag{1b}$$

We shall in the following gather together the properties of these matrices in the form of theorems.

Theorem 1. The lowest order of 4 matrices satisfying the conditions Eqs. (1a) and (1b) is 4×4.

The proof is given in the footnote following Eq. (17), Ch. 2.

Let us form the following 16 4×4 matrices Γ_A, $A = 1, ..., 16$,

$$I \quad (\text{unit matrix; } \Gamma_{16})$$

$$\gamma_1 \quad \gamma_2 \quad \gamma_3 \quad \gamma_4$$

$$i\gamma_1\gamma_2 \quad i\gamma_2\gamma_3 \quad i\gamma_3\gamma_4 \quad i\gamma_1\gamma_4 \quad i\gamma_2\gamma_4 \quad i\gamma_1\gamma_3$$

$$i\gamma_1\gamma_2\gamma_3 \quad i\gamma_2\gamma_3\gamma_4 \quad i\gamma_1\gamma_3\gamma_4 \quad i\gamma_1\gamma_2\gamma_4$$

$$\gamma_1\gamma_2\gamma_3\gamma_4 (\equiv \gamma_5) \tag{2}$$

Theorem 2. The Γ_A, $A = 1, ..., 16$, are hermitian.

The proof follows from Eqs. (1a) and (1b). For example

$$(i\gamma_1\gamma_2)^\dagger = -i\gamma_2^\dagger \gamma_1^\dagger = -i\gamma_2\gamma_1 = i\gamma_1\gamma_2, \quad \text{etc.}$$

Theorem 3.

$$\gamma_5\gamma_\mu + \gamma_\mu\gamma_5 = 0, \quad \mu = 1, ..., 4. \tag{3}$$

Theorm 4.

$$(\Gamma_A)^2 = I, \quad \text{i.e., the } \Gamma\text{'s are unitary.} \tag{4}$$

Theorem 5.

$$\Gamma_A \Gamma_B = \pm \Gamma_C, \quad \text{or} \quad \pm i \Gamma_C. \tag{5}$$

In view of Eqs. (1), any product $\Gamma_A \Gamma_B$ can be expressed as a product of at most 4 different $\gamma_1, \gamma_2, \gamma_3, \gamma_4$. But Eq. (2) contains all such products.

Theorem 6

$$\Gamma_A \Gamma_B + \Gamma_B \Gamma_A = 0, \quad \text{or} \quad \Gamma_A \Gamma_B - \Gamma_B \Gamma_A = 0. \tag{6}$$

Theorem 7.

$$\text{If} \quad \Gamma_A \Gamma_B = \Gamma_A \Gamma_C, \quad \text{then} \quad \Gamma_B = \Gamma_C. \tag{7}$$

If by hypothesis $\Gamma_A \Gamma_B = \Gamma_A \Gamma_C$, then

$$\Gamma_A \Gamma_A \Gamma_B = \Gamma_A \Gamma_A \Gamma_C.$$

$$\Gamma_B = \Gamma_C \quad \text{by Theorem 4.}$$

Theorem 8. Except for I, the trace of each Γ_A is zero.

Proof:
(i) From Eq. (1a),

$$\frac{1}{2}(\gamma_\mu \gamma_\mu + \gamma_\mu \gamma_\mu) = I \quad (\text{no summation over } \mu \text{ here});$$

$$\frac{1}{2}(\gamma_\nu \gamma_\mu \gamma_\mu + \gamma_\nu \gamma_\mu \gamma_\mu) = \gamma_\nu, \quad \nu \neq \mu$$

$$\frac{1}{2}(\gamma_\nu \gamma_\mu \gamma_\mu - \gamma_\mu \gamma_\nu \gamma_\mu) = \gamma_\nu \quad \text{by Eq. (1a)}$$

$$\frac{1}{2}(Tr(\gamma_\nu \gamma_\mu \gamma_\mu) - Tr(\gamma_\mu \gamma_\nu \gamma_\mu)) = Tr\gamma_\nu$$

But the lefthand side is zero by a theorem of the trace.

(ii)

$$\frac{1}{2}(\gamma_\mu \gamma_\nu + \gamma_\nu \gamma_\mu) = 0, \quad \mu \neq \nu;$$

$$\frac{1}{2}(Tr\gamma_\mu \gamma_\nu + Tr\gamma_\nu \gamma_\mu) = 0,$$

or

$$Tr(\gamma_\mu \gamma_\nu) = 0.$$

(iii) Take $\gamma_1\gamma_2\gamma_3\gamma_4\gamma_\mu$ where μ is among 1, 2, 3, 4. By three permutations of pair of indices, the product can be brought to the order $\gamma_\mu\gamma_1\gamma_2\gamma_3\gamma_4$. By Eq. (1a),

$$\gamma_1\gamma_2\gamma_3\gamma_4\gamma_\mu + \gamma_\mu\gamma_1\gamma_2\gamma_3\gamma_4 = 0,$$

$$Tr(\gamma_1\gamma_2\gamma_3\gamma_4\gamma_\mu) = 0.$$

By letting $\gamma_\mu = \gamma_1, \gamma_2, \gamma_3, \gamma_4$ in succession, one gets

$$Tr(\gamma_\mu\gamma_\nu\gamma_\sigma) = 0 \quad \text{for} \mu \neq \nu \neq \sigma \neq \mu.$$

(iv) From

$$\gamma_1\gamma_2\gamma_3\gamma_4 + \gamma_1\gamma_2\gamma_3\gamma_4 = \gamma_1\gamma_2\gamma_3\gamma_4 - \gamma_4\gamma_1\gamma_2\gamma_3$$

$$2Tr(\gamma_1\gamma_2\gamma_3\gamma_4) = Tr(\gamma_1\gamma_2\gamma_3\gamma_4) - Tr(\gamma_4\gamma_1\gamma_2\gamma_3) = 0.$$

Theorem 9. *The 16 Γ_A in Eq. (2) are linearly independent.*

If they are not linearly independent, i.e., if there exists a relation

$$\sum_{A=1}^{16} C_A\Gamma_A = 0, \tag{8}$$

then multiplying this by an arbitrary Γ_B and using Eq. (4), one gets

$$\sum_{A=1}^{16}{}' C_A\Gamma_A\Gamma_B + C_B I = 0, \tag{9}$$

where $\sum'$ indicates the exclusion of the term C_B from the summation, so that in the sum, Γ_B is different from Γ_A. By Theorems 5 and 8, $Tr(\Gamma_A\Gamma_B) = 0$ in each term of the sum $\sum'$. But $TrI = n$. Hence $C_B = 0$. As Γ_B is arbitrary among the 16 Γ_A 's , it follows that all $C_A = 0$ in Eq. (8) .

Theorem 10. *An arbitrary 4×4 matrix can be expressed as a linear combination of the 16 Γ_A 's ,*

$$X = \sum_{A=1}^{16} C_A\Gamma_A, \tag{10}$$

a 4×4 matrix has 16 (complex) numbers, and the 16 numbers C_A serve to represent them. The C_A are given by

$$C_A = \frac{1}{4}Tr(X\Gamma_A).\qquad(10a)$$

Theorem 11. If an arbitrary 4×4 matrix F commutes with γ_μ, $\mu = 1, ..., 4$,

$$F\gamma_\mu - \gamma_\mu F = 0, \quad \mu = 1, 2, 3, 4,$$

then

$$F\Gamma_A - \Gamma_A F = 0 \quad \text{for all } \Gamma_A.\qquad(11)$$

Theorem 12. Except for $\Gamma_{16} \equiv I$, for every Γ_A , there is at least one Γ_B such that

$$\Gamma_B\Gamma_A\Gamma_B = -\Gamma_A.\qquad(12)$$

The following table gives the Γ_B for each Γ_A satisfying Eq. (12). The numbering starts with $I = \Gamma_{16}$, $\gamma_1 = \Gamma_1$, $\gamma_2 = \Gamma_2$, etc.

A	2	3	4	5	6	7	8
B	3,4,5	4,5,2	3,5,2	2,3,4	2,3	3,4	4,5
	6,9,11	6,7,10	7,8,11	8,9,10	7,9,10,11	6,8,10,11	7,9,10,11
	13	11	15	12			

A	9	10	11	12	13	14	15	16
B	5,2	5,3	2,4	5	2	3	4	2,3,4,5
	6,8,10,11	6,7,8,9	6,7	8,9,10	6,9,11	6,7,10	7,8,11	12,13
	8,9			13,14,15	14,15,12	15,12,13	12,13,14	14,15

$$(12a)$$

Theorem 13. If F commutes with all Γ_A , F must be a multiple of the unit matrix I.

Proof: We write F in the form Eq. (10)

$$F = \sum_{A=1}^{16} C_A \Gamma_A$$
$$= C_B \Gamma_B + \sum_A{}' C_A \Gamma_A, \tag{13a}$$

where Γ_B is any one of the 16 Γ_A except I , and $\sum'$ indicates the exclusion of A = B . Let Γ_μ be one of the matrices that satisfy Eq. (12) (see (12a)) ,

$$\Gamma_\mu \Gamma_B \Gamma_\mu = -\Gamma_B, \quad \text{(no summation over } \mu \text{ here)}$$

They by Eq. (6) ,

$$\Gamma_\mu F \Gamma_\mu = -C_B \Gamma_B + \sum_A{}'(\pm) C_A \Gamma_A. \tag{13b}$$

By hypothesis, $\Gamma_\mu F \Gamma_\mu = \Gamma_\mu \Gamma_\mu F = F.$ Multiplying Eqs. (13a) and (13b) by Γ_B and taking the trace, one finds

$$C_B = -C_B, \quad C_B = 0.$$

Since Γ_B is arbitrary (except I) , hence

$$F = cI.$$

Theorem 14. Any two sets of γ_μ satisfying

$$\gamma_\mu \gamma_\nu + \gamma_\nu \gamma_\mu = 2\delta_{\mu\nu}, \quad \gamma'_\mu \gamma'_\nu + \gamma'_\nu \gamma'_\mu = 2\delta_{\mu\nu}, \tag{14a}$$

differ from each other only by a similarity transformation

$$\gamma'_\mu = S^{-1} \gamma_\mu S. \tag{14}$$

Proof: It is obvious that Eq. (14) is a sufficient condition for γ'_μ to satisfy Eq. (14a) if γ_μ's satisfy it.

To show the necessity part, let the Γ_A and the Γ'_A formed from the γ_μ, γ'_μ respectively be similarly ordered as in Eq. (2). By Theorem 5,

$$\Gamma_A \Gamma_B = \epsilon_{AB} \Gamma_C, \quad \Gamma'_A \Gamma'_B = \epsilon_{AB} \Gamma'_C$$

where

$$\epsilon = \pm 1, \pm i$$

Let F be an arbitrary nonzero 4×4 matrix, and define S by

$$S^{-1} = \sum_{B=1}^{16} \Gamma'_B F \Gamma_B.$$

Then

$$\Gamma'_A S^{-1} \Gamma_A = \sum_{B=1} \Gamma'_A \Gamma'_B F \Gamma_B \Gamma_A$$

$$= \sum_{C} \epsilon_{AB} \Gamma'_C F \frac{1}{\epsilon_{AB}} \Gamma_C = S^{-1},$$

$$\Gamma'_A S^{-1} = S^{-1} \Gamma_A,$$

$$\Gamma'_A = S^{-1} \Gamma_A S,$$

which includes Eq. (14).

Theorem 15. If in Theorem 14 the two sets of γ_μ, γ'_μ , in addition to satisfying Eq. (14a), are hermitian, then they differ by a unitary transformation,

$$\gamma'_\mu = U^\dagger \gamma_\mu U. \tag{15}$$

Theorem 16. The transformation matrix S in Eq. (14) is unique, up to a complex multiplicative constant.

Let S_1 and S_2 be two matrices satisfying Eq. (14).

$$\gamma'_\mu = S_1^{-1} \gamma_\mu S_1, \quad \gamma'_\mu = S_2^{-1} \gamma_\mu S_2.$$

Hence

$$\gamma'_\mu = S_1^{-1} (S_2 \gamma'_\mu S_2^{-1}) S_1$$

$$= (S_1^{-1} S_2) \gamma'_\mu (S_2^{-1} S_1)$$

$$= (S_1^{-1} S_2) \gamma'_\mu (S_1^{-1} S_2)^{-1},$$

$$\gamma'_\mu (S_1^{-1} S_2) = (S_1^{-1} S_2) \gamma'_\mu, \qquad \mu = 1, 2, 3, 4.$$

By Theorem 11, $\left(S_1^{-1}S_2\right)$ commutes with all Γ_A .

By Theorem 13,

$$S_1^{-1}S_2 = \text{const. } I.$$
$$S_2 = \text{const. } S_1.$$

Theorem 17. The Dirac equations[1]

$$\frac{\partial \Psi}{\partial t} + [(\alpha \cdot \nabla) + im_0\beta]\Psi = 0,$$
$$\frac{\partial \tilde{\Psi}^*}{\partial t} + (\nabla \tilde{\Psi}^* \cdot \alpha) - im_0 \tilde{\Psi}^*\beta = 0,$$
(16)

are invariant under the following unitary transformation

$$\alpha_\mu' = U^\dagger \alpha_\mu U, \qquad \beta' = U^\dagger \beta U,$$
$$\Psi' = U^\dagger \Psi, \qquad \tilde{\Psi}'^* = \tilde{\Psi}^* U.$$
(17)

The proof is obvious.

§.3.2. Helicity and neutrinos

Let **n** be a unit vector along the momentum **p** ,

$$\mathbf{n}(\xi, \eta, \varsigma)$$
(18)

where ξ, η, ς are the direction consines of **n**. Let $\sigma_x, \sigma_y, \sigma_z$ be the 2×2 Pauli matrices. Then the component of σ along **p** are

$$\begin{aligned}
\sigma_n &= (\sigma \cdot \mathbf{n}) \\
&= \sigma_x \xi + \sigma_y \eta + \sigma_z \varsigma, \\
&= \begin{pmatrix} \varsigma & \xi - i\eta \\ \xi + i\eta & -\varsigma \end{pmatrix},
\end{aligned}$$
(19a)

and

---
[1]
$$\frac{\partial \Psi}{\partial t} + [c(\alpha \cdot \nabla) + i\frac{m_0 c^2}{\hbar}\beta]\Psi = 0.$$
(16')

$$\sigma_n \text{ is hermitian,} \tag{19b}$$

$$(\sigma \cdot \mathbf{n})^2 = I. \tag{19c}$$

(1) Helicity, eigenvalues and eigenfunctions

The helicity operator (a 4×4 matrix) is defined as

$$\sigma_n = \begin{pmatrix} (\sigma \cdot \mathbf{n}) & 0 \\ 0 & (\sigma \cdot \mathbf{n}) \end{pmatrix}, \tag{20}$$

$$(\sigma_n)^2 = \begin{pmatrix} I & 0 \\ 0 & I \end{pmatrix}. \tag{21}$$

The eigenvalues and eigenfunctions of σ_n in Eq. (19a)

$$\sigma_n \begin{pmatrix} \psi_1 \\ \psi_2 \end{pmatrix} = h \begin{pmatrix} \psi_1 \\ \psi_2 \end{pmatrix} \tag{22}$$

are

$$h = \begin{cases} 1, & \frac{\psi_1}{\psi_2} = \frac{1+\varsigma}{\xi+i\eta} \left(= \frac{\xi-i\eta}{1-\varsigma} \right), \\ -1, & \frac{\psi_1}{\psi_2} = \frac{-\xi+i\eta}{1+\varsigma} \left(= \frac{-1+\varsigma}{\xi+i\eta} \right), \end{cases} \tag{23}$$

For the helicity σ_n in Eq. (20) ,

$$\sigma_n \begin{pmatrix} u_1 \\ u_2 \\ u_3 \\ u_4 \end{pmatrix} = h \begin{pmatrix} u_1 \\ u_2 \\ u_3 \\ u_4 \end{pmatrix}, \tag{24}$$

this equation decomposes into two equations

$$\sigma_n \begin{pmatrix} u_1 \\ u_2 \end{pmatrix} - h \begin{pmatrix} u_1 \\ u_2 \end{pmatrix} = 0, \quad \sigma_n \begin{pmatrix} u_3 \\ u_4 \end{pmatrix} - h \begin{pmatrix} u_3 \\ u_4 \end{pmatrix} = 0, \tag{25}$$

which are both identical with Eq. (22), so that

$$h = \pm 1 \text{ as in Eq. (23).} \tag{26}$$

Let the subscript u and ℓ denote "upper" and "lower" , and

$$\phi_u = \begin{pmatrix} u_1 \\ u_2 \end{pmatrix}, \quad \phi_\ell = \begin{pmatrix} u_3 \\ u_4 \end{pmatrix}, \tag{27}$$

each of which is a 2-component spinor. Eq. (25) can be written

$$\sigma_n \phi_u = h\phi_u, \qquad \sigma_n \phi_\ell = h\phi_\ell \qquad (28)$$

The normalized ϕ_u, ϕ_ℓ are, from Eq. (23) ,

$$h = 1, \quad \phi_{u+} = \phi_{\ell+} = \sqrt{\frac{1}{2(1+\varsigma)}} \begin{pmatrix} 1+\varsigma \\ \xi + i\eta \end{pmatrix},$$

$$h = -1, \quad \phi_{u-} = \phi_{\ell-} = \sqrt{\frac{1}{2(1+\varsigma)}} \begin{pmatrix} -\xi + i\eta \\ 1+\varsigma \end{pmatrix}, \qquad (29)$$

so that

$$\tilde{\phi}^*_{u+}\phi_{u+} = \tilde{\phi}^*_{u-}\phi_{u-} = \tilde{\phi}^*_{\ell+}\phi_{\ell+} = \tilde{\phi}^*_{\ell-}\phi_{\ell-},$$

$$\tilde{\phi}^*_{u+}\phi_{u-} = \tilde{\phi}^*_{\ell+}\phi_{\ell-} = 0.$$

We define the nomenclature:

$$h = \begin{cases} 1, & \text{positive helicity, right-handed,} \\ -1, & \text{negative helicity, left-handed.} \end{cases} \qquad (30)$$

For a free electron, it is seen from Eq. (88), Ch. 2, that the Hamiltonian commutes with σ_n ,

$$H\sigma_n - \sigma_n H = 0, \qquad (31)$$

so that H and the helicity have simultaneous eigenfunctions. Dirac's equation is[2]

$$((\boldsymbol{\alpha} \cdot \mathbf{p}) + m_0 \beta)\Psi = E\Psi. \qquad (32)$$

In terms of the spinors ϕ_u, ϕ_ℓ , we have

$$\Psi = \begin{pmatrix} \phi_u \\ \phi_\ell \end{pmatrix} = \begin{pmatrix} a_u \\ a_\ell \end{pmatrix} \exp(ip_\mu x_\mu), \qquad (33)$$

where

---
[2]

$$(c(\boldsymbol{\alpha} \cdot \mathbf{p}) + m_0 c^2)\Psi = E\Psi. \qquad (32')$$

$$p_\mu x_\mu = \mathbf{p} \cdot \mathbf{r} - Et. \tag{33a}$$

Using Eq. (19), Ch. 2, we obtain from Eq. (32)

$$
\begin{aligned}
(E - m_0)a_u - (\sigma \cdot \mathbf{p})a_\ell &= 0, \\
-(\sigma \cdot \mathbf{p})a_u + (E + m_0)a_\ell &= 0,
\end{aligned}
\tag{34}
$$

where σ are the 2×2 Pauli matrices, and $\mathbf{p}$ is the momentum but no longer an operator. The eigenvalues of E are

$$E = \pm\sqrt{\mathbf{p}^2 + m_0^2} = E_\pm, \tag{35}$$

and[3]

$$a_\ell = \frac{(\sigma \cdot \mathbf{p})}{E + m_0}a_u \quad \left(\text{or,} \quad \frac{E - m_0}{(\sigma \cdot \mathbf{p})}a_u\right). \tag{36}$$

For

$$E = \left\{\begin{matrix} E_+ \\ E_- \end{matrix}\right\}, \quad a_u = \left\{\begin{matrix} \text{large} \\ \text{small} \end{matrix}\right\} \quad \text{and} \quad a_\ell = \left\{\begin{matrix} \text{small} \\ \text{large} \end{matrix}\right\} \quad \text{components.} \tag{37}$$

This is what has been obtained in

$$h = \left\{\begin{matrix} +1 \\ -1 \end{matrix}\right\}, \qquad \frac{(\sigma \cdot \mathbf{p})}{|\,\mathbf{p}\,|}\Psi = \pm\Psi. \tag{38}$$

Let

$$a = \frac{|\,\mathbf{p}\,|}{|\,E\,| + m_0}, \qquad N^2 = \frac{|\,E\,| + m_0}{4\,|\,E\,|\,(1 + \varsigma)}. \tag{39}$$

The simultaneous eigenvalues and normalized eigenfunctions of H and σ_n are as follows,

---
[3]

$$a_\ell = \frac{c(\sigma \cdot \mathbf{p})}{E + m_0 c^2}a_u, \quad \text{etc.} \tag{36'}$$

$$E_+, \quad h = 1, \quad \Psi = N \begin{pmatrix} 1 + \varsigma \\ \xi + i\eta \\ a(1 + \varsigma) \\ a(\xi + i\eta) \end{pmatrix} \exp(ip_\mu x_\mu), \qquad (40a)$$

$$E_+, \quad h = -1, \quad \Psi = N \begin{pmatrix} -\xi + i\eta \\ 1 + \varsigma \\ -a(-\xi + i\eta) \\ -a(1 + \varsigma) \end{pmatrix} \exp(ip_\mu x_\mu), \qquad (40b)$$

$$E_-, \quad h = 1, \quad \Psi = N \begin{pmatrix} a(1 + \varsigma) \\ a(\xi + i\eta) \\ 1 + \varsigma \\ \xi + i\eta \end{pmatrix} \exp(-ip_\mu x_\mu), \qquad (40c)$$

$$E_-, \quad h = -1, \quad \Psi = N \begin{pmatrix} -a(-\xi + i\eta) \\ -a(1 + \varsigma) \\ -\xi + i\eta \\ 1 + \varsigma \end{pmatrix} \exp(-ip_\mu x_\mu). \qquad (40d)$$

On account of $h = \pm 1$, the E_+ and E_- state are each doubly degenerate.

(2) Neutrino, helicity and chirality

The Dirac equations Eq. (34) will simplify when the mass m_0 of the particle approaches zero. As in Eq. (18), let $\mathbf{n}$ be a unit vector along $\mathbf{p}$, and here

$$\mathbf{n} = \frac{1}{E}\mathbf{p}, \qquad (41)$$

so that

$$\mathbf{n} \;\uparrow\uparrow\; \mathbf{p} \quad \text{for } E_+,$$
$$\mathbf{n} \;\uparrow\downarrow\; \mathbf{p} \quad \text{for } E_-. \qquad (42)$$

Eq. (34) now become

$$\phi_u = (\sigma \cdot \mathbf{n})\phi_\ell,$$
$$\phi_\ell = (\sigma \cdot \mathbf{n})\phi_u, \qquad (43)$$
$$\sigma = \text{Pauli matrices.}$$

Eq. (33) becomes

$$\Psi = \begin{pmatrix} \phi_u \\ \phi_\ell \end{pmatrix} = \begin{pmatrix} \phi_u \\ (\sigma \cdot \mathbf{n})\phi_u \end{pmatrix} \tag{44a}$$

and

$$(\sigma \cdot \mathbf{n})\Psi = \begin{pmatrix} (\sigma \cdot \mathbf{n})\phi_u \\ (\sigma \cdot \mathbf{n})^2\phi_u \end{pmatrix} = \begin{pmatrix} \phi_\ell \\ \phi_u \end{pmatrix} \tag{44b}$$

so that $(\sigma \cdot \mathbf{n})$ interchanges ϕ_u and ϕ_ℓ.

consider $\gamma_5 = \gamma_1\gamma_2\gamma_3\gamma_4$ in Eq. (2).

$$\gamma_5 = -\begin{pmatrix} 0 & 0 & 1 & 0 \\ 0 & 0 & 0 & 1 \\ 1 & 0 & 0 & 0 \\ 0 & 1 & 0 & 0 \end{pmatrix} = -\begin{pmatrix} 0 & I \\ I & 0 \end{pmatrix} \tag{45a}$$

so that

$$\gamma_5 \begin{pmatrix} \phi_u \\ \phi_\ell \end{pmatrix} = -\begin{pmatrix} \phi_\ell \\ \phi_u \end{pmatrix} \tag{45b}$$

Hence

$$-\gamma_5 = (\sigma \cdot \mathbf{n}). \tag{46}$$

If we define two 2 component spinors α, β by

$$\alpha = \frac{1}{2}(\phi_u + \phi_\ell) = \frac{1}{2}(1 + \sigma \cdot \mathbf{n})\phi_u$$
$$\beta = \frac{1}{2}(\phi_u - \phi_\ell) = \frac{1}{2}(1 - \sigma \cdot \mathbf{n})\phi_u \tag{47}$$

then

$$(\sigma \cdot \mathbf{n})\alpha = \alpha, \qquad (\sigma \cdot \mathbf{n})\beta = -\beta, \tag{48}$$

or

$$(\sigma \cdot \mathbf{n})\begin{pmatrix} \alpha \\ \beta \end{pmatrix} = \begin{pmatrix} \alpha \\ -\beta \end{pmatrix}$$

i.e., α, β are the 2-component eigenfunctions of the helicity operator $(\sigma \cdot \mathbf{n})$ corresponding to the helicity $h = 1, -1$ respectively.

The above case of parallel and antiparallel σ and $\mathbf{p}$ comes about only for $m_0 = 0$, i.e., the particle travelling with the speed of light. For $m_0 \neq 0$, it is always possible to make a Lorentz transformation to a frame in which the particle is at rest, and the above relations do not hold.

The above theory, in which two-component wave functions for a particle of zero rest mass was in fact considered by H. Weyl (1929) before the neutrino was proposed by Pauli in 1931, has the following feature. Let P be the parity, or inversion, operator

$$P: \quad \mathbf{r} \rightarrow -\mathbf{r}. \tag{49}$$

The momentum $\mathbf{p}$ is a polar vector whereas the angular momentum σ is a pseudo (or, axial) vector ,

$$P\,\mathbf{p} = -\mathbf{p}\,P, \quad P\sigma = \sigma\,P,$$

$$P(\sigma \cdot \mathbf{n}) = -(\sigma \cdot \mathbf{n})P. \tag{50}$$

Applying P on the equation Eq. (48), one gets

$$-(\sigma \cdot \mathbf{n})(P\alpha) = (P\alpha)$$
$$-(\sigma \cdot \mathbf{n})(P\beta) = -(P\beta) \tag{51}$$

which differ from Eq. (48) by a negative sign, i.e., are not invariant under the inversion operation. It is for this reason Pauli rejected this theory (in his 1933 article in the Handbuch der Physik, 2nd edition).

This basis for the objection based on the violation of the "conservation of parity" was removed by the theory of T. D. Lee and C. N. Yang on the non-conservation of parity in weak interactions,[4] and its subsequent confirmation

---

[4] *Weak interactions are those that govern such processes as*

(i) β -decays,

$$n \rightarrow p + e^- + \bar{\nu}_e,$$

$$^{11}_{6}C \rightarrow {}^{11}_{5}B + e^+ + \nu_e, \tag{52}$$

(ii) $\pi - \mu$ decays ,

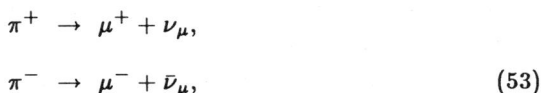

$$\pi^+ \rightarrow \mu^+ + \nu_\mu,$$

$$\pi^- \rightarrow \mu^- + \bar{\nu}_\mu, \tag{53}$$

by the experiments of C. S. Wu et al. and Lederman. The 2-component theory of neutrinos is revived by A. Salam, L. Landau, and Lee and Yang (1957).

The concept of helicity is related to the concept of neutrino and anti-neutrino. First of all, the electron e^- and the positron e^+ are called the 'particle' and the 'anti-particle' by convention. Once we start with this 'definition', the neutrino associated with e^- is an anti-neutrino denoted by $\bar{\nu}_e$, whereas the neutrino associated with e^+ is a neutrino, denoted by ν_e.[5] These two neutrinos are actually *different*. Experiments show that, with $\mathbf{s} = \frac{\sigma}{2}$,

the neutrino ν_e has negative helicity , $\mathbf{s}$ ↑ ↓ $\mathbf{p}$,

the anti-neutrino $\bar{\nu}_e$ has positive helicity, $\mathbf{s}$ ↑ ↑ $\mathbf{p}$. $\qquad(55)$

We shall introuce another related concept "chirality" here.

Let us transform the γ_μ in Eqs. (20a) and (20b), Ch. 2, by a unitary transformation

---

(iii) $\mu - e$ decay

$$\mu^- \; \rightarrow \; e^- + \bar{\nu}_e + \nu_\mu \qquad (54)$$

[5] *This naming of ν_e and $\bar{\nu}_e$ is not arbitrary. We introduce the concept of a quantum number called the "lepton number" L, such that a "particle" has $L = 1$, and an "anti-particle" $L = -1$. From the systematics of all elementary particle reactions, it has been found that the "selection rule"*

$$\triangle L = 0 \qquad (56)$$

holds. That e^+ has $L = -1$ follows from the pair-production process

$$h\nu \; \rightarrow \; e^+ + e^-.$$

It is to be noted that in Eq. (53), the neutrinos associated with the muons μ^-, μ^+ are denoted by $\bar{\nu}_\mu, \nu_\mu$. That they are distinct from the neutrinos associated with the e^-, e^+ is experimentally established by Lederman (1962) .
The notations in reactions (52), (53) and (54) all confirm the helicity relation (55) and lepton number conservation law (56).

$$\gamma'_\mu = U\gamma_\mu U^{-1}, \tag{57}$$

$$U = \frac{1}{\sqrt{2}}\begin{pmatrix} 1 & 1 \\ -1 & 1 \end{pmatrix}, \qquad U^\dagger = U^{-1} = \frac{1}{\sqrt{2}}\begin{pmatrix} 1 & -1 \\ 1 & 1 \end{pmatrix}, \tag{58}$$

(See Theorem 14 and 15 of Sect. 1) . It is found that

$$\gamma'_k = \gamma_k, \qquad k = 1, 2, 3,$$

$$\gamma'_4 = \begin{pmatrix} 0 & -1 \\ -1 & 0 \end{pmatrix}, \tag{58a}$$

$$\gamma'_5 = \begin{pmatrix} -1 & 0 \\ 0 & 1 \end{pmatrix}.$$

Thus

$$\gamma'_5\begin{pmatrix} \alpha \\ \beta \end{pmatrix} = \begin{pmatrix} -\alpha \\ \beta \end{pmatrix}. \tag{59}$$

The operator γ'_5 is called the chirality operator.[6] Comparison with the helicity $(\sigma \cdot n)$ in Eq. (48) shows that γ'_5 and $\sigma_n = (\sigma \cdot n)$ have opposite eigenvalues, i.e., γ'_5 and $-\sigma_n$ have the same property on the spinors $\begin{pmatrix} \alpha \\ \beta \end{pmatrix}$.

§.3.3. Charge conjugation

(1) Charge conjugate state Ψ_c

The Dirac equation for an electron (charge $-e$) in an electromagnetic field $(A, i\phi)$ is Eq. (24a),[7] Ch. 2.[8]

---

[6] *For a given Dirac spinor ψ as in Eq. (44a), the projected states $\psi_L = \frac{1}{2}(1 + \gamma_5)\psi$ and $\psi_R = \frac{1}{2}(1 - \gamma_5)\psi$ are referred to as "left-handed" and "right-handed" components of ψ , respectively. (Here γ_5 is given by Eq. (2).) Eq. (46) indicates that H and $\frac{1}{2}(1 \pm \gamma_5)$ have simultaneous eigenfunctions. (See Eqs. (31) and (40a)-(40d).) The components of definite chirality (or helicity) are basic entries in the construction of the Glashow-Salam-Weinberg $SU(2)_L \times U(1)$ electroweak theory. (See Chapter 13.)*

[7]

$$\left(\gamma_\mu(p_\mu + \frac{e}{c}A_\mu) - im_0 c\right)\Psi = 0. \tag{60'}$$

[8] *In the present section, we use the following notations: $\Psi_\mu, \Psi_\mu^\dagger, \Psi_c$ are column*

$$\left(\gamma_\mu(p_\mu + eA_\mu) - im_0\right)\Psi = 0, \tag{60}$$

$$p_k = \frac{1}{i}\frac{\partial}{\partial x_k}, \quad k = 1, 2, 3, \quad p_4 = \frac{1}{i}\frac{\partial}{\partial x_4}, \quad x_4 = it$$

$$A_k \text{ real}, \quad k = 1, 2, 3, \quad A_4 = i\phi = \text{imaginary}. \tag{61}$$

Taking the complex conjugate of Eq. (60) (not transpose), one gets

$$\left\{-\gamma_k^*(p_k - eA_k) + \gamma_4^*(p_4 - eA_4) + im_0\right\}\Psi^* = 0, \tag{62}$$

where p_k, p_4, A_k, A_4 have the same meaning as in Eq. (61).

For a particle of the same rest mass m_0 and charge $+e$, the Dirac equation is

$$\left(\gamma_\mu(p_\mu - eA_\mu) - im_0\right)\Psi_c = 0. \tag{63}$$

Ψ_c is defined as the *charge conjugate state* wave function . Note that Ψ_c is *not* the positron state wave function. See the following.

To obtain the relation between Ψ and Ψ_c , let C be a 4×4 matrix, and multiply Eq. (62) by C from the left, and get (using the summation convention for k)

$$C\left\{\gamma_k^*(p_k - eA_k) - \gamma_4^*(p_4 - eA_4) - im_0\right\}C^{-1}C\Psi^* = 0. \tag{64}$$

Comparison of Eqs. (63) and (64) shows that

$$\Psi_c \quad \text{and} \quad (\pm)C\Psi^* \tag{65}$$

satisfy the same equation if C satisfies the relations

$$C\gamma_k^*C^{-1} = \gamma_k, \quad k = 1, 2, 3, \quad -C\gamma_4^*C^{-1} = \gamma_4. \tag{66}$$

---

matrices, $\tilde{\Psi}_\mu, \tilde{\Psi}_\mu^*, \tilde{\Psi}_c, \quad \Psi^\dagger = \tilde{\Psi}^*, \quad \bar{\Psi} = \Psi^\dagger\gamma_4$ *are row matrices* .
Different notations are used in the literature. Thus in Pauli's article in Hand-buch der Physik , $i\tilde{\Psi}^*\gamma_4$ is denoted by $\Psi^\dagger$; in Davydov's Quantum Mechanics, $\tilde{\Psi}^*\gamma_4$ is denoted by $\bar{\Psi}$, as here.

These conditions do not suffice for the determination of the matrix C . On taking the transpose of Eq. (62), multiplying by γ_4 from the right and referring to Eqs. (20a)-(20b), Ch. 2, with

$$\tilde{\gamma}_\mu = \gamma_\mu^*, \qquad \gamma_4^* = \gamma_4,$$
$$\gamma_k \gamma_\mu = -\gamma_\mu \gamma_k, \qquad \gamma_k \gamma_\mu^* = -\gamma_\mu^* \gamma_k, \tag{67}$$
$$(\mu \neq k)$$

one gets

$$\tilde{\Psi}^* \gamma_4 \{ \gamma_k (p_k - eA_k) + \gamma_4 (p_4 - eA_4) + im_0 \} = 0. \tag{68}$$

Taking the transpose of Eq. (63), multiplying by $-\gamma_4 C^{-1}$ from the right, one gets

$$\tilde{\Psi}_c \gamma_4 \tilde{\gamma}_k C^{-1} (p_k - eA_k) - \tilde{\Psi}_c \gamma_4 \tilde{\gamma}_4 C^{-1} (p_4 - eA_4) + im_0 \tilde{\Psi}_c \gamma_4 C^{-1} = 0. \tag{69}$$

Comparison of Eqs. (68) and (69) (noting $\tilde{\gamma}_k C^{-1} = \gamma_k^* C^{-1} = C^{-1} \gamma_k$) shows that

$$\tilde{\Psi}^* \gamma_4 \quad \text{and} \quad (\pm) \tilde{\Psi}_c \gamma_4 C^{-1} \tag{70}$$

satisfy the same equation. From Eqs. (65) and (70), one gets (since $\gamma_4^* = \gamma_4$)

$$\gamma_4 C = \pm C \gamma_4. \tag{71}$$

To be consistent with Eq. (66)

$$\gamma_4 C = -C \gamma_4,$$

we may take

$$\tilde{C} = C = C^{-1}, \tag{72}$$

and in particular,

$$C = \gamma_2 = \begin{pmatrix} 0 & 0 & 0 & -1 \\ 0 & 0 & 1 & 0 \\ 0 & 1 & 0 & 0 \\ -1 & 0 & 0 & 0 \end{pmatrix}. \tag{73}$$

72

We finally take, from Eq. (65)

$$\Psi_c = C\Psi^*, \tag{74}$$

and from Eq. (70)

$$\tilde{\Psi}^*\gamma_4 = -\tilde{\Psi}_c\gamma_4 C^{-1},$$

which is identical with Eq. (74).[9]

Let us now apply Eq. (74) to the wave function Ψ of an electron (charge $-e$) in a positive energy E_+ state, helicity $\sigma_n = 1, p = |\mathbf{p}|$, namely Ψ in Eq. (40a) .

$$\Psi = N \begin{pmatrix} 1+\varsigma \\ \xi + i\eta \\ a(1+\varsigma) \\ a(\xi + i\eta) \end{pmatrix} \exp(ip_\mu x_\mu), \quad (a, N \text{ in Eq. (39)}). \tag{75}$$

From Eqs. (74) and (72) ,

$$\Psi_c = C\Psi^* = N \begin{pmatrix} -a(\xi - i\eta) \\ a(1+\varsigma) \\ \xi - i\eta \\ -(1+\varsigma) \end{pmatrix} \exp(-ip_\mu x_\mu), \tag{76}$$

Comparing this with Ψ in Eq. (40d) , one sees that Ψ_c is exactly the wave function of an electron e^- in a negative energy E_- state, helicity $\sigma_n = -1$, $p = -|\mathbf{p}|$ (except for a phase factor $e^{i\pi} = -1$) .

To sum up the above tedious calculations, it is found that the wave function Ψ_c of e^+ is the same as that of e^- in a negative energy E_- state.

---

9    $\gamma_4 C^{-1} = \gamma_4\gamma_2 = -\gamma_2\gamma_4.$

Thus, $\tilde{\Psi}^* = \tilde{\Psi}_c\gamma_2 = \tilde{\Psi}_c\tilde{\gamma}_2$, *so that* $\Psi^* = \gamma_2\Psi_c$ *or* $\Psi_c = \gamma_2\Psi^*.$

The following figure shows the above result

E_+ .... $p = |\mathbf{p}|$, $\sigma_n = 1$, $(\sigma \uparrow\uparrow \mathbf{p})$,

$m_0 c^2$ ————

0 ————

$-m_0 c^2$ ————

E_- .... $p = -|\mathbf{p}|$, $\sigma_n = -1$, $(\sigma \uparrow\downarrow \mathbf{p})$, $\Psi_c = \gamma_2 \Psi^*$.

$$\tag{77}$$

It is important to note that Ψ_c of e^+ , although it is the same as that of e^- in a E_- state, does *not* represent a positron. A positron is represented by a 'hole' in the sea of negative energy states. Thus the 'hole' has the following characteristics: charge $+e$, energe E_+ , momentum p . The 'hole' , or positron, is called an anti-electron .

Coming back to Eq. (74) , one sees that Ψ_c is the charge-conjugate state of Ψ . But from Eq. (74), one obtains (using Eqs. (72)-(73))

$$\Psi = C\Psi_c^*, \tag{78}$$

so that Ψ is also the charge-conjugate state of Ψ_c . But Ψ_c has just been seen to be the same as the function of e^- in a negative energy $E_- < 0$ state. Hence Eq. (78) shows that the charge conjugate state of e^- in a E_- state is the state of e^- in a E_+ state Ψ .

(2) Charge conjugate current

From Eqs. (1), (5), and (7) in Ch. 2., we write[10]

$$I_k = \tilde{\Psi}^* \alpha_k \Psi, \quad k = 1, 2, 3, \quad I_4 = i\tilde{\Psi}^* \Psi, \tag{79}$$

so that the equation of continuity Eq. (8), Ch. 2., now takes the form

$$\frac{\partial I_\mu}{\partial x_\mu} = 0. \tag{80}$$

---

[10]

$$I_k = c\tilde{\Psi}^* \alpha_k \Psi, \quad I_4 = ic\tilde{\Psi}^* \Psi. \tag{79'}$$

In the γ_μ -representation Eq. (20a)

$$\alpha_k = i\beta\gamma_k = i\gamma_4\gamma_k, \qquad \alpha_4 = \beta = \gamma_4,$$
$$I_k = i\tilde{\Psi}^*\gamma_4\gamma_k\Psi, \qquad I_4 = i\tilde{\Psi}^*\gamma_4\gamma_4\Psi. \tag{81}$$

We define the 4-electric current

$$J_\mu = -eI_\mu = -ie\tilde{\Psi}^*\gamma_4\gamma_\mu\Psi. \tag{82}$$

It will be shown in the following chapter that the I_μ defined in Eq. (81) trans-forms as a 4-vector under Lorentz transformations.

For the charge conjugate current, let us consider

$$\tilde{\Psi}_c^*\gamma_4\gamma_\mu\Psi_c. \tag{83}$$

From Eq. (74), we have, with $C = \gamma_2$, $\tilde{\gamma}_2^* = \gamma_2$,

$$\tilde{\Psi}_c^*\gamma_4\gamma_\mu\Psi_c = \tilde{\Psi}\gamma_2\gamma_4\gamma_\mu\gamma_2\Psi^*,$$
$$\tilde{\Psi}_c^*\gamma_4\gamma_k\Psi_c = \tilde{\Psi}\gamma_2\gamma_4\gamma_k\gamma_2\Psi^* = -\tilde{\Psi}\gamma_4\gamma_k^*\Psi^*, \quad \text{by Eq. (66)},$$
$$\tilde{\Psi}_c^*\gamma_4\gamma_4\Psi_c = \tilde{\Psi}\Psi^*.$$

Hence, if we define for the charge conjugate current

$$J_k^c = ie\{\tilde{\Psi}_c^*\gamma_4\gamma_k\Psi_c\}, \tag{84}$$

then

$$J_\mu^c = ie\tilde{\Psi}^*\gamma_4\gamma_\mu\Psi. \tag{84a}$$

Thus

$$J_\mu^c = -J_\mu. \tag{85}$$

§.3.4. Dirac equation in Majorana representation

In Chapter 2, Sect. 1, (ii), we have defined Majorana's representation of the Dirac matrices

$$\gamma_1^M = \begin{pmatrix} \sigma_z & 0 \\ 0 & \sigma_z \end{pmatrix}, \quad \gamma_2^M = \begin{pmatrix} 0 & -i\sigma_y \\ i\sigma_y & 0 \end{pmatrix},$$

$$\gamma_3^M = \begin{pmatrix} -\sigma_x & 0 \\ 0 & \sigma_x \end{pmatrix}, \quad \gamma_4^M = \begin{pmatrix} 0 & -\sigma_y \\ -\sigma_y & 0 \end{pmatrix}. \tag{86}$$

where $\sigma_x, \sigma_y, \sigma_z$ are the 2×2 Pauli matrices. (See Eq. (21) in Ch. 2.) In terms of the γ_μ^M matrices, the operator in Dirac's equation for a free electron[11]

$$(\gamma_\mu^M \frac{\partial}{\partial x_\mu} + m_0)\Psi = 0,$$

$$\gamma_\mu^M \frac{\partial}{\partial x_\mu} + m_0 = (\gamma_\mu^M \frac{\partial}{\partial x_\mu} + m_0)^*, \tag{87}$$

is an operator in real numbers. If we take as in Eq. (26) of Ch. 2,

$$\Psi_\mu = A_\mu \exp(i(\mathbf{p} \cdot \mathbf{r})), \quad A_\mu = \text{const.}, \tag{88}$$

as one solution for $p = |\mathbf{p}|$, $E = \sqrt{p^2 + m_0^2}$, then

$$\Psi_\mu^* = A_\mu^* \exp(-i(\mathbf{p} \cdot \mathbf{r})), \quad A_\mu^* = \text{const.}, \tag{89}$$

is also a solution for $p = -|\mathbf{p}|$, $E = -\sqrt{p^2 + m_0^2}$. As in Eqs. (40a)-(40d), for E_+ and $p = |\mathbf{p}|$, the double degeneracy is distinguished by the helicity $h = \pm 1$, and similarly for E_- and $p = -|\mathbf{p}|$.

The helicity operator

$$\sigma_n = (\sigma \cdot \mathbf{n}), \quad \mathbf{n} = \frac{\mathbf{p}}{|\mathbf{p}|}. \tag{90}$$

has been defined in Eq. (20) in terms of $(\sigma \cdot \mathbf{n})$ where $\sigma_x, \sigma_y, \sigma_z$ are the Pauli 2×2 matrices. But σ_n can also be expressed in terms of the 4×4 $\gamma_1, \gamma_2, \gamma_3$ matrices of Eq. (80e) of Ch. 2,

---

[11]

$$(\gamma_\mu^M \frac{\partial}{\partial x_\mu} + \frac{m_0 c}{\hbar})\Psi = 0. \tag{87'}$$

$$\sigma_j = -i\gamma_k\gamma_\ell, \quad \text{cyclic permutation of j, k, } \ell = 1, 2, 3. \tag{91}$$

In Majorana represention, we have

$$\sigma_j^M = -i\gamma_k^M\gamma_\ell^M. \tag{92}$$

As the $\gamma_k^M, k = 1, 2, 3$, are all real and symmetric, the σ_j^M are all pure imaginary

$$(\sigma_j^M)^* = -\sigma_j^M. \tag{93}$$

Hence

$$(\sigma_n^M)^* = -\sigma_n^M. \tag{93a}$$

From Eqs. (24), (25), and (30), we know the eigenvalues of σ_n to be ± 1 .

Let $\Psi_\mu^{(1)}, \Psi_\mu^{(2)}, \Psi_\mu^{(3)}, \Psi_\mu^{(4)}$ be solutions of the Dirac equation for the following states

$$\left\{ \begin{array}{c} \Psi_\mu^{(1)} \\ \Psi_\mu^{(2)} \end{array} \right\} \quad \text{are solutions for} \quad p = |\mathbf{p}|, \quad E_+, \quad \sigma_n^M = \left\{ \begin{array}{c} 1 \\ -1 \end{array} \right\}, \tag{94a}$$

$$\left\{ \begin{array}{c} \Psi_\mu^{(3)} \\ \Psi_\mu^{(4)} \end{array} \right\} \quad \text{are solutions for} \quad p = -|\mathbf{p}|, \quad E_-, \quad \sigma_n^M = \left\{ \begin{array}{c} -1 \\ 1 \end{array} \right\}. \tag{94b}$$

The eigenvalue equation for σ_n are

$$\begin{aligned} \sigma_n^M \Psi_\mu^{(1)} &= \Psi_\mu^{(1)}, \\ \sigma_n^M \Psi_\mu^{(2)} &= -\Psi_\mu^{(2)}, \\ \sigma_n^M \Psi_\mu^{*(1)} &= -\Psi_\mu^{*(1)}, \qquad \Psi^{*(1)} = \Psi^{(3)}; \\ \sigma_n^M \Psi_\mu^{*(2)} &= \Psi_\mu^{*(2)}, \qquad \Psi^{*(2)} = \Psi^{(4)}. \end{aligned} \tag{95}$$

Thus in the Majorana representation, Ψ_μ are separated into two pairs

$$\begin{pmatrix} \Psi^{(1)} \\ \Psi^{(2)} \end{pmatrix} \quad \text{and} \quad \begin{pmatrix} \Psi^{(3)} \\ \Psi^{(4)} \end{pmatrix} = \begin{pmatrix} \Psi^{*(1)} \\ \Psi^{*(2)} \end{pmatrix}.$$

Thus, if ψ_M is a solution of energy $E_\pm$ and Majorana spin $\sigma_n^M (= \pm 1)$, then ψ_M^* is also a Majorana spinor of energy $E_\mp$ and Majorana spin $-\sigma_n^M$. This aspect turns out to be of some importance in constructing theories for neutrinos.

References

Pauli, W., *Article in Handbuch der Physik Bd.* 24 (1933), or *Encycl. of Physics*, Vol. 5/1, (1958).

Good, R. H., *Reviews of Mod. Phys.* **27**, 187 (1955).

Eisele, J.A., *Modern Quantum Mechanics with Applications to Elementary Particle Physics* (Pergamon Press, New York, 1969).

Schiff, L.I., *Quantum Mechanics*, 3rd Edition (McGraw-Hill, New York, 1968).

Davydov, A. S., *Quantum Mechanics*, 2nd Edition (Pergamon Press, Oxford, 1965).

Exercises: *Chapter 3*

1. Theorem 1 states that the lowest order of 4 matrices satisfying the conditions (1a) and (1b) is 4 × 4. It is also of critical importance to ask whether the structure of the Dirac equation is *unique*. To this end, it is useful to examine Theorems 2-17 to see if Theorems 15-17 can be derived without any reference to the dimension of the γ matrices.

2. Demonstrate that the charge conjugation operator defined in Section 3.3 is unique up to a phase.

3. Assume that Ψ^M is a solution to the free Dirac equation in the Majorana representation. Obtain a solution to the free Dirac equation in the Pauli-Dirac representation, which can be expressed in terms of Ψ^M. Discuss the property of such solution under charge conjugation. You may now turn around the argument by showing a way to obtain the solution to the Dirac equation in the Majorana representation from that in the Pauli-Dirac representation.

Chapter 4.
Transformations of Dirac Equations

The properties of the Dirac relativistic equations of a charge particle under various transformtions such as gauge transformation of the electromagnetic fields, Lorentz transformations, space inversion will be the subject of our discussion in this chapter. The charge conjugation treated in the preceding chapter is another transformation.

§.4.1. Unitary transformations

Theorem: Under the unitary transformation

$$\alpha'_\mu = S^{-1}\alpha_\mu S, \quad \beta' = S^{-1}\beta S, \tag{1}$$

$$\Psi' = S^{-1}\Psi, \quad \Psi'^\dagger = \Psi^\dagger S, \tag{2}$$

the form of the Dirac equation is invariant.

We take the form Eq. (16) of Ch. 3,[1]

$$\frac{\partial \Psi'}{\partial t} + (\alpha' \cdot \nabla)\Psi' + im_0\beta'\Psi' = 0, \tag{3}$$

$$\frac{\partial \Psi'^\dagger}{\partial t} + (\nabla \Psi'^\dagger \cdot \alpha') - im_0\Psi'^\dagger \beta' = 0. \tag{4}$$

The theorem follows from the four equations (1)-(4).

§.4.2. Gauge transformations

Theorem: Under the transformation of the 4-potential[2]

$$A' = A + \nabla\chi, \quad \phi' = \phi - \frac{\partial\chi}{\partial t}, \tag{5}$$

$$\partial_\mu\partial_\mu\chi = 0 \text{ with } \chi \text{ a scalar function,}$$

1

$$\frac{\partial \Psi'}{\partial t} + c(\alpha' \cdot \nabla)\Psi' + \frac{im_0c^2}{\hbar}\beta'\Psi' = 0. \tag{3'}$$

2

$$\phi' = \phi - \frac{1}{c}\frac{\partial\chi}{\partial t}. \tag{5'}$$

and the simultaneous transformation[3]

$$\Psi' = \Psi \exp(-ie\chi), \tag{6}$$

the Dirac equation remains unchanged in form.[4]

We take the form Eqs. (24a) and (23), Ch. 2,

$$\{\gamma_\mu(\frac{1}{i}\,\frac{\partial}{\partial x_\mu} + eA'_\mu) - im_0\}\Psi' = 0. \tag{7}$$

The theorem follows from Eqs. (5), (6), and (7) .

§.4.3. Lorentz transformations

The Lorentz transformation

$$x' = \gamma(x - \beta t), \quad y' = y, \quad z' = z, \quad t' = \gamma(t - \beta x),$$

$$x = \gamma(x' + \beta t'), \quad y = y', \quad z = z', \quad t = \gamma(t' + \beta x'), \tag{8a}$$

$$\beta = \frac{v}{c}, \quad \gamma = \frac{1}{\sqrt{1 - \beta^2}}, \tag{8b}$$

3

$$\Psi' = \Psi \exp(-\frac{ie}{\hbar c}\chi). \tag{6'}$$

[4] *Eqs. (5) and (6) are referred to as "gauge transformation of the second kind". The invariance of the equation under such transformation is called "gauge invariance". It is now known that gauge field theories, i.e., field theories which respect gauge invariance, are of fundamental importance. However, it remains mysterious why theories with extra unphysical gauge degrees of freedom are so successful. Perhaps there is a much deeper meaning in connection with the gauge degree of freedom but any such meaning is yet to be found.*

will be put in the form[5],[6]

$$x'_\mu = a_{\mu\nu}x_\nu, \qquad x_4 = it,$$
$$x_\nu = x'_\mu a_{\mu\nu}, \qquad x'_4 = it', \tag{8c}$$

$$a_{\mu\rho}a_{\nu\rho} = \delta_{\mu\nu}, \tag{8d}$$

$$\det|\,a_{\mu\nu}\,| = 1, \tag{8e}$$

or

$$\begin{pmatrix} x'_1 \\ x'_2 \\ x'_3 \\ x'_4 \end{pmatrix} = \begin{pmatrix} \gamma & 0 & 0 & i\gamma\beta \\ 0 & 1 & 0 & 0 \\ 0 & 0 & 1 & 0 \\ -i\gamma\beta & 0 & 0 & \gamma \end{pmatrix} \begin{pmatrix} x_1 \\ x_2 \\ x_3 \\ x_4 \end{pmatrix}, \tag{8f}$$

$$(x_1, x_2, x_3, x_4) = (x'_1, x'_2, x'_3, x'_4) \begin{pmatrix} \gamma & 0 & 0 & i\gamma\beta \\ 0 & 1 & 0 & 0 \\ 0 & 0 & 1 & 0 \\ -i\gamma\beta & 0 & 0 & \gamma \end{pmatrix}. \tag{8g}$$

In the following, we shall use the notation[7]

$$\Psi^\dagger = i\tilde\Psi^*\gamma_4, \quad \text{or} \quad \bar\Psi = \tilde\Psi^*\gamma_4. \tag{9}$$

Dirac's equations are Eqs. (22a) and (24a), Ch. 2,

$$\left(\gamma_\mu \frac{\partial}{\partial x_\mu} + m_0\right)\Psi = 0, \tag{10}$$

---

[5] *Eq. (8a) states that the Jacobian of the transformation Eq. (8a), or Eqs. (8f) and (8g) is +1 . This condition restricts the transformations to proper Lorentz transformations. For space inversion and time reversal, the Jacobian is -1 . See Sect. 4 below.*

[6]

$$x'_\mu = a_{\mu\nu}x_\nu, \qquad x_4 = ict. \tag{8c'}$$

[7] *The notation $\Psi^\dagger$ here is that of Pauli in the Handbuch der Physik, Bd 24/1, (1933). $\Psi^\dagger, \tilde\Psi^*$ are row matrices. In Pauli's article, Ψ^* is a row matrix, so that Eq. (9) is $\Psi^\dagger = i\Psi^*\gamma_4$.*

$$(\gamma_\mu \Pi_\mu - im_0)\Psi = 0, \tag{11}$$

where[8]

$$\Pi_k = \frac{1}{i}\frac{\partial}{\partial x_k} + eA_k, \quad \Pi_4 = \frac{1}{i}\frac{\partial}{i\partial t} + ie\phi, \tag{12}$$

and

$$\Pi_\mu \Pi_\nu - \Pi_\nu \Pi_\mu = \frac{e}{i}F_{\mu\nu} \tag{13}$$

$$F_{\mu\nu} = \frac{\partial A_\nu}{\partial x_\mu} - \frac{\partial A_\mu}{\partial x_\nu} = \begin{pmatrix} 0 & H_z & -H_y & -iE_x \\ -H_z & 0 & H_x & -iE_y \\ H_y & -H_x & 0 & -iE_z \\ iE_x & iE_y & iE_z & 0 \end{pmatrix} = -F_{\nu\mu} \tag{14}$$

In terms of the $\Psi^\dagger$ in Eq. (9), we obtain from Eqs. (10) and (11)

$$\frac{\partial \Psi^\dagger}{\partial x_\mu}\gamma_\mu - m_0\Psi^\dagger = 0, \tag{10a}$$

$$\Pi_\mu^* \Psi^\dagger \gamma_\mu - im_0\Psi^\dagger = 0. \tag{11a}$$

We define[9] the 4-current s_μ

$$s_\mu = \Psi^\dagger \gamma_\mu \Psi \tag{15}$$

and the 4-divergence

$$\frac{\partial s_\mu}{\partial x_\mu} = 0. \tag{16}$$

On making the Lorentz transformation Eq. (8c)

---
[8]

$$\Pi_k = \frac{\hbar}{i}\frac{\partial}{\partial x_k} + \frac{e}{c}A_k, \quad \Pi_4 = \frac{\hbar}{i}\frac{\partial}{ic\partial t} + i\frac{e}{c}\phi. \tag{12'}$$

[9] *The proof that s_μ so defined is a 4-vector under Lorentz transformation will be given in the following. The J_μ in Eq. (82), Ch. 3, is in fact s_μ, except for a constant factor*

$$J_\mu = -es_\mu.$$

$$x'_\mu = a_{\mu\nu}x_\nu, \quad \text{or} \quad x_\nu = x'_\mu a_{\mu\nu}$$

and

$$\Psi' = S\Psi, \qquad \Psi^{\dagger'} = \Psi^\dagger S^{-1}, \tag{17}$$

then Eqs. (10) and (10a) become[10]

$$\gamma_\nu \frac{\partial}{\partial x'_\nu}\Psi' + m_0\Psi' = 0, \tag{18}$$

$$\frac{\partial \Psi^{\dagger'}}{\partial x'_\nu}\gamma_\nu - m_0\Psi^{\dagger'} = 0, \tag{19}$$

where

$$\gamma_\nu = (S^{-1}\gamma_\mu S)a_{\mu\nu}, \tag{20}$$

which can be written

$$S^{-1}\gamma_\mu S = a_{\mu\nu}\gamma_\nu. \tag{20a}$$

From Eqs. (9) and (17), we have

$$\Psi^{\dagger'} = i\tilde{\Psi}^*\gamma_4 S^{-1}. \tag{21}$$

We also have

$$\Psi^{\dagger'} = i\tilde{\Psi}^{*'}\gamma_4.$$

But from Eq. (17),

$$\tilde{\Psi}^{*'} = \tilde{\Psi}^* S^\dagger, \quad (S^\dagger = \tilde{S}^*).$$

Hence

$$\Psi^{\dagger'} = i\tilde{\Psi}^* S^\dagger \gamma_4. \tag{22}$$

10

$$(\gamma_\nu \frac{\partial}{\partial x'_\nu} + \frac{m_0 c}{\hbar})\Psi' = 0, \quad \text{etc.} \tag{18'}$$

From Eqs. (21) and (22) , we have the condition

$$S^\dagger \gamma_4 S = \gamma_4. \tag{23}$$

To insure the Lorentz invariance of the Dirac equations (10), (10a), (18), and (19), the conditions are Eqs. (20a) and (23a)

$$S^{-1}\gamma_\mu S = a_{\mu\nu}\gamma_\nu, \tag{20a}$$

$$S^\dagger \gamma_4 S = \gamma_4 \quad \text{(for proper Lorentz transformation).} \tag{23a}$$

We have to show that an S exists that satisfies these conditions.

Before we do that, we shall find the transformation properties of the 16 Γ_A's in Eq. (2), Ch. 3.

(i) I

$$\Psi^{\dagger'}\Psi' = \Psi^\dagger S^{-1}S\Psi = \Psi^\dagger\Psi, \quad \text{an invariant.}$$

(ii) $\gamma_1, \gamma_2, \gamma_3, \gamma_4$

$$\Psi^{\dagger'}\gamma_\mu\Psi' = \Psi^\dagger S^{-1}\gamma_\mu S\Psi = a_{\mu\nu}\Psi^\dagger\gamma_\nu\Psi, \quad \text{a 4-vector.}$$

(iii) $\gamma_1\gamma_2, \gamma_2\gamma_3, \gamma_3\gamma_4, \gamma_1\gamma_4, \gamma_2\gamma_4, \gamma_1\gamma_3$

$$M'_{\mu\nu} = \Psi^{\dagger'}\gamma_\mu\gamma_\nu\Psi' = \Psi^+ S^{-1}\gamma_\mu S S^{-1}\gamma_\nu S\Psi$$

$$= a_{\mu\rho}a_{\nu\sigma}M_{\rho\sigma}, \quad \text{an antisymmetric tensor (or, a 6-vector).} \tag{24}$$

(iv) $\gamma_1\gamma_2\gamma_3, \gamma_2\gamma_3\gamma_4, \gamma_1\gamma_3\gamma_4, \gamma_1\gamma_2\gamma_4$

$$M'_{\lambda\mu\nu} = \Psi^{\dagger'}S^{-1}\gamma_\lambda S S^{-1}\gamma_\mu S S^{-1}\gamma_\nu S\Psi$$

$$= a_{\lambda\rho}a_{\mu\sigma}a_{\nu\tau}M_{\rho\sigma\tau}, \quad \text{an antisymmetric tensor of the third rank.}$$

(v) $\gamma_1\gamma_2\gamma_3\gamma_4 \equiv \gamma_5$

$$\Psi^{\dagger'}\gamma_1\gamma_2\gamma_3\gamma_4\Psi' = \Psi^\dagger\gamma_1\gamma_2\gamma_3\gamma_4\Psi,$$

or

$$\Psi^{\dagger'}\gamma_5\Psi' = \Psi^{\dagger}\gamma_5\Psi.$$

These relations hold for proper Lorentz transformations (i.e., not involving space inversion or reflection, or time reversal).

§.4.4. Space inversion, charge conjugation, and time reversal

(1) Space inversion:

$$P: \quad \mathbf{r} \rightarrow -\mathbf{r} \tag{25}$$

The transformation equations are

$$x'_{\mu} = a_{\mu\nu}x_{\nu},$$

where

$$a_{\mu\nu} = \begin{pmatrix} -1 & 0 & 0 & 0 \\ 0 & -1 & 0 & 0 \\ 0 & 0 & -1 & 0 \\ 0 & 0 & 0 & 1 \end{pmatrix}. \tag{26}$$

The conditions Eqs. (20a) and (23) are then

$$S_p\gamma_k + \gamma_k S_p = 0, \quad k = 1, 2, 3;$$
$$S_p\gamma_4 - \gamma_4 S_p = 0, \tag{27}$$
$$S_p^{\dagger}\gamma_4 - \gamma_4 S_p^{-1} = 0.$$

These relations can be satisfied by choosing

$$S_p = \lambda\gamma_4, \quad \lambda = \text{constant.} \tag{28}$$

As two successive space inversions are equivalent to an identity transformation, one has, for $S_p(S_p\psi) = +\psi$,

$$(S_p)^2 = \lambda^2\gamma_4^2 = \lambda^2 = 1. \tag{29a}$$

But an inversion is equivalent to a reflection on a plane followed by a rotation through π about an axis perpendicular to that plane; two successive inversions are equivalent to a rotation through an angle 2π . It will be shown later that the S_r for rotation about an axis $j = 1,2,3$ (for x,y,z respectively) through an angle w is[11]

$$S_{rj} = \cos \frac{w}{2} + i\sigma_j \sin \frac{w}{2}.$$

and for $w = 2\pi$,

$$S_{2\pi} = -1.$$

Hence one also has

$$(S_p)^2 = S_{2\pi} = -1, \tag{29b}$$

which corresponds to $S_p(S_p\psi) = -\psi$.

From Eqs. (29a) and (29b) , one has 4 solutions[12]

$$\lambda = i, \; -i, \; 1, \; -1. \tag{30}$$

From Eq. (17), $\Psi' = S\Psi$ and Eq. (29), one has

$$\Psi' = \lambda\gamma_4\Psi \;\; \text{or} \;\; \psi'(\mathbf{r},t) = \lambda\gamma_4\psi(\mathbf{r},t), \tag{31}$$

or, on taking the complex conjugate,

$$\Psi^{*'} = \lambda^*\gamma_4^*\Psi^*, \tag{31a}$$

or,

$$\tilde{\Psi}^{*'} = \lambda^*\tilde{\Psi}^*\gamma_4, \quad \Psi^{\dagger'} = \lambda^*\Psi^\dagger\gamma_4.$$

The properties of the 16 Γ'_A in Eq. (2) of Ch. 3 under inversion are as follows,

$$\Psi^{\dagger'}\Psi' = \Psi^\dagger\Psi,$$

---

[11] See Eq. (58) below. For σ_j, $j = 1,2,3$, see Eq. (80e), Ch. 2.

[12] Cf. C. N. Yang and J. Tiomno, Phys. Rev. 79, 495 (1950).

$$\begin{cases} \Psi^{\dagger'}\gamma_k\Psi' = -\Psi^\dagger\gamma_k\Psi, & k = 1,2,3; \\ \Psi^{\dagger'}\gamma_4\Psi' = \Psi^\dagger\gamma_4\Psi, & \text{a vector.} \end{cases}$$

$$\begin{cases} \Psi^{\dagger'}\gamma_k\gamma_j\Psi' = \Psi^\dagger\gamma_k\gamma_j\Psi, \\ \Psi^{\dagger'}\gamma_k\gamma_4\Psi' = -\Psi^\dagger\gamma_k\gamma_4\Psi; \end{cases} \tag{32}$$

$$\Psi^{\dagger'}\gamma_1\gamma_2\gamma_3\Psi' = -\Psi^\dagger\gamma_1\gamma_2\gamma_3\Psi,$$

$$\Psi^{\dagger'}\gamma_k\gamma_j\gamma_4\Psi' = \Psi^\dagger\gamma_k\gamma_j\gamma_4\Psi \quad \text{an axial vector,}$$

$$\Psi^{\dagger'}\gamma_1\gamma_2\gamma_3\gamma_4\Psi' = -\Psi^\dagger\gamma_1\gamma_2\gamma_3\gamma_4\Psi, \quad \text{a pseudo-scalar.}$$

In Eqs. (44b) and (46), Ch. 3, it is seen that $-\gamma_5$ and the helicity operator $(\sigma \cdot \mathbf{n})$ have the same property in their operation on the two 2-component spinors ϕ_u, ϕ_l, namely, interchange of ϕ_u and ϕ_l. From the last relation above

$$\Psi^{\dagger'}\gamma_5\Psi' = -\Psi^\dagger\gamma_5\Psi$$

It is seen that both $\Psi^\dagger\gamma_5\Psi$ and $(\sigma \cdot \mathbf{n})$ change sign upon space inversion.

From Eqs. (73), (74), and (78), it is shown that

$$\Psi_c = \gamma_2\Psi^*, \quad \Psi^* = \gamma_2\Psi_c. \tag{33}$$

Upon space inversion

$$\Psi'_c = \gamma_2\Psi^{*'},$$

one has from Eq. (31a)

$$\begin{aligned} \Psi'_c &= \lambda^*\gamma_2\gamma_4^*\Psi^* \\ &= -\lambda^*\gamma_4\gamma_2\Psi^* \quad \text{by Eq. (66), Ch. 3} \\ &= -\lambda^*\gamma_4\Psi_c. \end{aligned} \tag{34}$$

On comparing with Eq. (31),

$$\Psi' = \lambda\gamma_4\Psi,$$

it is seen that

(i) if $\quad \lambda = \pm i, \quad \Psi_c \quad$ and $\quad \Psi \quad$ transform in the same way,

(ii) if $\quad \lambda = \pm 1, \quad \Psi_c \quad$ and $\quad \Psi \quad$ transform differently. $\tag{35}$

But

$$\tilde{\Psi}'_c \Psi' = -\lambda\lambda^* \tilde{\Psi}_c \gamma_4 \gamma_4 \Psi$$
$$= -\tilde{\Psi}_c \Psi, \tag{36}$$

showing that the intrinsic parity of Ψ (of e^-) and the intrinsic parity of Ψ_c (of the charge conjugate e^+) are always opposite to each other.[13]

The relationship between charge conjugation and parity operation is as follows.

In Eqs. (48) and (55), Ch. 3, it has been seen that for

$$\text{anti-neutrino,} \quad h = 1, \quad (\sigma \cdot \mathbf{n})\alpha = \alpha,$$
$$\text{neutrino,} \quad h = -1, \quad (\sigma \cdot \mathbf{n})\beta = -\beta, \tag{37a}$$

or,

$$(\sigma \cdot \mathbf{n})\begin{pmatrix} \alpha \\ \beta \end{pmatrix} = \begin{pmatrix} \alpha \\ -\beta \end{pmatrix}.$$

Now, upon parity (space inversion) operation P, Eq. (25), shows

$$P(\sigma \cdot \mathbf{n}) = -(\sigma \cdot \mathbf{n})P,$$
$$(\sigma \cdot \mathbf{n})(P\alpha) = -(P\alpha),$$
$$(\sigma \cdot \mathbf{n})(P\beta) = (P\beta).$$

Comparison between these relations and Eq. (37a) shows that the transformation property of $(P\,\alpha)$ (α an anti-neutrino state) is the same as β , a neutrino state, and $(P\,\beta)$ has the transformation property of α , i.e., a parity operation converts a neutrino into an anti-neutrino, and vice versa. This is expressed by the following expression

$$P\alpha \propto \beta,$$
$$P\beta \propto \alpha. \tag{37b}$$

Consider next the charge conjugation Eqs. (74) and (73), Ch. 3.

$$\Psi_c = \gamma_2 \Psi^*$$

---

[13] *By intrinsic parity, it is meant the parity apart from that pertaining to the orbital motion of a particle.*

$$\gamma_2 = \begin{pmatrix} 0 & \begin{pmatrix} 0 & -1 \\ 1 & 0 \end{pmatrix} \\ \begin{pmatrix} 0 & 1 \\ -1 & 0 \end{pmatrix} & 0 \end{pmatrix}.$$

Carrying out in succession the P and C operations and using $\alpha^* = \alpha$ and $\beta^* = \beta$, we obtain

$$CP\begin{pmatrix} \alpha \\ \beta \end{pmatrix} = CP(\sigma \cdot \mathbf{n})\begin{pmatrix} \alpha \\ -\beta \end{pmatrix} = -C(\sigma \cdot \mathbf{n})P\begin{pmatrix} \alpha \\ -\beta \end{pmatrix}$$

$$= -C(\sigma \cdot \mathbf{n})\begin{pmatrix} \beta \\ -\alpha \end{pmatrix} = C\begin{pmatrix} \beta \\ \alpha \end{pmatrix} \tag{37c}$$

$$= \gamma_2\begin{pmatrix} \beta \\ \alpha \end{pmatrix} = \begin{pmatrix} 0 & -1 \\ 1 & 0 \end{pmatrix}\left\{ \begin{matrix} \alpha \\ -\beta \end{matrix} \right\}.$$

In words, this states that the combined operation of parity P and charge

conjugation C on a $\left\{ \begin{matrix} antineutrino \\ neutrino \end{matrix} \right\}$ leads to a $\left\{ \begin{matrix} antineutrino \\ neutrino \end{matrix} \right\}$;

i.e., the nature of $\left\{ \begin{matrix} antineutrino \\ neutrino \end{matrix} \right\}$ is conserved under the combined CP operation.

(2) Time reversal:

$$T : t \rightarrow -t \tag{38}$$

Consider the Dirac equation for a particle of charge -e ,

$$\{\gamma \cdot (\mathbf{p} + e\mathbf{A}) + \gamma_4(-\frac{\partial}{\partial t} + ie\phi) - im_0\}\psi = 0. \tag{39a}$$

under the time reversal transformation, $t \rightarrow t' = -t$, we have $\mathbf{A}(-t) = -\mathbf{A}(t)$ and $\phi(-t) = \phi(t)$ so that

$$\{\gamma \cdot (\mathbf{p} - e\mathbf{A}(t)) + \gamma_4(\frac{\partial}{\partial t} + ie\phi) - im_0\}\psi(-t) = 0. \tag{39b}$$

Introduce

$$T\psi(-t) = \psi'(t). \tag{40}$$

We find, from Eq. (39b),

$$T(-i\gamma)T^{-1} \cdot \nabla\psi'(t) - T\gamma T^{-1} \cdot e\mathbf{A}(t)\psi'(t)$$
$$+ T\gamma_4 T^{-1}\frac{\partial}{\partial t'}\psi'(t) + Ti\gamma_4 T^{-1}e\phi(t)\psi'(t) \qquad (41)$$
$$- im_0\psi'(t) = 0.$$

In order that Eq. (41) reduces to Eq. (39a), we obtain

$$\begin{aligned} Ti\vec{\gamma}T^{-1} &= -i\vec{\gamma}, \\ T\gamma T^{-1} &= \gamma, \\ T\gamma_4 T^{-1} &= \gamma_4, \\ Ti\gamma_4 T^{-1} &= -i\gamma_4. \end{aligned} \qquad (42)$$

In the Pauli-Dirac representation of γ-matrices, we find

$$T = T^0 K, \qquad (43)$$

with K the complex conjugation operator and $T_0 = \lambda_T i\gamma_1\gamma_3$. (λ_T is a phase factor.) Or, we have

$$T\psi(\mathbf{x}, t) = \psi'(\mathbf{x}, -t) = \lambda_T i\gamma_1\gamma_3\psi^*(\mathbf{x}, t). \qquad (44)$$

§.4.5. The transformation matrix S

In Sect. 3, it is shown that to establish the Lorentz invariance of the form of the Dirac equations, one has only to show the existence of a transformation S that satisfies the conditions Eqs. (20) and (23a). In Sect. 4, we have the S for the special transformation of space inversion, or parity, operation P,

$$S_p = \lambda\gamma_4, \quad \lambda = i, -i, 1, -1. \qquad (33)$$

In the present section we shall obtain the S for other special Lorentz transformations.

(1) Infinitesimal Lorentz transformations
    Let the

$$a_{\mu\nu} = \delta_{\mu\nu} + \epsilon_{\mu\nu}, \qquad (45)$$

where

$$\epsilon_{\mu\nu} = -\epsilon_{\nu\mu}, \quad |\epsilon_{\mu\nu}| \ll 1. \tag{46}$$

The orthonormal condition Eq. (8d) is then

$$
\begin{aligned}
a_{\mu\nu}a_{\sigma\nu} &= (\delta_{\mu\nu} + \epsilon_{\mu\nu})(\delta_{\sigma\nu} + \epsilon_{\sigma\nu}) \\
&= \delta_{\mu\sigma} + \epsilon_{\sigma\mu} + \epsilon_{\mu\sigma} + O(\epsilon^2) \\
&= \delta_{\mu\sigma}
\end{aligned} \tag{47}
$$

Let $T^{\mu\nu}$ be a tensor, and $T^{\nu\mu}$ a tensor equal to $-T^{\mu\nu}$ and define a 4×4 matrix[14]

$$S = I + \frac{1}{2}\epsilon_{\mu\nu}T^{\mu\nu}, \tag{48a}$$

where I is a unit matrix, and

$$S^{-1} = I - \frac{1}{2}\epsilon_{\mu\nu}T^{\mu\nu}. \tag{48b}$$

The condition (20a) is then

$$(I - \frac{1}{2}\epsilon_{\mu\nu}T^{\mu\nu})\gamma_\rho(I + \frac{1}{2}\epsilon_{\alpha\beta}T^{\alpha\beta}) = (\delta_{\rho\sigma} + \epsilon_{\rho\sigma})\gamma_\sigma,$$

or

$$\gamma_\rho + \frac{1}{2}\epsilon_{\mu\nu}(\gamma_\rho T^{\mu\nu} - T^{\mu\nu}\gamma_\rho) = \gamma_\rho + \epsilon_{\rho\sigma}\gamma_\sigma$$

$$\frac{1}{2}\epsilon_{\mu\nu}(\gamma_\rho T^{\mu\nu} - T^{\mu\nu}\gamma_\rho) = \frac{1}{2}\epsilon_{\mu\nu}(\delta_{\mu\rho}\gamma_\nu - \delta_{\nu\rho}\gamma_\mu).$$

For this to hold for arbitrary $\epsilon_{\mu\nu}$, one must have

$$\gamma_\rho T^{\mu\nu} - T^{\mu\nu}\gamma_\rho = \delta_{\mu\rho}\gamma_\nu - \delta_{\nu\rho}\gamma_\mu. \tag{49}$$

The condition (23a) is then

$$
\begin{aligned}
S^\dagger &= \gamma_4 S^{-1}\gamma_4 \\
&= I - \frac{1}{2}\epsilon_{\mu\nu}\gamma_4 T^{\mu\nu}\gamma_4.
\end{aligned} \tag{50}
$$

---

[14] Note that μ, ν are not row-column indices of the tensor T. $T^{\mu\nu}$ is the name of the tensor itself. See Eq. (51) below.

From Eq. (48a),

$$S^\dagger = I + \frac{1}{2}\epsilon^*_{\mu\nu}T^{\dagger\mu\nu}. \tag{51}$$

Hence

$$\epsilon^*_{\mu\nu}T^{\dagger\mu\nu} = -\epsilon_{\mu\nu}\gamma_4 T^{\mu\nu}\gamma_4. \tag{52}$$

From Eq. (8f), one sees the generalized $a_{\mu\nu}$ to have the properties

$$\begin{aligned}
\epsilon_{ji} &= -\epsilon_{ij} = \text{real}, \quad i,j = 1,2,3 \\
\epsilon_{4j} &= -\epsilon_{j4} = \text{imaginary}, \quad j = 1,2,3.
\end{aligned} \tag{53}$$

From these and Eq. (52), one has

$$\begin{aligned}
T^{\dagger ij} &= -\gamma_4 T^{ij}\gamma_4, \\
T^{\dagger 4j} &= \gamma_4 T^{4j}\gamma_4.
\end{aligned} \tag{54}$$

It is seen that

$$T^{\mu\nu} = \frac{1}{2}\gamma_\mu\gamma_\nu \tag{55}$$

satisfies Eqs. (49) and (54). Thus

(i) Substituting Eq. (55) into (49), one gets

$$\frac{1}{2}\left(\gamma_\rho\gamma_\mu\gamma_\nu - \gamma_\mu\gamma_\nu\gamma_\rho\right) = \begin{cases} 0 & \text{for } \rho \neq \mu, \rho \neq \nu, \\ \gamma_\nu & \text{for } \rho = \mu, \rho \neq \nu, \\ -\gamma_\mu & \text{for } \rho \neq \mu, \rho = \nu, \end{cases}$$

$$\delta_{\mu\rho}\gamma_\nu - \delta_{\nu\rho}\gamma_\mu = \begin{cases} 0 & \text{for } \rho \neq \mu, \rho \neq \nu, \\ \gamma_\nu & \text{for } \rho = \mu, \rho \neq \nu, \\ -\gamma_\mu & \text{for } \rho \neq \mu, \rho = \nu. \end{cases} \quad \text{q.e.d.}$$

(ii)

$$T^{\dagger ij} = \frac{1}{2}\gamma_j^\dagger\gamma_i^\dagger = \frac{1}{2}\gamma_j\gamma_i,$$

$$-\gamma_4 T^{ij}\gamma_4 = -\frac{1}{2}\gamma_4\gamma_i\gamma_j\gamma_4 = -\frac{1}{2}\gamma_i\gamma_j = \frac{1}{2}\gamma_j\gamma_i,$$

$$\frac{1}{2}\gamma_4\gamma_4\gamma_j\gamma_4 = \frac{1}{2}\gamma_j\gamma_4 = T^{j4}. \quad \text{q.e.d.}$$

Hence

$$S = I + \frac{1}{4}\epsilon_{\mu\nu}\gamma_\mu\gamma_\nu, \tag{56}$$

insures the Lorentz invariance under infinitesimal Lorentz transformations.

(2) Finite rotations in 3-dimensional space

(i) Rotation about z-axis through an angle w

$$a_{\mu\nu} = \begin{pmatrix} cosw & sinw & 0 & 0 \\ -sinw & cosw & 0 & 0 \\ 0 & 0 & 1 & 0 \\ 0 & 0 & 0 & 1 \end{pmatrix} \tag{57}$$

For infinitesimal rotation $\triangle w$,

$$\epsilon_{12} = -\epsilon_{21} = \triangle w, \quad \text{all other } \epsilon_{\mu\nu} = 0.$$

$S_z(\triangle w)$ from Eq. (56) is

$$S_z(\triangle w) = I + \frac{1}{2}\triangle w\gamma_1\gamma_2$$

For finite rotation $w = \lim_{n\to\infty} n\triangle w$,

$$\begin{aligned} S_z(w) &= \lim_{n\to\infty} \Pi S(\triangle w) = \lim_{n\to\infty} (1 + \frac{1}{2}\triangle w\gamma_1\gamma_2)^n \\ &= \lim_{n\to\infty} (1 + \frac{w}{2n}\gamma_1\gamma_2)^n \\ &= e^{\frac{1}{2}w\gamma_1\gamma_2} \\ &= \{1 - \frac{1}{2!}(\frac{w}{2})^2 + \frac{1}{4!}(\frac{w}{2})^4 - ...\} + \gamma_1\gamma_2\{(\frac{w}{2}) - \frac{1}{3!}(\frac{w}{2})^3 + ...\} \\ &= I cos\frac{w}{2} + \gamma_1\gamma_2 sin\frac{w}{2} \tag{58} \\ &= I cos\frac{w}{2} + i \begin{pmatrix} 1 & 0 & 0 & 0 \\ 0 & -1 & 0 & 0 \\ 0 & 0 & 1 & 0 \\ 0 & 0 & 0 & -1 \end{pmatrix} sin\frac{w}{2} \\ &= \begin{pmatrix} e^{i\frac{w}{2}} & 0 & 0 & 0 \\ 0 & e^{-i\frac{w}{2}} & 0 & 0 \\ 0 & 0 & e^{i\frac{w}{2}} & 0 \\ 0 & 0 & 0 & e^{-i\frac{w}{2}} \end{pmatrix}. \tag{58a} \end{aligned}$$

(ii) Rotation about the x-axis through an angle w

$$a_{\mu\nu} = \begin{pmatrix} 1 & 0 & 0 & 0 \\ 0 & cosw & sinw & 0 \\ 0 & -sinw & cosw & 0 \\ 0 & 0 & 0 & 1 \end{pmatrix}, \tag{59}$$

$$S_x(w) = cos\frac{w}{2} + \gamma_2\gamma_3 sin\frac{w}{2} \tag{60}$$

$$= \begin{pmatrix} cos\frac{w}{2} & isin\frac{w}{2} & 0 & 0 \\ isin\frac{w}{2} & cos\frac{w}{2} & 0 & 0 \\ 0 & 0 & cos\frac{w}{2} & isin\frac{w}{2} \\ 0 & 0 & isin\frac{w}{2} & cos\frac{w}{2} \end{pmatrix}. \tag{60a}$$

(iii) Rotation about the y-axis through an angle w

$$a_{\mu\nu} = \begin{pmatrix} cosw & 0 & -sinw & 0 \\ 0 & 1 & 0 & 0 \\ sinw & 0 & cosw & 0 \\ 0 & 0 & 0 & 1 \end{pmatrix}, \tag{61}$$

$$S_y(w) = cos\frac{w}{2} + \gamma_3\gamma_1 sin\frac{w}{2} \tag{62}$$

$$= \begin{pmatrix} cos\frac{w}{2} & sin\frac{w}{2} & 0 & 0 \\ -sin\frac{w}{2} & cos\frac{w}{2} & 0 & 0 \\ 0 & 0 & cos\frac{w}{2} & sin\frac{w}{2} \\ 0 & 0 & -sin\frac{w}{2} & cos\frac{w}{2} \end{pmatrix}. \tag{62a}$$

(iv) A general rotation

Let $OX'Y'Z'$ be a rectangular coordinate system fixed in space, and $OXYZ$ be a rotating system.

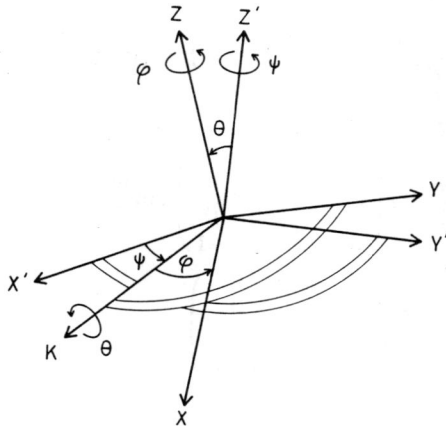

Initially, the two systems of axes coincide at the position indicated as $OX'Y'Z'$.

First rotate the $OXYZ$ system about OZ' through an angle ψ, thereby bringing OX from OX' to OK. Next rotate the $OXYZ$ system about OK through an angle ϑ, thereby bringing the OZ axis from OZ' to OZ in the figure. Finally rotate the $OXYZ$ system about the OZ axis through an angle φ, thereby bringing the OX axis at OK to OX in the figure. The angles ψ, ϑ, φ are the Euler angles.

Now rotations form a non-abelian group, so that the order of the rotations is relevant. If the three rotations above are performed in the order described, then it can be shown that the product

$$S(\varphi)S(\vartheta)S(\psi)$$

of the transformation matrices is given by

$$S(\varphi,\vartheta,\psi) = \begin{pmatrix} \alpha & \gamma & 0 & 0 \\ \beta & \delta & 0 & 0 \\ 0 & 0 & \alpha & \gamma \\ 0 & 0 & \beta & \delta \end{pmatrix} \tag{63}$$

where

$$\alpha = \cos\frac{\vartheta}{2}exp(-i\frac{\psi+\varphi}{2}), \quad \beta = -isin\frac{\vartheta}{2}exp(i\frac{\psi-\varphi}{2}),$$

$$\gamma = -isin\frac{\vartheta}{2}exp(-i\frac{\psi-\varphi}{2}), \quad \delta = \cos\frac{\vartheta}{2}exp(i\frac{\psi+\varphi}{2}), \tag{64}$$

are called the Cayley-Klein parameters. They satisfy the relation

$$\alpha\delta - \beta\gamma = 1. \tag{65}$$

The adjoint is

$$S^\dagger = \begin{pmatrix} \delta & -\gamma & 0 & 0 \\ -\beta & \alpha & 0 & 0 \\ 0 & 0 & \delta & -\gamma \\ 0 & 0 & -\beta & \alpha \end{pmatrix} \tag{66}$$

and it is seen that

$$SS^\dagger = S^\dagger S = 1 \tag{67}$$

so that S is unitary.

It is to be emphasized that in all the rotation matrices in Eqs. (58), (60), (62) and (63), the half-angles $\frac{w}{2}, \frac{\vartheta}{2}, \frac{\psi}{2}, \frac{\varphi}{2}$ appear.[15] The $\begin{pmatrix} \alpha & \gamma \\ \beta & \delta \end{pmatrix}$ is the transformation matrix for spinors — 2-component quantities. The appearance of spinors, as already referred to in Eq. (117) of Ch. 2 and Eqs. (27), (47) and (59) of Ch. 3, is characteristic of the Dirac equations.

(3) Special Lorentz transformation Eq. (8a)

For the special case of two inertial systems in uniform relative motion along their x-axes, the Lorentz transformation is Eq. (8a). We define an angle ϑ by

$$cosh\,\vartheta = \gamma, \quad sinh\,\vartheta = \beta\gamma,$$

$$\beta = \frac{v}{c} < 1, \quad \gamma = \frac{1}{\sqrt{1 - \beta^2}}, \tag{69}$$

so that

$$a_{\mu\nu} = \begin{pmatrix} cosh\,\vartheta & 0 & 0 & isinh\,\vartheta \\ 0 & 1 & 0 & 0 \\ 0 & 0 & 1 & 0 \\ -isinh\,\vartheta & 0 & 0 & cosh\,\vartheta \end{pmatrix} \tag{70}$$

By a generalization of the results Eqs. (57), (58), and (58a), one obtains

$$S(\vartheta) = I cosh\frac{\vartheta}{2} + i\gamma_1\gamma_4 sinh\frac{\vartheta}{2} \tag{71}$$

$$= \begin{pmatrix} cosh\frac{\vartheta}{2} & 0 & 0 & -sinh\frac{\vartheta}{2} \\ 0 & cosh\frac{\vartheta}{2} & -sinh\frac{\vartheta}{2} & 0 \\ 0 & -sinh\frac{\vartheta}{2} & cosh\frac{\vartheta}{2} & 0 \\ -sinh\frac{\vartheta}{2} & 0 & 0 & cosh\frac{\vartheta}{2} \end{pmatrix}. \tag{71a}$$

---

[15] *The expressions (60), (62) and (58) for the $S_x(w), S_y(w), S_z(w)$ can be expressed in terms of the $\sigma_1, \sigma_2, \sigma_3$ matrices by means of the relations (80e) of Ch. 2, namely*

$$S_x(w) = exp(i\frac{w}{2}\sigma_1),$$

$$S_y(w) = exp(i\frac{w}{2}\sigma_2), \tag{68}$$

$$S_z(w) = exp(i\frac{w}{2}\sigma_3).$$

For the inverse transformation, $S^{-1}(\vartheta)$ can be obtained from $S(\vartheta)$ by replacing the velocity v by $-v$ (i.e., β by $-\beta$, and ϑ by $-\vartheta$), and one sees from Eq. (71a) that

$$S^{-1}(\vartheta) \neq S^{\dagger}(\vartheta), \tag{72}$$

i.e., $S(\vartheta)$ for the special Lorentz transformation is not unitary.

The transformation of ψ under the special Lorentz transformation (8a) or (70) is, from Eq. (17),

$$\Psi' = S\Psi, \tag{73}$$

and with S given by Eq. (71a), one can write this transformation equation in the form

$$\begin{pmatrix} \Psi_1' \\ \Psi_2' \\ \Psi_3' \\ \Psi_4' \end{pmatrix} = \begin{pmatrix} c\Psi_1 - s\Psi_4 \\ c\Psi_2 - s\Psi_3 \\ -s\Psi_2 + c\Psi_3 \\ -s\Psi_1 + c\Psi_4 \end{pmatrix}, \quad c = \cosh\frac{\vartheta}{2}, \quad s = \sinh\frac{\vartheta}{2}, \tag{74}$$

which can be split up into a pair of equations

$$\begin{pmatrix} \Psi_1' \\ \Psi_4' \end{pmatrix} = \begin{pmatrix} c & -s \\ -s & c \end{pmatrix} \begin{pmatrix} \Psi_1 \\ \Psi_4 \end{pmatrix},$$
$$\begin{pmatrix} \Psi_2' \\ \Psi_3' \end{pmatrix} = \begin{pmatrix} c & -s \\ -s & c \end{pmatrix} \begin{pmatrix} \Psi_2 \\ \Psi_3 \end{pmatrix}. \tag{75}$$

The $\begin{pmatrix} \Psi_1 \\ \Psi_4 \end{pmatrix}$, $\begin{pmatrix} \Psi_2 \\ \Psi_3 \end{pmatrix}$ are two pairs of 2-components spinors (of rank 1); they transform according to Eq. (75), or[16]

$$\phi_1' = b_{11}\phi_1 + b_{12}\phi_2,$$
$$\phi_2' = b_{21}\phi_1 + b_{22}\phi_2, \tag{76a}$$

---

[16] *The transformation (71a), (or in the form (75)) and the transformation (63) involve the half-angles of the $a_{\mu\nu}$, are not the transformation laws of vectors or tensors. In the theory of groups, tensors do not constitute all the representations of the Lorentz group; the transformations of the ψ of the Dirac equation form another representation of the Lorentz group - namely the spinors, characterized by Eqs. (76a) and (76b) for spinors of the first rank.*

98

with

$$b_{11}b_{22} - b_{12}b_{21} = 1. \tag{76b}$$

References

Pauli, W., Article in *Handbuch der Physik*, Bd. 24 (1933), or in *Encycl. of Physics*, Vol. 5/1, Julius Springer (1958).

Bjorken, J. D. and Drell, S. D., *Relativistic Quantum Mechanics* (McGraw-Hill, N. Y. 1964).

Exercises: *Chapter 4*

1. In treating the Lorentz transformation of the Dirac equation in the present chapter, we have taken the view that the γ_μ's are constants (i.e. not subject to transformation) while the Dirac wave function $\psi(x)$ is transformed into $\psi'(x')$ (which is $S\psi(x)$). In fact, it is equivalent to assume that $\psi(x)$ is not transformed while γ_μ transforms like a Lorentz 4-vector,

$$\gamma'_\mu = a_{\mu\nu}\gamma_\nu.$$

   (i) Prove that the two views are indeed equivalent.

   (ii) Use the second view to discuss C, P, and T.

2. Choose a specific representation for γ-matrices such that the $T^{\mu\nu}$ in Eq. (55) assumes the following form:

$$T^{i4} = \frac{i}{2}\begin{pmatrix} \sigma_i & 0 \\ 0 & -\sigma_i \end{pmatrix},$$

   where σ_i $(i = 1, 2, 3)$ are three 2×2 matrices.

3. (i) Show that the effect of two successive rotations on the Dirac wave function depends on the order of the two operations.

   (ii) Try to generalize (i) to two arbitrary Lorentz transformations. Decide the commuting properties for these operators.

   (iii) Work out the effect of the combined CPT operation on the Dirac wave function.

Chapter 5.
Dirac Electron in an Electromagnetic Field

For convenience, let us collect together the Dirac equations and the notations of symbols here:[1]

$$x_\mu \; (x_1, x_2, x_3, x_4 = it),$$

$$A_\mu \; (A_1, A_2, A_3, A_4 = i\phi),$$

$$\text{charge of electron} = -e,$$

$$\Pi_k = \frac{1}{i}\frac{\partial}{\partial x_k} + eA_k, \quad k = 1,2,3,$$

$$\Pi_4 = \frac{1}{i}\frac{\partial}{i\partial t} + ie\phi. \tag{1}$$

$$\gamma_k = \begin{pmatrix} 0 & -i\sigma_k \\ i\sigma_k & 0 \end{pmatrix}, \quad k = x, y, z, \quad \sigma_k \quad \text{Pauli } 2 \times 2 \text{ matrices;}$$

$$\gamma_4 = \begin{pmatrix} I & 0 \\ 0 & -I \end{pmatrix}.$$

$$\gamma_1 = i\begin{pmatrix} 0 & 0 & 0 & -1 \\ 0 & 0 & -1 & 0 \\ 0 & 1 & 0 & 0 \\ 1 & 0 & 0 & 0 \end{pmatrix}, \quad \gamma_2 = \begin{pmatrix} 0 & 0 & 0 & -1 \\ 0 & 0 & 1 & 0 \\ 0 & 1 & 0 & 0 \\ -1 & 0 & 0 & 0 \end{pmatrix},$$

$$\gamma_3 = i\begin{pmatrix} 0 & 0 & -1 & 0 \\ 0 & 0 & 0 & 1 \\ 1 & 0 & 0 & 0 \\ 0 & -1 & 0 & 0 \end{pmatrix}, \quad \gamma_4 = \begin{pmatrix} 1 & 0 & 0 & 0 \\ 0 & 1 & 0 & 0 \\ 0 & 0 & -1 & 0 \\ 0 & 0 & 0 & -1 \end{pmatrix}. \tag{2}$$

$$\Psi^\dagger = i\tilde{\Psi}^*\gamma_4, \text{ a row matrix,}$$

$$\Psi, \text{ a column matrix.} \tag{3}$$

---

[1]
$$x_4 = ict, \quad \Pi_k = \frac{\hbar}{i}\frac{\partial}{\partial x_k} + \frac{e}{c}A_k, \quad \Pi_4 = \frac{\hbar}{i}\frac{1}{ic}\frac{\partial}{\partial t} + i\frac{e}{c}\phi \tag{1'}$$

The Dirac equations are[2]

$$
\begin{aligned}
\gamma_\mu \Pi_\mu \Psi - i m_0 \Psi &= 0, \\
\Pi_\mu^* \Psi^\dagger \gamma_\mu - i m_0 \Psi^\dagger &= 0.
\end{aligned}
\tag{4}
$$

These equations are covariant under the gauge transformation[3]

$$
A' = A + \nabla\chi, \quad \phi' = \phi - \frac{\partial\chi}{\partial t}, \qquad \chi = \text{a scalar function,}
\tag{5}
$$

$$
\Psi' = \Psi exp(-ie\chi).
\tag{6}
$$

On defining

$$
\Pi_\pm = \Pi_1 \pm i\Pi_2,
\tag{7}
$$

one can write Eq. (4) for Ψ in the explicit form[4]

$$
\begin{aligned}
(\frac{1}{i}\frac{\partial}{\partial t} - e\phi + m_0)\Psi_1 + \Pi_-\Psi_4 + \Pi_3\Psi_3 &= 0, \\
(\frac{1}{i}\frac{\partial}{\partial t} - e\phi + m_0)\Psi_2 + \Pi_+\Psi_3 - \Pi_3\Psi_4 &= 0, \\
(\frac{1}{i}\frac{\partial}{\partial t} - e\phi - m_0)\Psi_3 + \Pi_-\Psi_2 + \Pi_3\Psi_1 &= 0, \\
(\frac{1}{i}\frac{\partial}{\partial t} - e\phi - m_0)\Psi_4 + \Pi_+\Psi_1 - \Pi_3\Psi_2 &= 0,
\end{aligned}
\tag{8}
$$

2

$$
\gamma_\mu \Pi_\mu \Psi - i m_0 c \Psi = 0.
\tag{4'}
$$

3

$$
\phi' = \phi - \frac{1}{c}\frac{\partial\chi}{\partial t}.
\tag{5'}
$$

$$
\Psi' = \Psi exp(-\frac{ie}{\hbar c}\chi).
\tag{6'}
$$

4

$$
(\frac{\hbar}{i}\frac{1}{c}\frac{\partial}{\partial t} - \frac{e}{c}\phi + m_0 c)\Psi_1 + \Pi_-\Psi_4 + \Pi_3\Psi_3 = 0.
\tag{8'}
$$

which are the extension of the equations for a free electron [Eq. (25), Ch. 2] to the electron in an $(\mathbf{A}, i\phi)$ field.

§.5.1. Dirac equations in second-order form

One can apply the operator

$$(\gamma_\mu \Pi_\mu + im_0) = (\gamma_4 \Pi_4 + (\gamma \cdot \Pi) + im_0) \tag{9}$$

on Eq. (4), obtaining[5]

$$\{\Pi_4^2 + (\gamma \cdot \Pi)^2 + m_0^2 + \sum_k^3 (\gamma_k \gamma_4 \Pi_k \Pi_4 + \gamma_4 \gamma_k \Pi_4 \Pi_k)\}\Psi = 0. \tag{10}$$

On defining Π_0

$$\Pi_0 = -i\Pi_4 = (-\frac{1}{i}\frac{\partial}{\partial t} + e\phi), \tag{12}$$

---

[5] *This equation, on using Eqs. (13) and (14) of Ch. 4, can be put in the form*

$$\{\Pi_\mu \Pi_\mu + m_0^2 + \frac{e}{2i}\gamma_\mu \gamma_\nu F_{\mu\nu}\}\Psi = 0. \tag{10a}$$

From Eqs. (20c) and (80f) of Ch. 2, one has

$$\gamma_k = -i\beta\rho_1\sigma_k, \quad k = 1, 2, 3,$$

$$(\gamma \cdot \Pi)^2 = -(\beta\rho_1\sigma_j \Pi_j)(\beta\rho_1\sigma_k \Pi_k)$$
$$= \beta\sigma_j\rho_1\rho_1\beta\sigma_k \Pi_j \Pi_k \quad \text{(from Eq. (82), Ch. 2 and } \beta\rho_1 + \rho_1\beta = 0)$$
$$= \beta\sigma_j\beta\sigma_k \Pi_j \Pi_k = \sigma_j\sigma_k \Pi_j \Pi_k \quad \text{(from } \sigma_j\beta - \beta\sigma_j = 0)$$
$$= (\sigma \cdot \Pi)^2.$$
$$\gamma_k\gamma_4 = -i\beta\alpha_k\beta \quad \text{(using Eq. (20c), Ch. 2)}$$
$$= i\alpha_k = i\rho_1\sigma_k. \quad \text{(using Eq. (80f), Ch. 2)}$$

$$\tag{11}$$

Eq. (10) takes the form[6]

$$[-\Pi_0^2 + (\sigma \cdot \Pi)^2 + m_0^2 - \rho_1\{(\sigma \cdot \Pi)\Pi_0 - \Pi_0(\sigma \cdot \Pi)\}\Psi = 0. \qquad (13)$$

This equation can be rewritten as follows[7],[8]

6

$$[-\Pi_0^2 + (\sigma \cdot \Pi)^2 + m_0^2 c^2 - \rho_1\{(\sigma \cdot \Pi)\Pi_0 - \Pi_0(\sigma \cdot \Pi)\}\Psi = 0. \qquad (13')$$

[7] *If X is an arbitrary vector (operator) that commutes with σ , then on using Eq. (81), Ch. 2,*

$$\sigma_i\sigma_j + \sigma_j\sigma_i = 2\delta_{ij},$$

one has

$$(\sigma \cdot B)^2 = B^2 + i\sigma_1(X_2 X_3 - X_3 X_2) + i\sigma_2(X_3 X_1 - X_1 X_3) + i\sigma_3(X_1 X_2 - X_2 X_1). \qquad (15)$$

Hence

$$(\sigma \cdot \Pi)^2 = \sum_{k=1}^{3}(\frac{1}{i}\frac{\partial}{\partial x_k} + eA_k)^2 + e(\sigma \cdot \text{curl}A)$$
$$= \Pi^2 + e(\sigma \cdot \mathbf{H}), \quad \mathbf{H} = \text{magnetic field}, \qquad (16)$$

$$(\sigma \cdot \Pi)\Pi_0 - \Pi_0(\sigma \cdot \Pi) = \frac{e}{i}(\sigma \cdot (\nabla\phi + \frac{\partial \mathbf{A}}{\partial t}))$$
$$= -\frac{e}{i}(\sigma \cdot \mathbf{E}), \quad \mathbf{E} = \text{electric field}. \qquad (17)$$

Or, without using $\hbar = c = 1$,

$$(\sigma \cdot \Pi)^2 = \Pi^2 + \frac{e\hbar}{c}(\sigma \cdot \mathbf{H}), \qquad (16')$$

$$.... = -\frac{e\hbar}{ic}(\sigma \cdot \mathbf{E}). \qquad (17')$$

8

$$\{-(-\frac{\hbar}{i}\frac{\partial}{c\partial t} + \frac{e}{c}\phi)^2 + (\frac{\hbar}{i}\nabla + \frac{e}{c}A)^2 + m_0^2 c^2 + \frac{e\hbar}{c}(\sigma \cdot \mathbf{H}) - i\frac{e\hbar}{c}\rho_1(\sigma \cdot \mathbf{E})\}\Psi = 0. \quad (14')$$

$$\{-(-\frac{1}{i}\frac{\partial}{\partial t} + e\phi)^2 + (\frac{1}{i}\nabla + eA)^2 + m_0^2 + e(\sigma \cdot \mathbf{H}) - ie\rho_1(\sigma \cdot \mathbf{E})\}\Psi = 0. \quad (14)$$

On comparing Eq. (14) with the Klein-Gordon equation Eq. (14), Ch. 1 (in which e is now the charge of an electron $-e$), it is seen that the Dirac equation Eq. (14) has two extra terms[9]

$$\frac{e}{2m_0}(\sigma \cdot \mathbf{H})\Psi - i\frac{e}{2m_0}(\sigma \cdot \mathbf{E})\Psi. \quad (18)$$

The first term here with the operator

$$\mu_B(\sigma \cdot \mathbf{H}) = 2\mu_B(\mathbf{S} \cdot \mathbf{H}), \quad (19)$$

where $\mathbf{S} = \frac{1}{2}\sigma$ is the spin angular momentum operator. Eq. (91), Ch. 2, can immediately be interpreted as the interaction of the spin magnetic moment $\mu_B\mathbf{S}$ with the external magnetic field with the gyromagnetic ratio $g = 2$.

The second term in Eq. (18) is a pure imaginary and seems to have no physical meaning. But we shall see below that in the case of the hydrogenic atom it leads to the so-called "S-state correction".[10]

The above comparison between the Dirac equation Eq. (14) and the Klein-Gordon equation Eq. (14) of Ch. 1, shows that the latter applies to spinless particles, such as the pions $\pi^\pm, \pi^0$.

9

$$\frac{e\hbar}{2m_0c}(\sigma \cdot \mathbf{H})\Psi - i\frac{e\hbar}{2m_0c}(\sigma \cdot \mathbf{E})\Psi \quad (18')$$

[10] See Eq. (39), the last term containing $\frac{d\phi}{dr}$, and Eq. (57) in the following.

§.5.2. Dirac equation: approximate, iterated form

Let E_+ be the energy of an electron in a positive-energy state, and

$$w = E_+ - m_0, \qquad (20a)$$

and consider in this section the situation

$$|w| \ll m_0. \qquad (20b)$$

Let Ψ_k in Eq. (8) be

$$\Psi_k = \Phi_k \exp(-im_0 t), \qquad (21)$$

so that[11]

$$
\begin{aligned}
(\frac{1}{i}\frac{\partial}{\partial t} - e\phi + m_0)\Psi_k &= exp(-im_0 t)(\frac{1}{i}\frac{\partial}{\partial t} - e\phi)\Phi_k, \\
(\frac{1}{i}\frac{\partial}{\partial t} - e\phi - m_0)\Psi_k &= exp(-im_0 t)(\frac{1}{i}\frac{\partial}{\partial t} - e\phi - 2m_0)\Phi_k.
\end{aligned}
\qquad (22)
$$

It is seen from Eq. (8) that Ψ_1, Ψ_2 are the "big" components and Ψ_3, Ψ_4 the "small" components. Let us introduce the notations

$$\Phi \equiv \begin{pmatrix} \Phi_1 \\ \Phi_2 \end{pmatrix}, \qquad \varphi \equiv \begin{pmatrix} \varphi_1 \\ \varphi_2 \end{pmatrix} \equiv \begin{pmatrix} \Phi_3 \\ \Phi_4 \end{pmatrix}, \qquad (23)$$

$$|\varphi| \ll |\Phi|. \qquad (24)$$

By means of the 2×2 Pauli matrices

$$\sigma_x = \begin{pmatrix} 0 & 1 \\ 1 & 0 \end{pmatrix}, \quad \sigma_y = \begin{pmatrix} 0 & -i \\ i & 0 \end{pmatrix}, \quad \sigma_z = \begin{pmatrix} 1 & 0 \\ 0 & -1 \end{pmatrix}, \qquad (25)$$

one can express Eq. (8) in the form

---

[11]

$$(\frac{\hbar}{i}\frac{\partial}{\partial t} - \frac{e}{c}\phi + m_0 c)\Psi_k = exp(-\frac{i}{\hbar}m_0 c^2 t)(\frac{\hbar}{i}\frac{\partial}{\partial t} - \frac{e}{c}\phi)\Phi_k, \quad etc. \qquad (22')$$

$$\left(\frac{1}{i}\frac{\partial}{\partial t} - e\phi\right)\begin{pmatrix}\Phi_1\\\Phi_2\end{pmatrix} + (\sigma\cdot\Pi)\begin{pmatrix}\varphi_1\\\varphi_2\end{pmatrix} = 0, \tag{26a}$$

$$\left(\frac{1}{i}\frac{\partial}{\partial t} - e\phi - 2m_0\right)\begin{pmatrix}\varphi_1\\\varphi_2\end{pmatrix} + (\sigma\cdot\Pi)\begin{pmatrix}\Phi_1\\\Phi_2\end{pmatrix} = 0, \tag{26b}$$

or[12]

$$\left(-\frac{1}{i}\frac{\partial}{\partial t} + e\phi\right)\Phi - (\sigma\cdot\Pi)\varphi = 0, \tag{27a}$$

$$\left(-\frac{1}{i}\frac{\partial}{\partial t} + e\phi + 2m_0\right)\varphi - (\sigma\cdot\Pi)\Phi = 0. \tag{27b}$$

These equations are exact. The last equation[13]

$$\left\{1 + \frac{1}{2m_0}\left(-\frac{1}{i}\frac{\partial}{\partial t} + e\phi\right)\right\}\varphi = \frac{(\sigma\cdot\Pi)}{2m_0}\Phi, \tag{28}$$

shows that the second term on the left is of order $\left(\frac{v}{c}\right)^2$.

As an initial approximation, one may take

$$\varphi = \frac{(\sigma\cdot\Pi)}{2m_0}\Phi.$$

Now in the non-relativistic (Schrödinger) approximation, one has

$$\left(-\frac{1}{i}\frac{\partial}{\partial t} + e\phi\right)\Phi = \frac{\Pi^2}{2m_0}\Phi, \tag{29}$$

or on using Eq. (20a)

$$(w + e\phi)\Phi = \frac{\Pi^2}{2m_0}\Phi, \tag{30a}$$

12

$$\left(-\frac{\hbar}{ic}\frac{\partial}{\partial t} + \frac{e}{c}\phi\right)\Phi - (\sigma\cdot\Pi)\varphi = 0, \quad etc. \tag{27a'}$$

13

$$\left\{1 + \frac{1}{2m_0c^2}\left(-\frac{\hbar}{i}\frac{\partial}{\partial t} + e\phi\right)\right\}\varphi = \frac{(\sigma\cdot\Pi)}{2m_0c}\Phi. \tag{28'}$$

and the next approximation for φ in Eq. (28) is[14]

$$\varphi = \frac{1}{1 + \frac{\Pi^2}{4m_0^2}} \cdot \frac{(\sigma \cdot \Pi)}{2m_0} \Phi. \tag{30}$$

Let us define an operator F by

$$F \equiv [1 + \frac{\Pi^2}{4m_0^2}]^{-1} = [1 + \frac{w + e\phi}{2m_0}]^{-1}. \tag{31}$$

The equation Eq. (27a) can be approximated by

$$(-\frac{1}{i}\frac{\partial}{\partial t} + e\phi - \frac{1}{2m_0}(\sigma \cdot \Pi)F(\sigma \cdot \Pi))\Phi = 0. \tag{32}$$

Now

$$(\sigma \cdot \Pi)F(\sigma \cdot \Pi) = F(\sigma \cdot \Pi)^2 + (\sigma \cdot \Pi F)(\sigma \cdot \Pi). \tag{33}$$

From Eq. (30a)

$$\frac{\Pi^2}{4m_0^2} = \frac{w + e\phi}{2m_0}, \tag{34}$$

and for $w + e\phi \ll 2m_0$, as from Eq. (20b), one expands F in Eq. (31) and obtains, using Eq. (16),[15]

$$F(\sigma \cdot \Pi)^2 = (1 - \frac{\Pi^2}{4m_0^2})(\sigma \cdot \Pi)^2$$
$$= \Pi^2 - \frac{\Pi^4}{4m_0^2} + e(\sigma \cdot H) - \frac{\Pi^2 e}{4m_0^2}(\sigma \cdot H). \tag{35}$$

14
$$\varphi = \frac{1}{1 + \frac{\Pi^2}{4m_0^2 c^2}} \cdot \frac{(\sigma \cdot \Pi)}{2m_0 c} \Phi. \tag{30'}$$

15
$$F(\sigma \cdot \Pi)^2 = (1 - \frac{\Pi^2}{4m_0^2 c^2})(\sigma \cdot \Pi)^2$$
$$= \Pi^2 - \frac{\Pi^4}{4m_0^2 c^2} + \frac{e\hbar}{c}(\sigma \cdot H) - \frac{\Pi^2 e\hbar}{4m_0^2 c^2}(\sigma \cdot H). \tag{35'}$$

The calculation of the last term in Eq. (33) is long, but to the order $(\frac{v}{c})^2$, one may neglect the vector potential **A** and replace Π_k by p_k , i.e.,

$$(\sigma \cdot \Pi F)(\sigma \cdot \Pi) \rightarrow (\sigma \cdot \mathbf{p}F)(\sigma \cdot \mathbf{p}).$$

If the scalar potential ϕ is central field, then

$$(\sigma \cdot \mathbf{p}F) = (\sigma \cdot \frac{1}{i} \frac{d}{dr} F) = \frac{1}{ir} \frac{dF}{dr}(\sigma \cdot \mathbf{r}),$$

$$(\sigma \cdot \mathbf{p}F)(\sigma \cdot \mathbf{p}) = \frac{dF}{dr}(-\frac{\partial}{\partial r} + \frac{1}{r}(\mathbf{L} \cdot \sigma)), \tag{36}$$

where $\mathbf{L} = [\mathbf{r} \times \mathbf{p}]$ is the angular momentum, and

$$\frac{dF}{dr} = -[1 + \frac{w + e\phi}{2m_0}]^{-2} \cdot \frac{e}{2m_0} \frac{d\phi}{dr} \tag{37}$$

$$\cong -\frac{e}{2m_0} \frac{d\phi}{dr} \qquad \text{[From Eq. (34)]}. \tag{38}$$

Substituting Eqs. (35), (36), and (37) into (32), one obtains[16]

$$\{-\frac{1}{i} \frac{\partial}{\partial t} + e\phi - \frac{\Pi^2}{2m_0}(1 - \frac{\Pi^2}{4m_0^2}) - \frac{e}{2m_0}(1 - \frac{\Pi^2}{4m_0^2})(\sigma \cdot \mathbf{H})$$
$$+ [1 + \frac{w + e\phi}{2m_0}]^{-2} \frac{e}{4m_0^2} \frac{d\phi}{dr}[\frac{1}{r}(\mathbf{L} \cdot \sigma) - \frac{\partial}{\partial r}]\}\Phi = 0. \tag{39}$$

The first three terms are the Schrödinger non-relativistic theory Eq. (29), the term in Π^4 is the Sommerfeld correction (treated in Chapter 6, (VI-22) Vol. I); the term $\frac{e}{2m_0}(\sigma \cdot \mathbf{H})$ has been referred to in Eq. (19) already, (the term $\frac{\Pi^2}{4m_0^2}\mu_B(\sigma \cdot \mathbf{H})$ is a higher order correction to it, and can be neglected); the last two terms can be traced to the last term in Eq. (14); they are the spin-orbit interactions and will be examined in detail in the following section.

16

$$\{-\frac{\hbar}{i} \frac{\partial}{\partial t} + e\phi - \frac{\Pi^2}{2m_0}(1 - \frac{\Pi^2}{4m_0^2c^2}) - \frac{e\hbar}{2m_0c}(1 - \frac{\Pi^2}{4m_0^2c^2})(\sigma \cdot \mathbf{H})$$
$$+ [1 + \frac{w + e\phi}{2m_0c^2}]^{-2} \frac{e\hbar^2}{4m_0^2c^2} \frac{d\phi}{dr}[\frac{1}{r}(\mathbf{L} \cdot \sigma) - \frac{\partial}{\partial r}]\}\Phi = 0. \tag{39'}$$

Equation (39) has been obtained in approximate theories by Pauli and Darwin, before the Dirac theory from which Eq. (39) can be derived.

§.5.3. Hydrogenic atoms in Dirac's theory - approximate solution

For hydrogenic atoms,[17]

$$\phi(r) = \frac{Ze}{4\pi r}, \quad \frac{d\phi}{dr} = -\frac{Ze}{4\pi r^2};$$

$$A = 0, \quad \text{i.e.,} \quad H = 0;$$

$$S = \frac{1}{2}\sigma, \tag{40}$$

$$J = L + S = L + \frac{1}{2}\sigma,$$

$$2(L \cdot S) = J^2 - L^2 - S^2.$$

Let

$$\Phi(r,t) = \Psi(r)e^{-iwt}. \tag{41}$$

Eq. (39) leads to the stationary value problem[18]

$$\{\frac{p^2}{2m_0} - \frac{Ze^2}{4\pi r} - \frac{1}{2m_0}(w + \frac{Ze^2}{4\pi r})^2$$

$$+ \mu_B^2[1 + \frac{1}{2m_0}(w + \frac{Ze^2}{4\pi r})]^{-2}\frac{Z}{4\pi r^2}[\frac{2}{r}(L \cdot S) - \frac{\partial}{\partial r}]\}\Psi(r) = w\Psi(r). \tag{42}$$

17

$$\phi(r) = \frac{Ze}{r}, \text{ etc.} \tag{40'}$$

18

$$\{\frac{p^2}{2m_0} - \frac{Ze^2}{r} - \frac{1}{2m_0c^2}(w + \frac{Ze^2}{r})^2$$

$$+ \mu_B^2[1 + \frac{1}{2m_0c^2}(w + \frac{Ze^2}{r})]^{-2}\frac{Z}{r^2}[\frac{2}{r}(L \cdot S) - \frac{\partial}{\partial r}]\}\Psi(r) = w\Psi(r). \tag{42'}$$

The lefthand side may be regarded as an approximate Hamiltonian $\bar{H}$, and from Eq. (42) , it is seen that[19]

$$\bar{H}J^2 - J^2\bar{H} = 0,$$
$$\bar{H}S^2 - S^2\bar{H} = 0,$$
$$\bar{H}L^2 - L^2\bar{H} = 0.$$

(43)

The eigenvalues of $\bar{H}, J^2, S^2, L^2$ are

$$w, \quad j(j+1), \quad s(s+1), \quad \ell(\ell+1), \quad (s = \frac{1}{2}).$$

(44)

From Eq. (40), the eigenvalues of $2(L \cdot S)$ are given by

$$2(L \cdot S)\Psi = \{j(j+1) - \ell(\ell+1) - s(s+1)\}\Psi.$$

(45)

The Schrödinger approximation as obtained from Eq. (39) is

$$\{-\frac{1}{2m_0}(\frac{d^2}{dr^2} + \frac{2}{r}\frac{d}{dr} - \frac{\ell(\ell+1)}{r^2}) - e\phi\}R_0(r) = wR_0(r).$$

(46)

$R_0(r)$ is the Schrödinger wave function of Chapter 3, (III-139), Vol. I. The other terms may be treated by the perturbation method. Let

$$H_r = -\frac{1}{2m_0}(w + \frac{Ze^2}{4\pi r})^2,$$

(47)

$$H_{s.o.} = \mu_B^2[1 + \frac{w + \frac{Ze^2}{4\pi r}}{2m_0}]^{-2}\frac{2Z}{4\pi r^3}(L \cdot S),$$

(48a)

$$H_s = -\mu_B^2[1 + \frac{w + \frac{Ze^2}{4\pi r}}{2m_0}]^{-2}\frac{Z}{4\pi r^2}\frac{\partial}{\partial r}.$$

(48b)

---

[19] Note that $\mathbf{L}^2$ does not commute with the exact Hamiltonian H

$$i\frac{\partial}{\partial t}\Phi = H\Phi,$$

$$H = (\alpha \cdot \Pi) - e\phi + \beta m_0.$$

The term H_r is the Sommerfeld relativistic correction and has been calculated in Chapter 6, (VI-22), Vol. I. The result is[20]

$$< n, \ell | \, H_r \, | n, \ell > = -\frac{Z^4 \alpha^2}{n^4}\left(\frac{n}{\ell + \frac{1}{2}} - \frac{3}{4}\right), \qquad \alpha = \frac{e^2}{4\pi} \simeq \frac{1}{137}. \tag{49}$$

The term $H_{s.o.}$ is simplified for the approximation[21]

$$1 + \frac{w + e\phi}{2m_0} \simeq 1, \tag{50}$$

and is then the spin-orbit interaction already treated in Chapter 8, Sect. 1, (2), (VIII-11, 25a, b, 28), Vol. I.[22] For $\ell \neq 0$,

---

[20]

$$< n, \ell | \, H_r \, | n, \ell > = -\frac{Z^4 \alpha^2}{n^4}\left(\frac{n}{\ell + \frac{1}{2}} - \frac{3}{4}\right), \qquad \alpha = \frac{e^2}{\hbar c}. \tag{49'}$$

[21]

$$1 + \frac{w + e\phi}{2m_0 c^2} \simeq 1. \tag{50'}$$

[22] *The matrix element of $H_{s.o.}$ consists of two factors, namely,*

$$< n, \ell | \, \frac{Z}{r^3} \, | n, \ell >, \quad and \quad < j, m | \, \boldsymbol{\ell} \cdot \boldsymbol{s} \, | j, m > . \tag{52}$$

For S state, $\ell = 0$, one has the indeterminate result $\frac{0}{0}$, and one has to examine the situation more carefully. For $\ell = 0$, the matrix element $< n, 0 | \, \frac{Z}{r^3} \, | n, 0 >$

$$\propto \int_0^\infty R_{n,0}\frac{1}{r^3}R_{n,0}\, r^2 dr \quad \xrightarrow{(r \to 0)} \quad \int_0^\infty \frac{1}{r} dr, \tag{53}$$

is divergent at $r = 0$. But at $r = 0$, one must use the expression (48a), and then one finds the matrix element $< n, 0 | \, \frac{Z}{r^3} \, | n, 0 >$ to be finite. Hence for S state,

$$< n, 0 | \, H_{s.o.} \, | n, 0 > = 0. \tag{54}$$

$$< n, \ell|\; H_{s.o.}\; |n, \ell > = \frac{Z^4\alpha^2}{n^3}\frac{1}{\ell(\ell+\frac{1}{2})(\ell+1)} \cdot \begin{cases} \frac{\ell}{2}, & j = \ell + \frac{1}{2}, & \ell = 1, 2, \ldots \\ -\frac{\ell+1}{2}, & j = \ell - \frac{1}{2}, & \ell = 1, 2, \ldots \end{cases}$$

(51)

The term H_s in Eq. (48b) can also be calculated for the approximation Eq. (50). Then

$$< n, \ell|\; H_s\; |n, \ell > = -Z\mu_B^2 \int_0^\infty R_{n,\ell}\frac{1}{r^2}\frac{d}{dr}R_{n,\ell}r^2 dr$$
$$= \frac{1}{2}Z\mu_B^2|\; R_{n,\ell}(0)\;|^2,$$

(55)

which does not vanish only for S state $(\ell = 0)$, and[23]

$$< n, 0|\; H_s\; |n, 0 > = \frac{1}{2}Z\mu_B^2 \frac{4Z^3}{n^3 a^3}, \quad a_0 = \frac{1}{m_0 e^2}, \quad \text{Bohr radius};$$
$$= \frac{Z^4\alpha^2}{n^3}\; (in \quad \frac{m_0 e^4}{2} = \frac{e^2}{2a}).$$

(56)

which is called the S-state correction. This result may be combined together with Eq. (51) (which is valid for $\ell \neq 0$) , so that now one may write

$$< n, \ell|\; H_{s.o.} + H_s\; |n, \ell >$$
$$= \frac{Z^4\alpha^2}{n^3}\frac{1}{\ell(\ell+\frac{1}{2})(\ell+1)} \begin{cases} \frac{\ell}{2}, & j = \ell + \frac{1}{2}, \ell = 0, 1, .. \\ -\frac{\ell+1}{2}, & j = \ell - \frac{1}{2}, \ell = 1, 2, .. \end{cases}$$

(57)

The total contribution from $H_r, H_{s.o.}, H_s$ in Eqs. (47), (48a), and (48b) is then

$$< n, \ell|\; H_r + H_{s.o.} + H_s\; |n, \ell > = -\frac{Z^4\alpha^2}{n^4}(\frac{n}{j+\frac{1}{2}} - \frac{3}{4}).$$

(58)

This energy is degenerate for the two states

$$(n, \ell = j - \frac{1}{2}) \quad \text{and} \quad (n, \ell' = j + \frac{1}{2}),$$

(59)

---

[23]

$$a_0 = \frac{\hbar^2}{m_0 e^2}, \quad \alpha = \frac{e^2}{\hbar c},$$
$$= \frac{Z^4\alpha^2}{n^3}\; (\text{ in units of} \quad \frac{m_0 e^4}{2\hbar^2} = \frac{e^2}{2a_0}).$$

(56′)

for example

$$^2S_{\frac{1}{2}} \quad \text{and} \quad ^2P_{\frac{1}{2}},$$
$$^2P_{\frac{3}{2}} \quad \text{and} \quad ^2D_{\frac{3}{2}}, \quad \text{etc.} \tag{59a}$$

Let us find the wave functions of the approximate Hamiltonian $\bar{H}$ in Eq. (42).

In Eq. (43), we have seen that J^2, S^2, L^2 commute with $\bar{H}$. In addition to these,

$$J_z = L_z + S_z \tag{60}$$

also commutes with $\bar{H}$

$$\bar{H}J_z - J_z\bar{H} = 0, \tag{61}$$

and J_z has simultaneous eigenvalues m with J^2, S^2, L^2 ,

$$-J \leq m \leq J. \tag{62}$$

But L_z and S_z separately do not commute with $\bar{H}$ so that m_ℓ, m_s are not "exact" quantum numbers.

A state is now defined by the quantum numbers

$$n, j, \ell, s = \frac{1}{2} \quad \text{and} \quad m. \tag{63}$$

A state having n, ℓ, j, m is a combination

$$\Psi = R_{n,\ell}(r)[aY_{\ell,m_\ell} + bY_{\ell,m_\ell'}], \tag{64}$$

where

$$m_\ell + \frac{1}{2} = m_\ell' - \frac{1}{2} = m. \tag{64a}$$

The ratio $\frac{a}{b}$ is determined by the requirement that Ψ of Eq. (64) be also an eigenfunction of $(L \cdot S)$. The theory has already been given in Chapter 8, Sect. 1, (3) (VIII-31a, 31b), Vol. I. Thus for
$j = \ell + \frac{1}{2}, \quad (\ell \cdot s) = \frac{\ell}{2}$,

$$\Phi_{n,\ell,j,m} = \begin{pmatrix} \Phi_1 \\ \Phi_2 \end{pmatrix} = \frac{1}{\sqrt{2\ell+1}} R_{n,\ell}(r) \begin{pmatrix} \sqrt{\ell+m+\frac{1}{2}} & Y_{\ell,m-\frac{1}{2}} \\ -\sqrt{\ell-m+\frac{1}{2}} & Y_{\ell,m+\frac{1}{2}} \end{pmatrix}, \tag{65a}$$

$$j = \ell - \tfrac{1}{2}, \quad (\boldsymbol{\ell} \cdot \boldsymbol{s}) = -\tfrac{\ell+1}{2} ,$$

$$\Phi_{n,\ell,j,m} = \begin{pmatrix} \Phi_1 \\ \Phi_2 \end{pmatrix} = \frac{1}{\sqrt{2\ell+1}} R_{n,\ell}(r) \begin{pmatrix} \sqrt{\ell - m + \tfrac{1}{2}} & Y_{\ell, m-\frac{1}{2}} \\ \sqrt{\ell + m + \tfrac{1}{2}} & Y_{\ell, m+\frac{1}{2}} \end{pmatrix}. \tag{65b}$$

§.5.4. Hydrogenic atoms in Dirac's theory - Exact solution

For hydrogenic atoms, we have[24]

$$A = 0, \qquad \phi(r) = \frac{Ze}{4\pi r},$$

$$p_{\pm} \equiv \frac{1}{i}\left(\frac{\partial}{\partial x} \pm i \frac{\partial}{\partial y}\right)$$
$$= \frac{1}{i} e^{\pm i\varphi}\left[\sin\vartheta \frac{\partial}{\partial r} + \cos\vartheta \frac{1}{r}\frac{\partial}{\partial \vartheta} \pm \frac{i}{r\sin\vartheta}\frac{\partial}{\partial \varphi}\right],$$
$$p_3 = \frac{1}{i}\frac{\partial}{\partial z}$$
$$= \frac{1}{i}\left[\cos\vartheta \frac{\partial}{\partial r} - \sin\vartheta \frac{\partial}{r\partial\vartheta}\right]. \tag{66}$$

$$E = w + m_0 = w + E_0, \qquad w < 0 \quad \text{for bound states.}$$

The exact equations are, from Eq. (8),

$$-\left(E + \frac{Ze^2}{4\pi r} - E_0\right)\Phi_1 + p_-\Phi_4 + p_3\Phi_3 = 0,$$
$$-\left(E + \frac{Ze^2}{4\pi r} - E_0\right)\Phi_2 + p_+\Phi_3 - p_3\Phi_4 = 0,$$
$$-\left(E + \frac{Ze^2}{4\pi r} + E_0\right)\Phi_3 + p_-\Phi_2 + p_3\Phi_1 = 0,$$
$$-\left(E + \frac{Ze^2}{4\pi r} + E_0\right)\Phi_4 + p_+\Phi_1 - p_3\Phi_2 = 0. \tag{67}$$

For hydrogenic atoms in their stationary states,

$$w + \frac{Ze^2}{4\pi r} \ll w + \frac{Ze^2}{4\pi r} + 2m_0. \tag{68}$$

---

[24] *In ordinary units, $\phi(r) = \frac{Ze}{r}$ in (67), (68), (73), (74), (77), (78), (80) and m_0 is $m_0 c^2$ in (68).*

Thus Φ_1, Φ_2 are the big components and Φ_3, Φ_4 the small components (for positive energy $E_+ > 0$ states).

The functions $\binom{\Phi_1}{\Phi_2}$ and $\binom{\Phi_3}{\Phi_4}$ are coupled in Eq. (67) so that the Schrödinger radial wave functions $R_{n,\ell}(r)$ are not solutions. From Eq. (65a), we may write

$$
\begin{cases}
j = \ell + \tfrac{1}{2}, & \Phi_1(r) = g(r)\sqrt{\dfrac{\ell+m+\tfrac{1}{2}}{2\ell+1}}\, Y_{\ell, m-\tfrac{1}{2}}(\vartheta, \varphi), \\
m & \Phi_2(r) = -g(r)\sqrt{\dfrac{\ell-m+\tfrac{1}{2}}{2\ell+1}}\, Y_{\ell, m+\tfrac{1}{2}}(\vartheta, \varphi)
\end{cases}
\tag{69}
$$

Substituting these into the last two equations in (67), one gets[25]

$$
(E + \frac{Ze^2}{4\pi r} + E_0)\Phi_3 = \frac{1}{i}\sqrt{\frac{\ell - m + \tfrac{3}{2}}{2\ell+3}}\Big(\frac{dg}{dr} - \ell\frac{g}{r}\Big) Y_{\ell+1, m-\tfrac{1}{2}}(\vartheta, \varphi),
$$

$$
(E + \frac{Ze^2}{4\pi r} + E_0)\Phi_4 = \frac{1}{i}\sqrt{\frac{\ell + m + \tfrac{3}{2}}{2\ell+3}}\Big(\frac{dg}{dr} - \ell\frac{g}{r}\Big) Y_{\ell+1, m+\tfrac{1}{2}}(\vartheta, \varphi).
\tag{70}
$$

---

[25] *Use is made of Eq. (66) and the following relations*

$$
\frac{\partial}{\partial z}(gY_{\ell,m}) = \sqrt{\frac{(\ell + m + 1)(\ell - m + 1)}{(2\ell+1)(2\ell+3)}}\, Y_{\ell+1,m}\Big(\frac{dg}{dr} - \ell\frac{g}{r}\Big)
$$

$$
+ \sqrt{\frac{(\ell + m)(\ell - m)}{(2\ell-1)(2\ell+1)}}\, Y_{\ell-1,m}\Big(\frac{dg}{dr} + (\ell+1)\frac{g}{r}\Big),
$$

$$
(\frac{\partial}{\partial x} + i\frac{\partial}{\partial y})(gY_{\ell,m}) = \sqrt{\frac{(\ell + m + 1)(\ell + m + 2)}{(2\ell+1)(2\ell+3)}}\, Y_{\ell+1,m+1}\Big(\frac{dg}{dr} - \ell\frac{g}{r}\Big)
$$

$$
- \sqrt{\frac{(\ell - m - 1)(\ell - m)}{(2\ell-1)(2\ell+1)}}\, Y_{\ell-1,m+1}\Big(\frac{dg}{dr} + (\ell+1)\frac{g}{r}\Big), \tag{71}
$$

$$
(\frac{\partial}{\partial x} - i\frac{\partial}{\partial y})(gY_{\ell,m}) = -\sqrt{\frac{(\ell - m + 1)(\ell - m + 2)}{(2\ell+1)(2\ell+3)}}\, Y_{\ell+1,m-1}\Big(\frac{dg}{dr} - \ell\frac{g}{r}\Big)
$$

$$
+ \sqrt{\frac{(\ell + m - 1)(\ell + m)}{(2\ell-1)(2\ell+1)}}\, Y_{\ell-1,m-1}\Big(\frac{dg}{dr} + (\ell+1)\frac{g}{r}\Big).
$$

If we introduce a function f(r) by letting

$$\Phi_3 = -i\sqrt{\frac{\ell - m + \frac{3}{2}}{2\ell + 3}} f(r) Y_{\ell+1, m-\frac{1}{2}},$$

$$\Phi_4 = -i\sqrt{\frac{\ell + m + \frac{3}{2}}{2\ell + 3}} f(r) Y_{\ell+1, m+\frac{1}{2}}, \tag{72}$$

then Eq. (70) becomes

$$(E + \frac{Ze^2}{4\pi r} + E_0) f(r) = \frac{dg}{dr} - \ell\frac{g}{r}. \tag{73}$$

Substituting this into the first two equations of (67) yields the same equation

$$(E - \frac{Ze^2}{4\pi r} - E_0) g(r) = -\frac{df}{dr} - (\ell + 2)\frac{f}{r}. \tag{74}$$

Similarly, from Eq. (65b), let

$$\begin{cases} j = \ell - \frac{1}{2} & \Phi_1 = g(r)\sqrt{\frac{\ell - m + \frac{1}{2}}{2\ell + 1}} Y_{\ell, m-\frac{1}{2}}(\vartheta, \varphi), \\ m & \Phi_2 = g(r)\sqrt{\frac{\ell + m + \frac{1}{2}}{2\ell + 1}} Y_{\ell, m+\frac{1}{2}}(\vartheta, \varphi), \end{cases} \tag{75}$$

and by a procedure similar to Eq. (72),

$$\Phi_3 = -i\sqrt{\frac{\ell + m - \frac{1}{2}}{2\ell - 1}} f(r) Y_{\ell-1, m-\frac{1}{2}},$$

$$\Phi_4 = i\sqrt{\frac{\ell - m - \frac{1}{2}}{2\ell - 1}} f(r) Y_{\ell-1, m+\frac{1}{2}}, \tag{76}$$

and corresponding to Eqs. (73) and (74)

$$(E + \frac{Ze^2}{4\pi r} + E_0) f(r) = \frac{dg}{dr} + (\ell + 1)\frac{g}{r}, \tag{77}$$

$$(E + \frac{Ze^2}{4\pi r} - E_0) g(r) = -\frac{df}{dr} + (\ell - 1)\frac{f}{r}. \tag{78}$$

The 4 equation (73), (74), (77) and (78) can be expressed in the form of a pair of equations by the following notation:[26]

---

[26] For $j = \ell - \frac{1}{2}$, ℓ must not be 0; otherwise $\Phi_1, \Phi_2, \Phi_3, \Phi_4$ all vanish, as seen from Eqs. (75) and (76).

$$j = \begin{cases} \ell + \frac{1}{2} : \\ \ell - \frac{1}{2} : \end{cases} \qquad \xi = \begin{cases} -(j + \frac{1}{2}) = -(\ell + 1) \\ (j + \frac{1}{2}) = \ell \end{cases} \tag{79}$$

$$\frac{df}{dr} + (1 - \xi)\frac{f}{r} + (E + \frac{Ze^2}{4\pi r} - E_0)g(r) = 0,$$
$$\frac{dg}{dr} + (1 + \xi)\frac{g}{r} - (E + \frac{Ze^2}{4\pi r} + E_0)f(r) = 0. \tag{80}$$

Introducing

$$G(r) \equiv rg(r), \qquad F(r) \equiv rf(r), \tag{81}$$

one has from Eq. (80),[27]

$$\frac{dF}{dr} - \xi\frac{F}{r} = [m_0(1 - \epsilon) - \frac{\beta}{r}]G,$$
$$\frac{dG}{dr} + \xi\frac{G}{r} = [m_0(1 + \epsilon) + \frac{\beta}{r}]F. \tag{82}$$

$$\epsilon \equiv \frac{E}{m_0} < 1, \qquad \alpha = \frac{e^2}{4\pi}(\simeq \frac{1}{137.037}), \qquad \beta \equiv Z\alpha. \tag{83}$$

On introducing a dimensionless length ρ

$$\rho = 2m_0\sqrt{1 - \epsilon^2}\, r, \tag{84}$$

and a pair of functions $u(\rho)$, $v(\rho)$,

$$F(\rho) = \sqrt{1 - \epsilon}\, e^{-\frac{1}{2}\rho}\rho^\gamma(u - v),$$
$$G(\rho) = \sqrt{1 + \epsilon}\, e^{-\frac{1}{2}\rho}\rho^\gamma(u + v), \tag{85}$$

27

$$\frac{dF}{dr} - \xi\frac{F}{r} = [\frac{m_0 c}{\hbar}(1 - \epsilon) - \frac{\beta}{r}]G,$$
$$\frac{dG}{dr} + \xi\frac{G}{r} = [\frac{m_0 c}{\hbar}(1 + \epsilon) + \frac{\beta}{r}]F. \tag{82'}$$

$$\epsilon = \frac{E}{m_0 c^2} < 1, \qquad \alpha = \frac{e^2}{\hbar c}(\simeq \frac{1}{137.037}), \qquad \beta = Z\alpha. \tag{83'}$$

one gets for $u(\rho)$, $v(\rho)$

$$\frac{du}{d\rho} - [1 - \frac{1}{\rho}(\gamma + \frac{\beta\epsilon}{\sqrt{1-\epsilon^2}})]u + \frac{1}{\rho}(\xi + \frac{\beta}{\sqrt{1-\epsilon^2}})v = 0,$$
$$\frac{dv}{d\rho} + \frac{1}{\rho}(\gamma - \frac{\beta\epsilon}{\sqrt{1-\epsilon^2}})v + \frac{1}{\rho}(\xi - \frac{\beta}{\sqrt{1-\epsilon^2}})u = 0. \tag{86}$$

Let

$$u = \sum_{s=0} c_s \rho^s, \qquad v = \sum_{s=0} d_s \rho^s. \tag{87}$$

Then one gets

$$-c_{s-1} + (s + \gamma + \frac{\beta\epsilon}{\sqrt{1-\epsilon^2}})c_s + (\xi + \frac{\beta}{\sqrt{1-\epsilon^2}})d_s = 0,$$

$$(\xi - \frac{\beta}{\sqrt{1-\epsilon^2}})c_s + (s + \gamma - \frac{\beta\epsilon}{\sqrt{1-\epsilon^2}})d_s = 0. \tag{88}$$

The indicial equation is (for $s = 0$)

$$\begin{vmatrix} \gamma + \frac{\beta\epsilon}{\sqrt{1-\epsilon^2}} & \xi + \frac{\beta}{\sqrt{1-\epsilon^2}} \\ \xi - \frac{\beta}{\sqrt{1-\epsilon^2}} & \gamma - \frac{\beta\epsilon}{\sqrt{1-\epsilon^2}} \end{vmatrix} = 0, \tag{89}$$

$$\gamma = \sqrt{\xi^2 - Z^2\alpha^2}$$
$$= \text{real} \quad \text{for } \xi^2 > (\frac{Z}{137})^2. \tag{90}$$

To obtain the eigenvalues of Eq. (82) or (86), on eliminating d_s between the two equations in (88), one gets (using Eq. (90a)),

$$c_s = \frac{s + \gamma - \frac{\beta\epsilon}{\sqrt{1-\epsilon^2}}}{(s+\gamma)^2 - \gamma^2} c_{s-1}. \tag{91}$$

For the series in $u(\rho)$, $v(\rho)$ to terminate at a certain power, i.e.,

$$c'_n = 0, \qquad n' = \text{a positive integer},$$

one has

$$n' + \gamma - \frac{\beta\epsilon}{\sqrt{1-\epsilon^2}} = 0. \tag{92}$$

From this expression, it is seen that only for[28]

$$\epsilon = \frac{E}{m_0} = \frac{w + m_0}{m_0} < 1,$$

i.e.,

$$w < 0, \tag{93}$$

will n' be real. The condition that $n' =$ positive integer determines the eigenvalues ϵ, or w .

From Eq. (79)

$$\xi^2 = (j + \frac{1}{2})^2.$$

Let us introudce a positive integer n

$$n = n' + |\xi| = n' + j + \frac{1}{2}. \tag{94}$$

$|\xi| = \ell + 1$, and n' can be identified with the radial quantum number n_r in Sommerfeld's theory (I-53), Vol. I. Thus n is the principal quantum number.

From Eq. (92), one obtains[29]

$$\frac{w + m_0}{m_0} = [1 + \frac{Z^2 \alpha^2}{[n - j - \frac{1}{2} + \sqrt{(j + \frac{1}{2})^2 - Z^2 \alpha^2}]^2}]^{-\frac{1}{2}} \tag{95}$$

which is the exact eigenvalue formula from Dirac's theory.

For $Z^2 \alpha^2 \ll 1$, one obtains, on expanding[30]

28

$$\epsilon = \frac{E}{m_0 c^2} = \frac{w + m_0 c^2}{m_0 c^2} < 1. \tag{93'}$$

29

$$\frac{w + m_0 c^2}{m_0 c^2} = [1 + \frac{Z^2 \alpha^2}{[n - j - \frac{1}{2} + \sqrt{(j + \frac{1}{2})^2 - Z^2 \alpha^2}]^2}]^{-\frac{1}{2}}. \tag{95'}$$

[30] *If multiplied by $R\hbar c$, $R = $ Rydberg constant, the above result is in c.g.s. units.*

$$w = -\frac{Z^2}{n^2} - \frac{Z^4\alpha^2}{n^4}\left\{\begin{matrix}\frac{n}{\ell+1} - \frac{3}{4}\\[4pt]\frac{n}{\ell} - \frac{3}{4}\end{matrix}\right. + \dots \quad \text{for} \quad j = \left\{\begin{matrix}\ell + \frac{1}{2}\\[4pt]\ell - \frac{1}{2}\end{matrix}\right.,$$

$$\text{in units of} \quad \frac{e^2}{2a_0}. \tag{95a}$$

To obtain the wave functions, one has from the recursion relation Eq. (91) together with (92),

$$c_s = -\frac{n' - s}{s(2\gamma + s)}c_{s-1}$$
$$= (-1)^s \frac{(n'-1)\dots(n'-s)}{s!(2\gamma+1)\dots(2\gamma+s)}c_0. \tag{96}$$

Thus $u(\rho)$ is the confluent hypergeometric series

$$u(\rho) = c_0\, F(-n'+1, 2\gamma+1; \rho), \tag{97}$$

which has been defined in Eq. (III-147), Vol. I.

From Eqs. (88) and (92), one has

$$d_s = -\frac{-\xi + \beta/\sqrt{1-\epsilon^2}}{n'-s}c_s, \tag{98}$$

so that from the second equation in Eq. (88) again,

$$\frac{d_s}{c_s} = \frac{d_0}{c_0}\frac{n'}{n'-s},$$

$$d_s = (-1)^s \frac{n'(n'-1)\dots(n'-s+1)}{s!(2\gamma+1)\dots(2\gamma+s)}d_0, \quad \text{from Eq. (96)}.$$

Thus $v(\rho)$ is

$$v(\rho) = d_0\, F(-n', 2\gamma+1; \rho). \tag{99}$$

From Eq. (98),

$$\frac{d_0}{c_0} = -\frac{-\xi + \dfrac{Z\alpha}{\sqrt{1-\epsilon^2}}}{n'}.$$

We shall take

$$c_0 = -C \frac{n'}{\sqrt{-\xi + Z\alpha/\sqrt{1-\epsilon^2}}},$$

$$d_0 = C\sqrt{-\xi + \frac{Z\alpha}{\sqrt{1-\epsilon^2}}}, \tag{100}$$

where the constant C is to be determined from the normalization condition for the $\Phi_1, \Phi_2, \Phi_3, \Phi_4$ of Eq. (67)

$$\int \{ |\Phi_1|^2 + |\Phi_2|^2 + |\Phi_3|^2 + |\Phi_4|^2 \} dr = 1. \tag{101}$$

After a lengthy calculation, one obtains

$$
\begin{aligned}
f(r) &= -\sqrt{1-\epsilon}\, W(r) \{ n' F(-n'+1, 2\gamma+1; \frac{2Z}{Na_0}r) \\
&\quad + (N-\xi) F(-n', 2\gamma+1; \frac{2Z}{Na_0}r) \}, \\
g(r) &= -\sqrt{1+\epsilon}\, W(r) \{ -n' F(-n'+1, 2\gamma+1; \frac{2Z}{Na_0}r) \\
&\quad + (N-\xi) F(-n', 2\gamma+1; \frac{2Z}{Na_0}r) \}, \\
N^2 &= n^2 - 2n'(\xi - \sqrt{\xi^2 - Z^2\alpha^2}), \\
W(r) &= \sqrt{\frac{\Gamma(2\gamma+n'+1)}{\Gamma(2\gamma+1)\sqrt{n'!}}} \frac{1}{\sqrt{4N(N-\xi)}} \left(\frac{2Z}{Na_0} \right)^{\gamma+\frac{1}{2}} e^{-Zr/Na_0}.
\end{aligned}
\tag{102}
$$

§.5.5. Dirac equation and many-body features

In Chapter 2, Sect. 3, it has been pointed out that the negative energy states in the Dirac theory - in fact in any relativistic theory - bring forth a consequence of basic importance, namely, the Dirac equation cannot strictly be regarded as theory of *one* particle, but is bound in a fundamental way with an infinite sea of particles in the negative energy states. That the negative energy states cannot be simply ignored as in the classical theory is first shown by Klein's study (Chap. 2, Sect. 3, (3)) and is then experimentally shown by the discovery of the positron and of the process of pair production. (Chap. 2, Sect. 3, (4)).

Then, in the mid and later 1940's, the development of quantum electro-dynamics - the theory of the quantized coupled electron and electromagnetic field - and the experimental discovery of the Lamb shift and the "anomaly" of the gyromagnetic ratio g of the electron by Kusch points without doubt to the many-body or field nature of the relativistic theory of the electron.

This many-body nature of the Dirac equation can be brought out in a very elementary way without using the method of quantum electrodynamics. We see in Section 4, Eq. (95), that for a (hypothetical) heavy enough atom

$$Z > \frac{(j + \frac{1}{2})}{\alpha} \tag{103}$$

i.e.,

$$Z > 137, \tag{103}$$

imaginary numbers appear in the energy formula. Let us trace this a little way back to the equation (82) for the radial wave functions, namely,

$$\begin{aligned}
\frac{dF}{dr} - \xi\frac{F}{r} &= [\frac{1}{\lambda}(1 - \epsilon) - \frac{Z\alpha}{r}]G, \\
\frac{dG}{dr} + \xi\frac{G}{r} &= [\frac{1}{\lambda}(1 + \epsilon) + \frac{Z\alpha}{r}]F,
\end{aligned} \tag{104}$$

where

$$\lambda = \frac{\hbar}{m_0 c} = \frac{1}{2\pi} \times \text{Compton wave length of the electron.}$$

Introducing ρ as in Eq. (84), eliminating the small component $F(r)$ between the two equations and writing

$$G(\rho) = e^{-\rho/2}W(\rho), \tag{105}$$

one obtains

$$\begin{aligned}
(\frac{1}{2}\sqrt{\frac{1+\epsilon}{1-\epsilon}} + \frac{Z\alpha}{\rho})(\frac{d^2W}{d\rho^2} - \frac{dW}{d\rho} + \frac{Z\alpha\epsilon}{\sqrt{1-\epsilon^2}}\frac{W}{\rho} + \frac{Z^2\alpha^2 - \xi^2}{\rho^2}W) \\
+ \frac{Z\alpha}{\rho^2}(\frac{dW}{d\rho} - W) - \frac{1}{2}\sqrt{\frac{1+\epsilon}{1-\epsilon}}\frac{\xi}{\rho^2}W(\rho) = 0.
\end{aligned} \tag{106}$$

Setting

$$W(\rho) = \rho^\gamma \sum_{s=0} a_s\rho^s, \tag{107}$$

one gets the indicial equation

$$\gamma^2 - \xi^2 + Z^2\alpha^2 = 0. \tag{108}$$

If

$$Z > \frac{\xi}{\alpha} = 137\xi, \tag{109}$$

γ is imaginary, and it is easy to see from the recursion formula for the a_s's that the series does not teminate into a polynomial but approaches

$$W(\rho) \simeq e^\rho, \tag{110}$$

so that $G(\rho)$ approaches $e^{\frac{1}{2}\rho}$ and is not quadratically integrable. Thus no stable states exist for $Z > 137$, in contradistinction to the non-relativistic Schrödinger theory.

The interpretation of this result is as follows. As Z becomes large, the electric field near the nucleus is intense and mixes up the positive and negative energy states so that the many-body properties become increasingly important, rendering the one-particle concept invalid. In this connection, it is interesting to see that the fine structure constant α, which is the coupling strength parameter in quantum field theory, plays the role of a critical value for the breaking down of the one-particle theory.

§.5.6. "Extra moments" to the Dirac equation?

In Eq. (10a), it has been seen that the Dirac equation can be put in the form[31] (summation convention)

$$[\Pi_\mu^2 + m_0^2 + \frac{e}{2i}\gamma_\mu\gamma_\nu F_{\mu\nu}]\Psi = 0. \tag{111}$$

In Eq. (14), it is seen that

$$\frac{e}{2i}\gamma_\mu\gamma_\nu F_{\mu\nu} = e(\sigma \cdot H) - ie\rho_1(\sigma \cdot E), \tag{112}$$

31

$$[\Pi_\mu^2 + m_0^2 c^2 + \frac{\hbar e}{2ic}\gamma_\mu\gamma_\nu F_{\mu\nu}]\Psi = 0. \tag{111'}$$

124

representing a moment σ .

But the Dirac equation Eq. (4)

$$(\gamma_\mu \Pi_\mu - im_0)\Psi = 0$$

can be extended by two extra terms without destroying the Lorentz covariance,[32]

$$
\begin{aligned}
(\gamma_\mu \Pi_\mu - im_0)\Psi = {} & g_1 \left(\frac{e}{4m_0}\right)\gamma_\mu \gamma_\nu F_{\mu\nu} \Psi \\
& - g_2 e \left(\frac{1}{m_0}\right)^2 \gamma_\mu \partial_\nu \partial_\nu A_\mu \Psi,
\end{aligned}
\tag{113}
$$

where g_1, g_2 are two dimensionless constants; $\gamma_\mu \gamma_\nu F_{\mu\nu}$ is a scalar under Lorentz transformations;

$$\partial_\nu \partial_\nu A_\mu = -4\pi j_\mu,\tag{114}$$

is a 4-vector, so that $\gamma_\mu \partial_\nu \partial_\nu A_\mu$ is also a scalar. The g_2 term in Eq. (113) may be regarded as representing a replacement of the A_μ in Eq. (4) by

$$A_\mu + g_2 \left(\frac{1}{m_0}\right)^2 \partial_\nu \partial_\nu A_\mu.\tag{115}$$

We shall in the following consider only the additional term g_1 (i.e., disregarding the g_2 term).

Denote by

$$G = g_1 \left(\frac{e}{4m_0}\right)\gamma_\mu \gamma_\nu F_{\mu\nu}\tag{116}$$

and carry out a calculation parallel to that from Eq. (9) to (14), one obtains the following equation[33]

32

$$
\begin{aligned}
(\gamma_\mu \Pi_\mu - im_0 c)\Psi = {} & g_1 \left(\frac{e\hbar}{4m_0 c^2}\right)\gamma_\mu \gamma_\nu F_{\mu\nu} \Psi \\
& - g_2 \frac{e}{c} \left(\frac{\hbar}{m_0 c}\right)^2 \gamma_\mu \partial_\nu \partial_\nu A_\mu \Psi.
\end{aligned}
\tag{113$'$}
$$

33

$$[\Pi^2 + m_0^2 c^2 + (1 + g_1)\frac{e\hbar}{2ic}\gamma_\mu \gamma_\nu F_{\mu\nu}] = [G^2 + \gamma_\sigma \Pi_\sigma G - G\gamma_\sigma \Pi_\sigma].\tag{117$'$}$$

$$[\Pi^2 + m_0^2 + (1 + g_1)\frac{e}{2i}\gamma_\mu\gamma_\nu F_{\mu\nu}] = [G^2 + \gamma_\sigma\Pi_\sigma G - G\gamma_\sigma\Pi_\sigma]. \tag{117}$$

The extra magnetic moment arising from the term G (and g_1) is suggested by Pauli. So far there seems to have been no experimental evidence for such an extra moment in the case of an electron.[34,35]

---

[34] *There seems to be a story in connection with this Pauli moment. In 1947, G. Breit (at Yale University) and I. I. Rabi (Columbia University) discussed the possibility of detecting, by means of the atomic beam facilities at Columbia, any extra magnetic moment - deviation from the value g = 2 of Dirac's theory. In the careful measurements of the value of g by P. Kusch (1947) and the analysis by H. M. Foley, the value*

$$g = 2(1 + \frac{\alpha}{2\pi} + ...), \tag{118}$$

was found, which is in excellent agreement with the theoretical value just predicted by J. Schwinger (at Harvard University) on his covariant treatment of quantum electrodynamics. The Lamb shift (1947) and this "g-2" discovery are most important as direct supports for the then new developments of quantum electrodynamics.

[35] *Another modification of the Dirac equation has been examined, by G. Feinberg, Phys. Rev. 112, 1637 (1958); E. F. Salpeter, ibid. 112, 1642 (1958), namely*

$$(\gamma_\mu\Pi_\mu - im_0)\Psi = \xi\frac{e}{4m_0}\gamma_5\gamma_\mu\gamma_\nu F_{\mu\nu}\Psi. \tag{113a}$$

The extra term containing the pseudo-scalar γ_5 corresponds to a parity non-conserving electric dipole moment. ξ is a numerical parameter. This term leads to terms of the form

$$\xi\mu_B(\sigma\cdot\mathbf{E}_{in}), \quad \xi\mu_B(\sigma\cdot\mathbf{E}_{ext}),$$
$$-\xi\frac{\mu_B}{2m_0}(\sigma\cdot[\mathbf{H}\times\mathbf{p}]),$$
$$i\xi\frac{\mu_B}{2m_0}(\mathbf{p}\cdot\mathbf{H}).$$

$\mathbf{E}_{in}$ is the internal electric field due to the nucleus, and $\mathbf{E}_{ext}$ the external field. Of the various terms, only the $\xi\mu_B(\sigma \cdot \mathbf{E}_{in})$, where $\mathbf{E}_{in} = \frac{Ze}{4\pi r^2}$, is important, and for the ground state is $^2S_{\frac{1}{2}}$ of hydrogen,

$$E(1s\,^2S_{\frac{1}{2}}) = -\frac{1}{2}\xi^2\alpha^2\,(Rydberg), \qquad \alpha = \frac{e^2}{4\pi} \simeq \frac{1}{137}$$
$$= -8.7 \times 10^4\xi^2 MC/sec.$$
$$= -2.9\xi^2 cm^{-1},$$

which is a downward shift (opposite to the Lamb shift). If the optical method for the Lyman α-line can detect at best a shift of $\simeq 0.01cm^{-1}$, it is not sensitive enough for an upper limit for ξ . Salpeter shows that for the $2s\,^2S_{\frac{1}{2}}$ state, a considerably lower value can be concluded for the upper limit of $\xi(\xi < 0.004)$.

Exercises: *Chapter 5*

1. Solve the free Dirac equation in a spherical cavity of radius R subject to the linear boundary condition proposed in the MIT bag model:

$$(\gamma_j r_j / r)\psi(\mathbf{r}) = \psi(\mathbf{r}) \qquad \text{at } r = R.$$

Show that the eigenvalue ω is given by

$$\omega = \frac{1}{R}\sqrt{x^2 + (mR)^2},$$

with x determined by

$$\tan x = \frac{x}{1 - mR - \sqrt{x^2 + (mR)^2}}.$$

For $m = 0$, one finds $x = 2.0428$ as the lowest eigenvalue. Find the next eigenvalue if you are familiar with the numerical method. Otherwise, find the forms for the first few excited states.

2. Solve the problem of a Dirac particle moving in a confining potential:

$$(1 + \gamma_4)\frac{kr^2}{4} + (a + b\gamma_4).$$

(i) Fix the parameters such that the ground-state solution will be of the form:

$$\psi(\mathbf{r}, s) = \begin{pmatrix} u(r) \\ i\sigma \cdot rv(r) \end{pmatrix} \chi_s,$$

with χ_s a two-component Pauli spinor and $v(r)/u(r) = $ constant.
(ii) Fix the parameters such that the mean square radius is 0.517 fm^2 for the ground state (a value for the nucleon in the MIT bag).

3. (i) Solve the problem of an electron moving in a constant magnetic field in the z direction.
(ii) Solve the problem of a neutron as a neutral Dirac particle moving in a constant magnetic field. (Note that the neutron has an anomalous magnetic moment of $-1.913\ n.m.$)
(iii) Compare the above two cases. Could you confine neutrons in a suitably designed magnetic field?

128

Part B. Introduction to Quantum Field Theory

Chapter 6.
Classical Field Theory

§.6.1. Introduction

In classical physics, "particles" and "fields" are two different concepts; they deal with different phenomena; they are characterized by discreteness and continuity respectively. A system may contain a vast number of particles - such as the molecules in a gas, but as long as they are denumerable, the basic theory is classical dynamics. But if the variables of a system are not denumerable - such as the electric field intensity or the velocity field of a fluid, the basic theory is the so-called field theory.

The fundamental theory in classical dynamics is the Newtonian dynamics, which can be expressed in different mathematical forms, such as the Lagrangian form, the Hamiltonian form, and the variational form.

The phenomena and the laws of fields are usually expressed in partial differential equations - such as those of fluid dynamics and of electromagnetic fields. But these equations are also expressed in the form of Lagrange and Hamilton equations, and these latter equations can in turn be derived from a Lagrangian density function via a variational principle.

In classical physics, electromagnetic fields are regarded as continuous functions - nondenumerably infinite numbers of variables. If all electromagnetic phenomena, including the propagation of light, are continuous, then the descriptions by the Maxwell equations, or by means of variational equations, are equivalent.

But since the quantum theory of radiation of Einstein in 1905, direct experiments such as the photo-electric effect and the Compton effect, give strong evidence for the particle properties of electromagnetic radiation. This leads us to seek a mathematical theory to describe the quantum properties of continuous fields - analogous to the quantization of atomic and molecular phenomena in quantum mechanics. To extend quantum mechanics to the quantized properties of fields, the theory is called the theory of quantized fields.

There are various kinds of fields. One kind is the "classical fields", such as the electromagnetic field. The concept of the "quantum" of radiation originates with Einstein's theory of the photon in 1905; but the formal mathematical method of "quantization" of particle dynamics begins with Bohr's theory of the hydrogen atom in 1913, in the form of the quantum conditions, as generalized by Sommerfeld and Wilson,

$$\oint p_k dq_k = n_k h. \tag{1}$$

This "old quantum theory" has amazingly successful results for the hydrogen atom, but then has met with very basic difficulties in other systems. Then quantum mechanics was born, and the quantization condition can be formulated in the form of the commutation relation

$$pq - qp = \frac{\hbar}{i}. \tag{2}$$

On this relation, quantum mechanics in the form of the matrix theory and the Schrödinger theory has been established. Quantum mechanics represents both the particle and the wave properties of what in classical physics have been described as "particles" and "waves", mutually exclusive properties now joined by the Einstein-de Broglie relations

$$E = h\nu, \qquad p = \frac{h}{\lambda}. \tag{3}$$

Quantum mechanics has successfully treated all known phenomena in atomic, molecular, solid state, and nuclear physics.

Now we wish to go from particles back to the older problem - the quantum theory of the electromagnetic field, to extend the method of quantization for particle dynamics (coordinates and momenta, angular momenta and energies) to the continuous fields.

The method consists of the following program:

(1) Express the laws of electromagnetic field, namely, Maxwell's field equations, in terms of the 4-potential $A_1, A_2, A_3, A_4 = i\phi$, in the form of the Hamiltonian equations in classical dynamics,

(2) define the field variables (such as the 4-potential) and their canonical conjugates,

(3) quantize the field variables by relations similar to the relation (2).

This program, when applied to the so-called "free field" (such as electromagnetic field free from the charge density ρ and current density j), can be readily carried out, and in fact the quantized field is represented by the photons.

The electromagnetic field having a 4-potential is not the simplest classical field. If we regard the Klein-Gordon equation[1]

$$(\nabla^2 - \frac{\partial^2}{\partial t^2})\Psi = (\frac{1}{\lambda_c})^2\Psi. \qquad \lambda_c = \frac{1}{m}, \tag{4}$$

as the equation of a classical scalar Ψ field, the quantization process is even simpler. We shall see that the quantized field of Eq. (4) is appropriate for the description of the π mesons (pions, π^0, π^+, π^-).

But there are other kinds of fields than the above-mentioned classical fields. We must return to the quantization of particle systems.

In the last four chapters, we have treated the Dirac relativistic wave equation of the electron. It has the following successes: it is Lorentz covariant; it contains the electron spin with the gyromagnetic ratio $g = 2$; it predicts the anti-particle positron which is experimentally discovered; it gives the correct fine structure of the hydrogenic atom levels.

As already mentioned in the introductory chapter (Ch. 0), there are "difficulties" associated with the Dirac equation. The first one has to do with the question of extending the theory for a single electron to a system of many electrons, since it is not clear if the negative-energy spectra associated with different electrons can be treated in a consistent manner. Another "difficulty" is of an even more basic nature, namely, a theory started out to represent a single electron ends up being inseparably bound with a many-body effect on account of the infinite sea of electrons in negative energy states. Using Dirac's "hole" picture, we see that, in the presence of a strong electric field, an electron does not exist as a lone electron but is a system containing infinitely many (and not constant in number) electrons and positrons. Thus, strictly speaking, the single-electron theory is valid only in the sense of a limiting case of a free electron and must be replaced by a theory of "many-electrons".

Just as the representation of "photons" is the electromagnetic field, the appropriate representation of a system of an infinitely large and variable number of "particles" is a "field". Thus, we see that, starting from the particle point of view, one seeks a theory of particles that satisfies the relativity principle and finds that the appropriate theory is a many-body theory, and the appropriate representation of a many-body system is "field".

---

[1] λ_c *is the Compton wave length. In ordinary units, it is $\hbar/(mc)$.*

Therefore, the problem is then to formulate a many-electron field. One may start from the Dirac equation for a free electron:

$$(\gamma_\mu \frac{1}{i} \frac{\partial}{\partial x_\mu} - im_0)\Psi = 0, \tag{5}$$

and regard it as the equation of a classical field Ψ - in this case a 4-component field. Then treat Ψ_j as the field variables (similarly to the 4-potential A_μ of the electromagnetic field), and their canonical conjugates. Then introduce the quantization conition in the form of the commutation relations (2), and as expected, one obtains the electron as the particles of the quantized field. Such a quantization procedure is called "second quantization". The quantized Ψ possesses both the wave and the particle properties - just as the quantized electromagnetic field possesses the photon and the wave property.

Unfortunately, the story did not end here. The quantum theory of a pure electron field presents no difficulties but, for the system of the electron coupled to the electromagnetic field, a relativistic quantum field theory, known as quantum electrodynamics or QED, presents serious deep-rooted difficulties, namely the persistent presence of infinities in the results of calculations of physical quantities.

The theory of quantized electromagnetic fields begins with the work of Dirac in 1927, followed immediately by the work of Jordan and Wigner, and by Fermi in 1930. A breakthrough came in the mid 1940's with the work of Tomonaga in Japan and Schwinger, Feynman, and Dyson in the United States. Although infinities still remain, the theory succeeds in "subtracting" them away in a consistent way so that finite results can be obtained, which have been found to be in excellent agreement with the observed Lamb shifts and the "g anomaly". The decade from the mid 1940's to the mid 1950's is a period of fervent studies of quantum electrodynamics, both in further calculations on this "renormalization" theory and in attempts to rid the theory of the infinities (not just to isolate and bury them, so to speak).

During the last two decades, physicists have arrived at a successful description of elementary particles and the nature of their interactions and their unification — the so-called "Standard Model" which consists of quantum chromodynamics (QCD) for describing strong interactions among quarks and gluons and the Glashow-Salam-Weinberg (GSW) theory for a unified description of electromagnetic and weak interactions. Both QCD and the GSW electroweak theory are constructed using QED as a prototype theory. Till now, however, a satisfactory answer is yet to be found toward the question why QED and more generally the Standard Model, in which infinities appear but get subtracted away in a certain way, have been so successful.

In this part of the book, we thus wish to introduce in a pedagogic manner basic elements associated with QED, ranging from free classical fields and their quantization, to calculations of scattering matrix, or S-matrix, elements, and to concepts of "regularization" and "renormalization". A primary objective of our presentation is to set up the framework in order to introduce in the third part of this book basic ingredients associated with the Standard Model. Another objective is to stimulate further thoughts, or developments, concerning the treatment (or "renormalization") of infinities in QED.

§.6.2. Classical field equation

In classical dynamics, the state of a system of N particles is defined by $3N$ pairs of generalized coordinates q_k and their conjugate momenta p_k.

In classical field theories, a field is described by a continuous function of real variables $\phi(\mathbf{r}, t)$, and n fields by n such functions $\phi_\alpha(\mathbf{r}, t), \alpha = 1, 2, 3, ..., n$. Each field is described by partial differential equations of independent variables $\mathbf{r}, t$, called the field equations. Thus in classical electromagnetic field theory, ther are four fields for $(A_1, A_2, A_3, A_4 = i\phi)$, forming a vector field with four components.

If the Klein-Gordon equation is regarded as an equation for a classical field Ψ, then the field is a scalar field, i.e., a Ψ having only one component.

The equations of motion of particles in classical dynamics can be obtained from a variational principle. We may extend this view to a field in the following manner.

A field is regarded as a dynamical system having a non-denumerably infinite number of degrees of freedom, whose generalized coordinates and conjugate momenta are the components $\varphi_\alpha(\mathbf{r}, t)$ and $\dot{\varphi}_\alpha = \frac{\partial \phi_\alpha}{\partial t}$. Here the $\mathbf{r}$ are *not* the coordinates, but are just independent parameters. Let us imagine a volume V in the 3-dimentional space $\mathbf{r}$ be divided in small cells, $\triangle V^{(s)}$ at $(\mathbf{r}_s, t)$, and let $\triangle V^{(s)}$ be smaller and smaller. At a point x_s at time t, $\varphi_\alpha^{(s)}$ is almost constant within $\triangle V^{(s)}$. We shall choose as "coordinates" for the field φ the average values of φ_α at various x_s

$$\varphi_\alpha^{(s)}(x_s, t), \qquad \alpha = 1, 2, 3, \ldots \quad s = 1, 2, 3, \ldots \qquad (6)$$

and s is denumerable. Now let $\triangle V^{(s)}$ approach zero in the limit, and the degree of freedom of the field becomes non-denumerably infinite. In this sense, we can extend the variational method of obtaining the Lagrange and Hamiltonian equations in classical dynamics to obtain the field equations.

Let us introduce the notations

$$\partial_\mu = \frac{\partial}{\partial x_\mu}, \qquad \mu = 1, 2, 3, 4$$

$$d^3x = dx_1 dx_2 dx_3$$

$$d^4x = dx_1 dx_2 dx_3 dt \quad \text{(scalar, invariant)},$$

and define the Lagrangian density $\mathcal{L}$ and Lagrangian L

$$\mathcal{L} = \mathcal{L}(\varphi_\alpha, \partial_\mu \varphi_\alpha), \qquad (7)$$

$$L = \int_{V_3} \mathcal{L}(\varphi_\alpha, \partial_\mu \varphi_\alpha) d^3x, \qquad (8)$$

$$\text{Action function} = S = \int_{t_1}^{t_2} L dt = \int_{V_4} \mathcal{L}(\varphi_\alpha, \partial_\mu \varphi_\alpha) d^4x, \qquad (9)$$

where V_4 is a 4-dimensional volume $V_4 = V_3 \times (t_2 - t_1)$. d^4x is scalar (invariant) under Lorentz transformations. If $\mathcal{L}$ is a scalar, then S is also an invariant. (Conversely, if S is an invariant, then $\mathcal{L}$ is an invariant; but L is not an invariant.) Let $\delta\varphi_\alpha$ be variations such that

$$\delta\varphi_\alpha = \text{arbitrary within} \quad V_4, \qquad (10)$$

$$\delta\varphi_\alpha = 0 \quad \text{on the boundary surface} \quad S \quad \text{of} \quad V_4,$$

and formulate the variational principle

$$\delta S = \delta \int_{V_4} \mathcal{L} d^4x = 0. \qquad (11)$$

From

$$\delta \mathcal{L} = \frac{\partial \mathcal{L}}{\partial \varphi_\alpha} \delta \varphi_\alpha + \frac{\partial \mathcal{L}}{\partial (\partial_\mu \varphi_\alpha)} \delta (\partial_\mu \varphi_\alpha)$$

and on integrating by parts,[2]

$$\delta S = \int_{V_4} \{ \frac{\partial \mathcal{L}}{\partial \varphi_\alpha} - \partial_\mu (\frac{\partial \mathcal{L}}{\partial (\partial_\mu \varphi_\alpha)}) \} \delta \varphi_\alpha d^4 x \tag{12}$$

from which one obtains the Euler equation

$$\frac{\partial \mathcal{L}}{\partial \varphi_\alpha} - \partial_\mu (\frac{\partial \mathcal{L}}{\partial (\partial_\mu \varphi_\alpha)}) = 0, \tag{13}$$

or

$$\frac{\partial \mathcal{L}}{\partial \varphi_\alpha} - \partial_k \frac{\partial \mathcal{L}}{\partial (\partial_k \varphi_\alpha)} - \partial_t \frac{\partial \mathcal{L}}{\partial (\partial_t \varphi_\alpha)} = 0, \tag{13a}$$

which is a second order differential equation if $\mathcal{L}$ is a function of the first order partial derivatives $\partial_\mu \varphi_\alpha$.

We introduce the $\varphi_\alpha^{(s)}$ of Eq. (6) and replace the integral L in Eq. (8) by

$$L = \lim_{\Delta V^{(s)} \to 0} \sum_s \mathcal{L}(\varphi_\alpha^{(s)}, \partial_\mu \varphi_\alpha^{(s)}) \Delta V^{(s)} \tag{14}$$

and take $\varphi_\alpha^{(s)}$ as the coordinates (like the q_k for particles), and define their conjugate variables p_α and conjugate momentum densities $\pi_\alpha^{(s)}$ by

$$p_\alpha = \frac{\delta \mathcal{L}}{\delta \dot{\varphi}_\alpha^{(s)}} \Delta V^{(s)} \tag{15}$$

$$\pi_\alpha^{(s)} = \frac{p_\alpha}{\Delta V^{(s)}} = \frac{\partial \mathcal{L}}{\partial \dot{\varphi}_\alpha^{(s)}} \tag{16}$$

---

[2] *The integral*

$$\int_{V_4} \partial_\mu (\frac{\partial \mathcal{L}}{\partial (\partial_\mu \varphi_\alpha)} \delta \varphi_\alpha) d^4 x = \int_S \frac{\partial \mathcal{L}}{\partial (\partial_\mu \varphi_\alpha)} \delta \varphi_\alpha dS = 0,$$

where the integral of a 4-divergence can be transformed into a surface integral over the boundary $\sum$ of V_4 , and $\delta \varphi_\alpha = 0$ on $\sum$ according to Eq. (10).

As in classical dynamics, we define the Hamiltonian density $\mathcal{H}$ and the Hamiltonian H[3]

$$\mathcal{H}(\varphi_\alpha, \pi_\alpha, \partial_k \varphi_\alpha) = \pi_\alpha^{(s)} \dot{\varphi}_\alpha^{(s)} - \mathcal{L} \tag{17}$$

$$
\begin{aligned}
H &= \lim_{\triangle V^{(s)} \to 0} \sum_s \mathcal{H} \triangle V^{(s)} \\
&= \int_{V_3} \mathcal{H} d^3 x.
\end{aligned}
\tag{18}
$$

On carrying out the variation δH of the integral (18) and the integration by parts, one obtains

$$\delta H = \int_{V_3} \{[\frac{\partial \mathcal{H}}{\partial \varphi_\alpha} - \partial_k \frac{\partial \mathcal{H}}{\partial(\partial_k \varphi_\alpha)}]\delta\varphi_\alpha + \frac{\partial \mathcal{H}}{\partial \pi_\alpha}\delta\pi_\alpha\}d^3 x, \tag{19}$$

$$\frac{\delta H}{\delta \varphi_\alpha} = \frac{\partial \mathcal{H}}{\partial \varphi_\alpha} - \partial_k \frac{\partial \mathcal{H}}{\partial(\partial_k \varphi_\alpha)}, \tag{20}$$

$$\frac{\delta H}{\delta \pi_\alpha} = \frac{\partial \mathcal{H}}{\partial \pi_\alpha}. \tag{21}$$

From Eq. (17), one obtains

$$\frac{\partial \mathcal{H}}{\partial \varphi_\alpha} - \partial_k \frac{\partial \mathcal{H}}{\partial(\partial_k \varphi_\alpha)} = -\frac{\partial \mathcal{L}}{\partial \varphi_\alpha} + \partial_k \frac{\partial \mathcal{L}}{\partial(\partial_k \varphi_\alpha)}, \tag{22}$$

$$\frac{\partial \mathcal{H}}{\partial \pi_\alpha} = \dot{\varphi}_\alpha. \tag{23}$$

From Eqs. (21), (16), (13), (22), and (20), one obtains

$$\dot{\varphi}_\alpha = \frac{\delta H}{\delta \pi_\alpha}, \qquad \dot{\pi}_\alpha = -\frac{\delta H}{\delta \varphi_\alpha}, \tag{24}$$

which are the field equations in canonical form.

From Eqs. (18), (17), (16), and (13), it can be shown that

$$\frac{dH}{dt} = -\int_{V_3} \frac{\partial \mathcal{L}}{\partial t} d^3 x \tag{25}$$

---

[3] The $\mathcal{H}$ defined by the Legendre transformation Eq. (17) is a function of $\partial_k \varphi_\alpha^{(s)}$, $k = 1, 2, 3$, but not of $\partial_t \varphi_\alpha^{(s)} = \dot{\varphi}_\alpha^{(s)}$.

so that if $\frac{\partial \mathcal{L}}{\partial t} = 0$, then $\frac{dH}{dt} = 0$, i.e., H is a constant of motion.

If $\mathcal{L}$ is replaced by

$$\mathcal{L}' = \mathcal{L} + \partial_\mu g_\mu(\varphi_\alpha), \tag{26}$$

where g_μ is an arbitrary 4-vector function of φ_α so that $\partial_\mu g_\mu$ is a 4-divergence, then the Euler-Lagrange equation obtained from

$$\delta \int \mathcal{L}' d^4 x = 0, \tag{27}$$

is the same as Eq. (13).[4]

§.6.3. Noether's Theorem

The Lagrangian density $\mathcal{L}$ as specified by Eq. (7) contains all the information concerning dynamical invariants such as energy-momentum 4-vector and angular momentum tensor, just like in the case of classical dynamics for a system of particles where constants of motion are contained in the Lagrangian. This aspect is summarized by the well-known Noether's theorem.

Noether's Theorem: To every continuous transformation of field functions and simultaneously coordinates which depends on n_0 continuous parameters and which ensures that the variation of the action is zero, there corresponds n_0 dynamical invariants which are combinations of field functions and their derivatives that are conserved in time.

Proof:

We define the continuous transformation in question as follows:

$$x_\mu \rightarrow x'_\mu = x_\mu + \delta x_\mu = x_\mu + X_{\mu\alpha}\delta w_\alpha, \tag{28a}$$

$$\phi_i(x) \rightarrow \phi'_i(x') = \phi_i(x) + \delta\phi_i(x) = \phi_i(x) + \Psi_{i\alpha}\delta w_\alpha, \tag{28b}$$

where δw_α $(\alpha = 1, 2, ..., n_0)$ are n_0 infinitesimal parameters, $X_{\mu\alpha}$ specify the variation in coordinates, and $\Psi_{i\alpha}$ characterize the change in field functions $\phi_i(x)$.

---

[4] $g_\mu(\varphi_\alpha)$ *have only to satisfy the condition that* $g_\mu = 0$ *on the boundary surface* $\sum$ *of* V_4.

The theorem states that, if the corresponding variation in the action S is zero,

$$\delta S = \delta \int \mathcal{L}(x) d^4 x \equiv \int \mathcal{L}'(x') d^4 x' - \int \mathcal{L}(x) d^4 x = 0, \tag{29}$$

then we must be able to find n_0 dynamical quantities $\Theta_{\mu\alpha}$ such that

$$\partial_\mu \Theta_{\alpha\mu} = 0 \quad (\alpha = 1, 2, ..., n_0). \tag{30}$$

The procedure to find the expression for $\Theta_{\alpha\mu}$ is described below: Define the variation in the form of the field function,

$$\bar{\delta}\phi_i(x) \equiv \phi_i'(x) - \phi_i(x), \tag{31}$$

or,

$$\begin{aligned}\bar{\delta}\phi_i(x) &= \delta\phi_i(x) - (\partial_\mu \phi_i)\delta x_\mu \\ &= (\Psi_{i\alpha} - (\partial_\mu \phi_i)X_{\mu\alpha})\delta w_\alpha.\end{aligned} \tag{32}$$

We have

$$\mathcal{L}'(x') \equiv \mathcal{L}(\phi_i'(x'), \partial_\mu' \phi_i'(x')) = \mathcal{L}(x) + \delta\mathcal{L}(x), \tag{33a}$$

with

$$\delta\mathcal{L} = \frac{\partial\mathcal{L}}{\partial\phi_i}\delta\phi_i + \frac{\partial\mathcal{L}}{\partial(\partial_\mu\phi_i)}\delta(\partial_\mu\phi_i) = \bar{\delta}\mathcal{L}(x) + \frac{d\mathcal{L}}{dx_\mu}\delta x_\mu, \tag{33b}$$

$$\bar{\delta}\mathcal{L}(x) = \frac{\partial\mathcal{L}}{\partial\phi_i}\bar{\delta}\phi_i + \frac{\partial\mathcal{L}}{\partial(\partial_\mu\phi_i)}\bar{\delta}(\partial_\mu\phi_i). \tag{33c}$$

Eq. (29) becomes

$$\delta S = \int (\bar{\delta}\mathcal{L}(x) + \frac{d\mathcal{L}}{dx_\mu}\delta x_\mu) d^4 x + \int \mathcal{L}(x) d^4 x' - \int \mathcal{L}(x) d^4 x. \tag{34}$$

Noting that

$$d^4 x' = dx_1' dx_2' dx_3' dt' = \frac{\partial(x_1', x_2', x_3', t')}{\partial(x_1, x_2, x_3, t)} d^4 x \cong (1 + \frac{\partial\delta x_\mu}{\partial x_\mu}) d^4 x, \tag{35}$$

we obtain

$$\delta S = \int \{\bar{\delta}\mathcal{L}(x) + \frac{d}{dx_\mu}(\mathcal{L}(x)\delta x_\mu)\}d^4x. \tag{36}$$

Using Eq. (13), we find from Eq. (33c)

$$\bar{\delta}\mathcal{L}(x) = \partial_\mu\{\frac{\partial\mathcal{L}}{\partial(\partial_\mu\phi_i)}\bar{\delta}\phi_i\}, \tag{37}$$

so that Eq. (36) becomes

$$\begin{aligned}\delta S &= \int \frac{d}{dx_\mu}\{\frac{\partial\mathcal{L}}{\partial(\partial_\mu\phi_i)}\bar{\delta}\phi_i + \mathcal{L}(x)\delta x_\mu\}d^4x \\ &\equiv -\int (\partial_\mu\Theta_{\alpha\mu}(x))d^4x\delta w_\alpha,\end{aligned} \tag{38}$$

with

$$\Theta_{\alpha\mu} = -\frac{\partial\mathcal{L}}{\partial(\partial_\mu\phi_i)}(\Psi_{i\alpha} - (\partial_\nu\phi_i)X_{\nu\alpha}) - \mathcal{L}(x)X_{\mu\alpha}. \tag{39}$$

Since δw_α is arbitrary, we obtain from Eq. (29)

$$\partial_\mu\Theta_{\alpha\mu}(x) = 0. \tag{30'}$$

This completes the proof of Noether's theorem.

(1) The Energy-Momentum Tensor $T_{\mu\nu}$.

Consider Noether's theorem in the case of translation:

$$x_\mu \to x'_\mu = x_\mu + \delta a_\mu = x_\mu + \delta_{\mu\nu}\delta w_\nu, \qquad \phi_i(x) \to \phi'_i(x') = \phi_i(x). \tag{40}$$

140

We obtain the energy-momentum tensor,[5,6,7,8]

$$T_{\mu\nu} = \partial_\mu \varphi_\alpha \frac{\partial \mathcal{L}}{\partial(\partial_\nu \varphi_\alpha)} - \delta_{\mu\nu} \mathcal{L}$$

$$\neq T_{\nu\mu}.$$

(41)

$T_{jk}, \quad j,k = 1,2,3, \quad$ is the stress tensor, (42)

$$\frac{1}{i} T_{j4} = \partial_j \varphi_\alpha \frac{\partial \mathcal{L}}{\partial \dot{\varphi}_\alpha} = \pi_\alpha \partial_j \varphi_\alpha = P_j,$$

$P_j, j = 1,2,3,$ form the momentum vector,

(43)

5

$$\frac{1}{ic} T_{j4} = \partial_j \varphi_\alpha \frac{\partial \mathcal{L}}{\partial \dot{\varphi}_\alpha} = \pi_\alpha \partial_j \varphi_\alpha = P_j.$$

(43')

6

$$ic\, T_{4j} = \dot{\varphi}_\alpha \frac{\partial \mathcal{L}}{\partial(\partial_j \varphi_\alpha)} = S_j.$$

(44')

7

$$c\, T_{44} = \dot{\varphi}_\alpha \frac{\partial \mathcal{L}}{\partial \dot{\varphi}_\alpha} - \mathcal{L} = \mathcal{H}.$$

(45')

[8]$T_{\mu\nu}$ defined in Eq. (41) is not symmetric in general. In Eqs. (26) and (27), it is known that adding a 4-divergence $\partial_\mu g_\mu$ to $\mathcal{L}$ leaves the field equations unchanged. But if one adds to $\mathcal{L}$ a $\partial_\mu g_\mu$ satisfying the relations

$$\{\partial_\nu \varphi_\alpha \frac{\partial}{\partial(\partial_\mu \varphi_\alpha)} - \partial_\mu \varphi_\alpha \frac{\partial}{\partial(\partial_\nu \varphi_\alpha)}\}(\mathcal{L} + \partial_\beta g_\beta) = 0,$$

(46)

and

$$\partial_\beta \left(\frac{\partial g_\beta}{\partial \varphi_\alpha} - \partial_\mu \frac{\partial g_\beta}{\partial(\partial_\mu \varphi_\alpha)} \right) = 0,$$

(47)

then one can make

$$T_{\mu\nu} = T_{\nu\mu}.$$

(48)

$$i T_{4j} = \dot{\varphi}_\alpha \frac{\partial \mathcal{L}}{\partial(\partial_j \varphi_\alpha)} = S_j,$$

$$(44)$$

$S_j, \quad j = 1, 2, 3,$ from energy flux density vector.

$$T_{44} = \dot{\varphi}_\alpha \frac{\partial \mathcal{L}}{\partial \dot{\varphi}_\alpha} - \mathcal{L} = \mathcal{H}, \qquad \text{the energy density.} \qquad (45)$$

If $\mathcal{L}$ does not explicitly depend on x_μ , one has from Eq. (41)

$$\partial_\nu T_{\mu\nu} = -\left(\frac{\partial \mathcal{L}}{\partial \varphi_\alpha} - \partial_\nu \frac{\partial \mathcal{L}}{\partial(\partial_\nu \varphi_\alpha)}\right)\partial_\mu \varphi_\alpha. \qquad (49)$$

Hence we have, as a special case of Noether's theorem,

Theorem: If $\mathcal{L}$ satisfies the Lagrange equation Eq. (13), then

$$\partial_\nu T_{\mu\nu} = 0. \qquad (50)$$

Theorem: If Eq. (50) holds, then $\mathcal{L}$ satisfies the Lagrange equation Eq. (13).

The physical meaning of Eq. (50) is seen as follows. On integrating over a volume V_3 , one has

$$\begin{aligned}
0 &= \int_{V_3} \partial_\nu T_{\mu\nu} d^3 x \\
&= \int_{V_3} (\partial_j T_{\mu j} + \partial_4 T_{\mu 4}) d^3 x \\
&= \int_S (T_{\mu j} dS_j) + \frac{d}{dx_4} \int_{V_3} T_{\mu 4} d^3 x, \quad \mu = 1, 2, 3, 4.
\end{aligned} \qquad (50a)$$

The first term is a surface integral and vanishes if $T_{\mu j} = 0$ on the surface S of V_3 . Then the above equation gives, from Eqs. (43) and (45),

$$\frac{d}{dt} \int_{V_3} P_j d^3 x = 0, \quad j = 1, 2, 3, \qquad (51)$$

$$\frac{d}{dt} \int_{V_3} \mathcal{H} d^3 x = 0, \qquad (52)$$

which are the momentum and energy conservation relations.

Again from Eq. (50), one has[9]

$$\frac{\partial}{\partial x_j} T_{4j} + \frac{1}{i}\frac{\partial}{\partial t} T_{44} = 0, \tag{53}$$

$$\frac{\partial}{\partial x_j} T_{kj} + \frac{1}{i}\frac{\partial}{\partial t} T_{k4} = 0, \tag{54}$$

which are, on using Eqs. (44), (43), and (45),

$$div\, S + \frac{\partial}{\partial t} \mathcal{H} = 0, \tag{53a}$$

$$\frac{\partial}{\partial x_j} T_{kj} + \frac{\partial}{\partial t} P_k = 0. \tag{54a}$$

Eq. (53a) is the continuity equation (energy conservation). If the field is an electromagnetic field, then S is the Poynting vector.

The $T_{\mu\nu}$ tensor is[10]

$$T_{\mu\nu} = \begin{pmatrix} T_{11} & T_{12} & T_{13} & iP_1 \\ T_{21} & T_{22} & T_{23} & iP_2 \\ T_{31} & T_{32} & T_{33} & iP_3 \\ \frac{1}{i}S_1 & \frac{1}{i}S_2 & \frac{1}{i}S_3 & \mathcal{H} \end{pmatrix}. \tag{55}$$

If $\mathcal{L}$ is an explicit function of x_μ, then Eq. (50) does not hold, and energy and momentum are not conserved, the field interacting with external sources.

(2) The Angular-Momentum Tensor.[11]

9

$$\frac{\partial}{\partial x_j} T_{4j} + \frac{1}{ic}\frac{\partial}{\partial t} T_{44} = 0. \tag{53'}$$

$$\frac{\partial}{\partial x_j} T_{kj} + \frac{1}{ic}\frac{\partial}{\partial t} T_{k4} = 0. \tag{54'}$$

10

$$T_{\mu\nu} = \begin{pmatrix} T_{11} & T_{12} & T_{13} & icP_1 \\ T_{21} & T_{22} & T_{23} & icP_2 \\ T_{31} & T_{32} & T_{33} & icP_3 \\ \frac{1}{ic}S_1 & \frac{1}{ic}S_2 & \frac{1}{ic}S_3 & \mathcal{H} \end{pmatrix}. \tag{55'}$$

[11] *Rotations in space and Lorentz transformations (boosts) yield*

$$x_\mu \rightarrow x'_\mu = x_\mu + x_\nu \delta w_{\mu\nu} = x_\mu + X_{\mu\alpha\beta}\delta w_{\alpha\beta}, \qquad (\alpha < \beta) \tag{63}$$

The angular momentum density $\mathcal{M}_{jk}$ is defined by[12]

$$
\begin{aligned}
\mathcal{M}_{jk} &= x_j P_k - x_k P_j \\
&= \frac{1}{i}(x_j T_{k4} - x_k T_{j4}),
\end{aligned}
\tag{56}
$$

and the total angular momentum of the field in volume V_3 is

$$
M_{jk} = \int_{V_3} \mathcal{M}_{jk} d^3 x.
\tag{57}
$$

Let us define a $4 \times 4 \times 4$ tensor $m_{\mu\nu\rho}$ by

$$
m_{\mu\nu\rho} = \frac{1}{i}(x_\mu T_{\nu\rho} - x_\nu T_{\mu\rho}),
\tag{58}
$$

so that

$$
m_{jk4} = \mathcal{M}_{jk}.
\tag{59}
$$

Then

$$
\begin{aligned}
\partial_\rho m_{\mu\nu\rho} &= \frac{1}{i}(T_{\nu\mu} + x_\mu \partial_\rho T_{\nu\rho} - T_{\mu\nu} - x_\nu \partial_\rho T_{\mu\rho}) \\
&= \frac{1}{i}(T_{\nu\mu} - T_{\mu\nu}) \quad \text{if Eq. (50) holds,} \\
&= 0 \qquad \text{if Eq. (48) holds.}
\end{aligned}
$$

$$
\tag{60}
$$
$$
\tag{61}
$$

with $\delta w_{\nu\mu} = -\delta w_{\mu\nu}$ and $X_{\mu\alpha\beta} = x_\beta \delta_{\mu\alpha} - x_\alpha \delta_{\mu\beta}$. The corresponding change in the field function is, with $\alpha < \beta$,

$$
\phi_i(x) \rightarrow \phi_i'(x') = \phi_i(x) + \delta\phi_i(x) = \phi_i(x) + A_{ij\alpha\beta}\phi_j(x)\delta w_{\alpha\beta}.
\tag{64}
$$

Using Noether's theorem, we obtain the angular momentum tensor,

$$
M_{\alpha\beta\mu} = (x_\beta T_{\alpha\mu} - x_\alpha T_{\beta\mu}) - \frac{\partial \mathcal{L}}{\partial(\partial_\mu \phi_i)} A_{ij\alpha\beta}\phi_j(x).
\tag{65}
$$

For a scalar field, $A_{ij\alpha\beta} = 0$ so that Eq. (58) and Eq. (65) are related. For a vector field, we have

$$
A_{ij\alpha\beta} = \delta_{i\alpha}\delta_{j\beta} - \delta_{i\beta}\delta_{j\alpha},
\tag{66}
$$

so that Eq. (58) should be modified. So is the case for the Dirac field.

---

[12] *Note that, in ordinary units, i stands for ic in Eqs. (56), (58), and (60).*

Integrating this over V_3 , one has

$$0 = \int_V \partial_j m_{\mu\nu j} d^3 x + \frac{1}{i} \int_V \frac{\partial}{\partial t} m_{\mu\nu 4} d^3 x$$
$$= \int_S m_{\mu\nu j} dS + \frac{1}{i} \frac{d}{dt} \int_V M_{\mu\nu} d^3 x,$$

where we have written, as extension of Eq. (59),

$$M_{\mu\nu} \equiv m_{\mu\nu 4}.$$

The surface integral vanishes if $m_{\mu\nu j} = 0$ on S. Thus

$$\frac{d}{dt} M_{\mu\nu} = \frac{d}{dt} \int_V M_{\mu\nu} d^3 x = 0, \tag{62}$$

which includes Eq. (57). Thus if $T_{\mu\nu} = T_{\nu\mu}$, one has the conservation of the generalized angular momentum $M_{\mu\nu}$.

§.6.4. The Klein-Gordon Field in Lagrangian Form.

(1) The simplest free field is a scalar, real field satisfying the Klein-Gordon equation

$$(\partial_\alpha \partial_\alpha - \mu^2)\phi(x_\nu) = 0,$$
$$\mu = \frac{m_0 c}{\hbar} = \frac{2\pi}{\lambda_c}. \tag{67}$$

A Lagrangian density leading to this equation is[13,14]

---

[13] *At any given instant,*

$$\frac{\partial \mathcal{L}}{\partial \phi} = -\mu^2 \phi,$$

$$\frac{\partial \mathcal{L}}{\partial(\partial_k \phi)} = -\partial_k \phi, \quad k = 1, 2, 3, \qquad \frac{\partial \mathcal{L}}{\partial \dot\phi} = \pi \quad \text{from Eq. (16).} \tag{69}$$

From Eq. (13), one obtains Eq. (67).
[14]

$$\mathcal{L}(\phi, \partial_\nu \phi) = -\frac{c^2}{2}(\partial_\nu \phi \partial_\nu \phi + \mu^2 \phi^2). \tag{68'}$$

$$\mathcal{L}(\phi, \partial_\nu \phi) = -\frac{1}{2}(\partial_\nu \phi \partial_\nu \phi + \mu^2 \phi^2). \tag{68}$$

$\mathcal{L}$ is Lorentz invariant for a scalar or a pseudo-scalar ϕ, since it is quadratic in ϕ.

The Hamiltonian density is, from Eq. (17) is[15]

$$\begin{aligned}
\mathcal{H}(\phi, \pi, \partial_k \phi) &= \pi^2 + \frac{1}{2}(\partial_\nu \phi \partial_\nu \phi + \mu^2 \phi^2) \\
&= \frac{1}{2}\{\pi^2 + \mu^2 \phi^2 + (\nabla \phi)^2\}.
\end{aligned} \tag{70}$$

The canonical energy-momentum tensor is, from Eq. (28),[16]

$$T_{\mu\nu} = -\partial_\mu \phi \partial_\nu \phi - \delta_{\mu\nu} \mathcal{L} = T_{\nu\mu}. \tag{71}$$

In analogy with Eqs. (61) and (62), one obtains the angular momentum conservation equation

$$\frac{dM_{\mu\nu}}{dt} = \frac{1}{i} \frac{d}{dt} \int_{V_3} (x_\mu T_{\nu 4} - x_\nu T_{\mu 4}) d^3 x = 0. \tag{72}$$

From Eqs. (69)-(71), it is seen that

$$T_{44} = \mathcal{H}. \tag{73}$$

From Eqs. (43), (44), (53a), and (54a), one obtains

$$S_j = i T_{4j} = -\dot{\phi} \partial_j \phi = -\pi \partial_j \phi,$$

---

[15]

$$\begin{aligned}
\mathcal{H}(\phi, \pi, \partial_k \phi) &= \pi^2 + \frac{c^2}{2}(\partial_\nu \phi \partial_\nu \phi + \mu^2 \phi^2) \\
&= \frac{1}{2}\{\pi^2 + c^2 \mu^2 \phi^2 + c^2 (\nabla \phi)^2\}.
\end{aligned} \tag{70'}$$

[16]

$$T_{\mu\nu} = -c^2 \partial_\mu \phi \partial_\nu \phi - \delta_{\mu\nu} \mathcal{L} = T_{\nu\mu}. \tag{71'}$$

146

$$P_j = \frac{1}{i}T_{j4} = \dot{\phi}\partial_j\phi = \pi\partial_j\phi,$$

$$= \frac{1}{2}(\pi\partial_j\phi + (\partial_j\phi)\pi), (P \quad \text{hermitian}), \tag{74}$$

$$\partial_t \mathcal{H} + \nabla \cdot S = 0,$$

$$\partial_t P_k + \partial_j T_{kj} = 0. \tag{75}$$

(2) Another Klein-Gordon field is a complex scalar field $\Psi(x_\nu)$ which may be regarded as made up of two independent real scalar fields $\psi_1(x_\nu), \psi_2(x_\nu)$:

$$\Psi(x_\nu) = \frac{1}{\sqrt{2}}(\psi_1(x_\nu) + i\psi_2(x_\nu)),$$

$$\Psi^*(x_\nu) = \frac{1}{\sqrt{2}}(\psi_1(x_\nu) - i\psi_2(x_\nu)), \tag{76}$$

so that Ψ and Ψ^* are also independent. $\psi_1(x), \psi_2(x)$ are hermitian, while Ψ, Ψ^* are not hermitian operators.

The Lagrangian density $\mathcal{L}(\psi_1, \psi_2, \partial_\nu\psi_1, \partial_\nu\psi_2)$ is[17]

$$\mathcal{L} = -(\partial_\nu\Psi\partial_\nu\Psi^* + \mu^2\Psi\Psi^*)$$

$$= -\frac{1}{2}(\partial_\nu\psi_j\partial_\nu\psi_j + \mu^2\psi_j\psi_j), \quad \text{(summed over } j = 1, 2). \tag{77}$$

The Lagrange equations[18]

---
[17]

$$\mathcal{L} = -c^2(\partial_\nu\Psi\partial_\nu\Psi^* + \mu^2\Psi\Psi^*)$$

$$= -\frac{c^2}{2}(\partial_\nu\psi_j\partial_\nu\psi_j + \mu^2\psi_j\psi_j). \tag{77'}$$

[18] *The Lagrange equations*

$$\frac{\partial\mathcal{L}}{\partial\psi_j} - \partial_\nu\frac{\partial\mathcal{L}}{\partial(\partial_\nu\psi_j)} = 0, \qquad j = 1, 2; \tag{80a}$$

give

$$(\partial_\alpha\partial_\alpha - \mu^2)\begin{pmatrix}\psi_1\\\psi_2\end{pmatrix} = 0,$$

$$\Pi_j = \frac{\partial\mathcal{L}}{\partial\dot{\psi}_j} = \dot{\psi}_j(x_\nu),$$

$$\frac{\partial \mathcal{L}}{\partial \Psi} - \partial_\nu \frac{\partial \mathcal{L}}{\partial(\partial_\nu \Psi)} = 0, \qquad (\Psi = \Psi, \Psi^*), \tag{78}$$

give the Klein-Gordon equations

$$(\partial_\alpha \partial_\alpha - \mu^2)\begin{pmatrix} \Psi \\ \Psi^* \end{pmatrix} = 0, \qquad \mu = \frac{m_0 c}{\hbar}.$$

The conjugate momenta $\Pi(x_\nu), \Pi^*(x_\nu)$ are

$$\begin{aligned} \Pi &= \frac{\partial \mathcal{L}}{\partial \dot{\Psi}} = \dot{\Psi}^*(x_\nu), \\ \Pi^* &= \frac{\partial \mathcal{L}}{\partial \dot{\Psi}^*} = \dot{\Psi}(x_\nu). \end{aligned} \tag{79}$$

Π, Π^* are non-hermitian operators.

The Hamiltonian density $\mathcal{H}$ and Hamiltonian H are

$$\begin{aligned} \mathcal{H} &= \Pi \dot{\Psi} + \Pi^* \dot{\Psi}^* - \mathcal{L} \\ &= \Pi \Pi^* + (\nabla \Psi \cdot \nabla \Psi^* + \mu^2 \Psi \Psi^*), \\ H &= \int_V \mathcal{H} d^3 x. \end{aligned} \tag{81}$$

The energy-momentum tensor $T_{\mu\nu}$ is[19]

$$\begin{aligned} T_{\mu\nu} &= -(\partial_\mu \Psi \partial_\nu \Psi^* + \partial_\mu \Psi^* \partial_\nu \Psi) - \delta_{\mu\nu} \mathcal{L} \\ &= T_{\nu\mu}. \end{aligned} \tag{82}$$

It is seen from the definition (77) that $\mathcal{L}$ is a real scalar (invariant under proper Lorentz transformations). $\mathcal{L}$ is invariant under the following gauge transformation (of the first kind):

$$\Pi = \frac{1}{\sqrt{2}}(\Pi_1 - i\Pi_2), \qquad \Pi^* = \frac{1}{\sqrt{2}}(\Pi_1 + i\Pi_2), \tag{80b}$$

Π_1, Π_2 are hermitian operators.

19

$$T_{\mu\nu} = -c^2(\partial_\mu \Psi \partial_\nu \Psi^* + \partial_\mu \Psi^* \partial_\nu \Psi) - \delta_{\mu\nu} \mathcal{L}. \tag{82'}$$

$$\Psi' = e^{i\alpha}\Psi, \qquad\qquad \Psi'^* = e^{-i\alpha}\Psi^*, \qquad\qquad (83)$$

where α is an arbitrary real constant,

$$\delta\mathcal{L} = 0. \qquad\qquad (84)$$

For infinitesimal transformations

$$\Psi(x_\nu) \to \Psi'(x_\nu) = (1 + i\alpha)\Psi(x_\nu),$$

$$\Psi^*(x_\nu) \to \Psi'^*(x_\nu) = (i - i\alpha)\Psi^*(x_\nu),$$

i.e.,

$$\delta\Psi(x_\nu) = i\alpha\Psi, \qquad\qquad \delta\Psi^* = -i\alpha\Psi^*, \qquad\qquad (85)$$

$$\delta\mathcal{L} = \left(\frac{\partial\mathcal{L}}{\partial\Psi} - \partial_\nu\frac{\partial\mathcal{L}}{\partial(\partial_\nu\Psi)}\right)\delta\Psi + \left(\frac{\partial\mathcal{L}}{\partial\Psi^*} - \partial_\nu\frac{\partial\mathcal{L}}{\partial(\partial_\nu\Psi^*)}\right)\delta\Psi^*$$
$$+ \partial_\nu\left(\frac{\partial\mathcal{L}}{\partial(\partial_\nu\Psi)}\delta\Psi + \frac{\partial\mathcal{L}}{\partial(\partial_\nu\Psi^*)}\delta\Psi^*\right) = 0.$$

From the Lagrange equations (78) and (85), one has[20]

---

[20] *Derivation of a conserved current $j_\mu(x)$ is just an application of Noether's theorem. Under the gauge transformation of the first kind, we have*

$$x_\mu \to x'_\mu = x_\mu \qquad or \qquad X_{\mu\alpha} = 0, \qquad\qquad (90a)$$

and, from Eq. (85),

$$\Psi_{j\alpha} = i\psi \quad for \quad j = \psi$$
$$\qquad - i\psi^* \quad for \quad j = \psi^*. \qquad\qquad (90b)$$

Thus, the Noether's current is

$$\theta_\mu = -\frac{\partial\mathcal{L}}{\partial(\partial_\mu\psi)}\cdot i\psi - \frac{\partial\mathcal{L}}{\partial(\partial_\mu\psi^*)}(-i\psi^*)$$
$$= i\{(\partial_\mu\psi^*)\psi - \psi^*(\partial_\mu\psi)\}, \qquad\qquad (91)$$

which is proportional to the current given by Eq. (87).

$$\partial_\nu \Big(\frac{\partial \mathcal{L}}{\partial(\partial_\nu \Psi)}\Psi - \frac{\partial \mathcal{L}}{\partial(\partial_\nu \Psi^*)}\Psi^*\Big) = 0,$$

or

$$\partial_\nu\big((\partial_\nu \Psi^*)\Psi - (\partial_\nu \Psi)\Psi^*\big) = 0. \tag{86}$$

one may define a 4-current density[21]

$$j_\nu = ie\{(\partial_\nu \Psi^*)\Psi - (\partial_\nu \Psi)\Psi^*\}, \tag{87}$$

so that (86) becomes an equation of continuity

$$\partial_\nu j_\nu = 0, \tag{86a}$$

or[22]

$$\begin{aligned} j_k &= ie\big((\partial_k \Psi^*)\Psi - (\partial_k \Psi)\Psi^*\big), \\ j_4 &= e(\Psi\dot{\Psi}^* - \Psi^*\dot{\Psi}) = i\rho \\ \nabla \cdot \mathbf{j} &+ \frac{\partial \rho}{\partial t} = 0. \end{aligned} \tag{88}$$

The total charge of the field is

$$Q = \int_V \rho d^3x = -ie \int_V (\Psi\dot{\Psi}^* - \Psi^*\dot{\Psi})d^3x. \tag{89}$$

This we shall see is the electric charge operator.

§.6.5. The Electromagnetic Field in Lagrangian Form.

The fields φ_α in the preceding sections are the 4-potentials

$$\varphi_\alpha = A_1, A_2, A_3, A_4 = i\phi$$

21

$$j_\nu = i\frac{e}{\hbar}c^2\{(\partial_\nu \Psi^*)\Psi - (\partial_\nu \Psi)\Psi^*\}. \tag{87'}$$

22

$$j_k = i\frac{ec^2}{\hbar}\big((\partial_k \Psi^*)\Psi - (\partial_k \Psi)\Psi^*\big), \qquad j_4 = \frac{ec}{\hbar}(\Psi\dot{\Psi}^* - \Psi^*\dot{\Psi}) = ic\rho. \tag{88'}$$

150

The electric and magnetic field are

$$\mathbf{E} = -\nabla\phi - \frac{\partial}{\partial t}\mathbf{A}, \qquad \mathbf{B} = \nabla \times \mathbf{A}. \tag{92}$$

Ampère's law and Coulomb's laws are, for free fields,

$$\nabla \times \mathbf{B} = \frac{\partial}{\partial t}\mathbf{E}, \qquad \nabla \cdot \mathbf{E} = 0. \tag{93}$$

The Lorentz relation (Lorentz gauge) is

$$\partial_\mu A_\mu = 0. \tag{94}$$

The skew-symmetric (antisymmetric) electromagnetic field tensor $F_{\mu\nu}$ is defined by

$$F_{\mu\nu} \equiv \partial_\mu A_\nu - \partial_\nu A_\mu = -F_{\nu\mu}$$
$$= \begin{pmatrix} 0 & B_3 & -B_2 & -iE_1 \\ & 0 & B_1 & -iE_2 \\ & & 0 & -iE_3 \\ & & & 0 \end{pmatrix}. \tag{95}$$

The Lagrangian density $\mathcal{L}$ is chosen to be

$$\mathcal{L} = -\frac{1}{4}[F_{\mu\nu}F_{\mu\nu} + 2(\partial_\mu A_\mu)^2]$$
$$= \frac{1}{2}(\mathbf{E}\cdot\mathbf{E} - \mathbf{B}\cdot\mathbf{B}) - \frac{1}{2}(\nabla\cdot\mathbf{A} + \frac{\partial\phi}{\partial t})^2. \tag{96}$$

Under the gauge transformation

$$A_\mu \to A_\mu + \partial_\mu\chi, \qquad \chi = \text{a scalar function}, \tag{97}$$

with

$$\partial_\mu\partial_\mu\chi = 0, \tag{98}$$

both the Lorentz relation (94) and the Lagrangian density $\mathcal{L}$ are invariant.

In any arbitrary Lorentz frame, it is always possible to choose a $\Lambda(x_\mu)$ satisfying (98) to make

$$A_4 = i\phi + \frac{\partial}{\partial x_4}\Lambda = 0, \tag{99}$$

for a source-free field, so that the Lorentz relation (94) reduces to

$$\partial_k A_k = 0 \qquad (i.e., \quad \nabla \cdot \mathbf{A} = 0). \tag{100}$$

The Euler equation (13) now gives the field equation for A_μ ,

$$\partial_\mu \partial_\mu A_\nu = 0, \tag{101}$$

or

$$\partial_\mu^2 A_\nu = 0, \qquad \nu = 1, 2, 3, 4. \tag{101a}$$

The canonical conjugates π_μ of the A_μ are, from Eqs. (16), (96a), and (95),

$$\pi_\mu = \frac{\partial \mathcal{L}}{\partial \dot{A}_\mu} = i(F_{4\mu} + \partial_\nu A_\nu \delta_{\mu 4}), \tag{102}$$

$$\pi_k = -E_k, \qquad \pi_4 = i\partial_\nu A_\nu. \tag{102a}$$

If one introduces the Lorentz relation (94), then

$$\pi_4 = 0. \tag{103}$$

The Hamiltonian density $\mathcal{H}$ is

$$\begin{aligned}
\mathcal{H} &= \pi_\mu \dot{A}_\mu - \mathcal{L} \\
&= \pi_\mu \dot{A}_\mu + \frac{1}{4}[F_{\mu\nu} F_{\mu\nu} + 2(\partial_\alpha A_\alpha)^2].
\end{aligned} \tag{104}$$

From Eqs. (92) and (102),

$$\begin{aligned}
\dot{A}_k &= -E_k - \partial_k \phi = \pi_k + i\partial_k A_4, \\
\dot{A}_4 &= \pi_4 - i\partial_k A_k,
\end{aligned} \tag{105}$$

$$\begin{aligned}
\mathcal{H} &= \pi_\mu \pi_\mu + i(\pi_k \partial_k A_4 - \pi_4 \partial_k A_k) + \frac{1}{4}[F_{\mu\nu} F_{\mu\nu} + 2(\partial_\mu A_\mu)^2] \\
&= \frac{1}{2}(\mathbf{E} \cdot \mathbf{E} + \mathbf{B} \cdot \mathbf{B}) + i(\pi_k \partial_k A_4 - \pi_4 \partial_k A_k) + \frac{1}{2}\pi_4^2.
\end{aligned} \tag{104a}$$

As

$$\begin{aligned}
i\int_V \Pi_k \partial_k A_4 d^3 x &= \int_V (\mathbf{E} \cdot \nabla \phi) d^3 x \\
&= -\int \phi(\nabla \cdot \mathbf{E}) d^3 x \\
&= 0, \qquad \text{(See Eq. (93))},
\end{aligned}$$

and on using Lorentz gauge (94), $\pi_4 = 0$ (Eq. (103)) and $\partial_k A_k = 0$ from Eq. (100), one has for the Hamiltonian H

$$
\begin{aligned}
H &= \int_V \mathcal{H} d^3 x \\
&= \int_V \frac{1}{2} (\mathbf{E} \cdot \mathbf{E} + \mathbf{B} \cdot \mathbf{B}) d^3 x.
\end{aligned}
\tag{106}
$$

The electromagnetic field tensor $T_{\mu\nu}$ is, from Eqs. (41), (95), (96a), and (94)

$$
\begin{aligned}
T_{\mu\nu} &= (\partial_\mu A_\alpha) F_{\alpha\nu} - \delta_{\mu\nu} \mathcal{L} \\
&= F_{\mu\alpha} F_{\alpha\nu} + (\partial_\alpha A_\mu) F_{\alpha\nu} - \delta_{\mu\nu} \mathcal{L}.
\end{aligned}
\tag{107}
$$

It can be shown that the second term on the right contributes nothing to the momentum or energy conservation relations in Eqs. (50)-(52).[23] Hence one has

$$
T_{\mu\nu} = F_{\mu\alpha} F_{\alpha\nu} - \delta_{\mu\nu} \mathcal{L}
\tag{108}
$$

or

---

[23] *Taking the 4-divergence and integrating over the volume V_3 of the field, and by repeated application of the relation*

$$
\partial_\mu F_{\nu\mu} = \partial_\mu \partial_\nu A_\mu - \partial_\mu^2 A_\nu = \partial_\nu \partial_\mu A_\mu = 0,
$$

one obtains

$$
\begin{aligned}
\int_V \partial_\nu (\partial_\alpha A_\mu F_{\alpha\nu}) d^3 x &= \int_V \partial_\nu (\partial_\alpha A_\mu) F_{\alpha\nu} d^3 x \\
&= \int_V \partial_k (\partial_\alpha A_\mu) F_{\alpha k} d^3 x + \int_V \partial_4 (\partial_\alpha A_\mu) F_{\alpha 4} d^3 x \\
&= -\int_V \partial_\alpha A_\mu \partial_k F_{\alpha k} d^3 x - \int_V \partial_\alpha A_\mu \partial_4 F_{\alpha 4} d^3 x + \int_V \partial_4 ((\partial_\alpha A_\mu) F_{\alpha 4}) d^3 x \\
&= -\int \partial_\alpha A_\mu \partial_\nu F_{\alpha\nu} d^3 x + \partial_4 \int \partial_\alpha (A_\mu F_{\alpha 4}) d^3 x - \partial_4 \int A_\mu \partial_\alpha F_{\alpha 4} d^3 x \\
&= \partial_4 \int \partial_k (A_\mu F_{k4}) d^3 x = -\partial_4 \int_s A_\mu F_{4k} \cdot dS = 0.
\end{aligned}
$$

$$T_{\mu\nu} = \begin{pmatrix} E_1E_1 + B_1B_1 - \mathcal{H} & E_1E_2 + B_1B_2 & E_1E_3 + B_1B_3 & iP_1 \\ E_2E_1 + B_2B_1 & E_2E_2 + B_2B_2 - \mathcal{H} & E_2E_3 + B_2B_3 & iP_2 \\ E_3E_1 + B_3B_1 & E_3E_2 + B_3B_2 & E_3E_3 + B_3B_3 - \mathcal{H} & iP_3 \\ \frac{1}{i}S_1 & \frac{1}{i}S_2 & \frac{1}{i}S_3 & \mathcal{H} \end{pmatrix},$$

(109)

where

$$\mathcal{H} = \frac{1}{2}(\mathbf{E} \cdot \mathbf{E} + \mathbf{B} \cdot \mathbf{B}),$$

(110)

$$S = [\mathbf{E} \times \mathbf{B}], \qquad \text{the Poynting vector (energy flux density)},$$

$$P = -S, \qquad \text{momentum density}.$$

§.6.6. The Dirac Field in Lagrangian Form

We shall regard the Dirac relativistic equations of the electron, not as the quantum mechanical (wave) equation of a particle, but as a "classical" field equation - the spinor field $\Psi(x)$ (see Chapter 2, 3). We shall quantize the Dirac free field, i.e., the field given by the equations (22a), Ch. 2.

To obtain the Dirac equations as the equations of a classical field, one starts with the Lagrangian density[24],[25]

$$\mathcal{L} = -\frac{1}{2}\{\Psi^\dagger\gamma_4[\gamma_\mu\partial_\mu + m_0]\Psi + [(\partial_\mu\Psi)^\dagger\gamma_\mu + m_0\Psi^\dagger]\gamma_4\Psi\}.$$

(111)

The Euler equations (13)

$$\frac{\partial\mathcal{L}}{\partial\Psi_\alpha^\dagger} - \partial_\mu\frac{\partial\mathcal{L}}{\partial(\partial_\mu\Psi_\alpha^\dagger)} = 0,$$

$$\frac{\partial\mathcal{L}}{\partial\Psi_\alpha} - \partial_\mu\frac{\partial\mathcal{L}}{\partial(\partial_\mu\Psi_\alpha)} = 0, \qquad \alpha = 1, 2, 3, 4,$$

---

[24] *Here $\Psi^\dagger$ is a row matrix Ψ_1^*, Ψ_2^*, Ψ_3^*, Ψ_4^*, i.e., $\Psi^\dagger = \tilde{\Psi}^*$.*
[25]

$$\mathcal{L} = -\frac{\hbar c}{2}\{\Psi^\dagger\gamma_4[\gamma_\mu\partial_\mu + \frac{m_0c}{\hbar}]\Psi + [(\partial_\mu\Psi)^\dagger\gamma_\mu + \frac{m_0c}{\hbar}\Psi^\dagger]\gamma_4\Psi\}.$$

(111')

154

give the equations[26]

$$(\gamma_\mu \partial_\mu + m_0)\Psi = 0, \tag{112}$$

$$(\partial_\mu \Psi)^\dagger \gamma_\mu + m_0 \Psi^\dagger = 0. \tag{113}$$

Eq. (113) is, since $\partial_4^* = -\partial_4$,

$$\partial_j \Psi^\dagger \gamma_j - \partial_4 \Psi^\dagger \gamma_4 + m_0 \Psi^\dagger = 0. \tag{113a}$$

It is seen from Eqs. (112) and (113) that for Ψ, $\Psi^\dagger$ satisfying the Dirac equations.

The tensor component T_{44} (Eq. (45)) is[27]

$$T_{44} = \frac{\partial \mathcal{L}}{\partial \dot{\Psi}} \dot{\Psi} + \dot{\Psi}^\dagger \frac{\partial \mathcal{L}}{\partial \dot{\Psi}^\dagger} - \mathcal{L}, \tag{114}$$

and the Hamiltonian density $\mathcal{H}$ is (using $\mathcal{L} = 0$)[28]

$$\mathcal{H} = \frac{i}{2}(\Psi^\dagger \dot{\Psi} - \dot{\Psi}^\dagger \Psi). \tag{115}$$

The energy H is

$$H = \int_V \mathcal{H} d^3 x. \tag{116}$$

The electric charge 4-current J_μ is, from Eq. (82), Ch. 3,[29,30]

------

[26]
$$(\gamma_\mu \partial_\mu + \frac{m_0 c}{\hbar})\Psi = 0. \tag{112'}$$

[27]
$$c T_{44} = \frac{\partial \mathcal{L}}{\partial \dot{\Psi}} \dot{\Psi} + \dot{\Psi}^\dagger \frac{\partial \mathcal{L}}{\partial \dot{\Psi}^\dagger} - \mathcal{L}. \tag{114'}$$

[28]
$$\mathcal{H} = \frac{i\hbar}{2}(\Psi^\dagger \dot{\Psi} - \dot{\Psi}^\dagger \Psi). \tag{115'}$$

[29] *J_μ can also be obtained from Noether's theorem.*
[30]
$$J_\mu = -ice\Psi^\dagger \gamma_4 \gamma_\mu \Psi, \qquad J_4 = ic\rho. \tag{117'}$$

$$J_\mu = -ie\Psi^\dagger \gamma_4 \gamma_\mu \Psi,$$
$$J_4 = i\rho, \qquad \rho = e\Psi^\dagger \Psi, \qquad (117)$$

and the total charge Q is

$$Q = \int_V \rho d^3 x. \qquad (118)$$

Under the transformation specified by Eq. (64) (which is to be compared with Eqs. (8c), (45), and (46) in Ch. 4), we find, with $\mu < \nu$,

$$\psi(x) \to \psi'(x') = (1 + \frac{1}{2}\delta w_{\mu\nu}\gamma_\mu \gamma_\nu)\psi(x), \qquad (119)$$

which yields

$$A_{\psi\psi\mu\nu} = \frac{1}{2}\gamma_\mu \gamma_\nu, \qquad A_{\psi\bar\psi\mu\nu} = 0. \qquad (120a)$$

Analogously, we have

$$A_{\bar\psi\bar\psi\mu\nu} = -\frac{1}{2}\gamma_\mu \gamma_\nu, \qquad A_{\bar\psi\psi\mu\nu} = 0. \qquad (120b)$$

Thus, we obtain from Eq. (65) the angular momentum tensor

$$M_{\mu\alpha\beta} = (x_\beta T_{\mu\alpha} - x_\alpha T_{\mu\beta}) + \frac{1}{4}\bar\psi\{\gamma_\mu \gamma_\alpha \gamma_\beta + \gamma_\alpha \gamma_\beta \gamma_\mu\}\psi. \qquad (121)$$

This completes our treatment of the Dirac field in Lagrangian form.

Appendix

Electromagnetic Fields

For convenient reference, we collect in this appendix together some formulas relevant for the electromagnetic field. As in §.6.5., we adopt the m.k.s.a. unit system (instead of $\hbar = c = 1$).

Faraday's law

$$\nabla \times \mathbf{E} = -\frac{\partial \mathbf{B}}{\partial t} \tag{A1}$$

Amperè's law

$$\nabla \times \mathbf{H} = \mathbf{j} + \frac{\partial \mathbf{D}}{\partial t}, \quad j = \rho \mathbf{v}, \tag{A2}$$

Coulomb's law

$$\nabla \cdot \mathbf{B} = 0, \tag{A3}$$

Coulomb's law

$$\nabla \cdot \mathbf{D} = \rho, \tag{A4}$$

$$\mathbf{B} = \mu_0 \mathbf{H} \, (\mu_0 = 4\pi \times 10^{-17} \, \text{henry/m}),$$

$$\mathbf{D} = \epsilon_0 \mathbf{E} \, (\epsilon_0 = \frac{1}{36\pi} \times 10^{-9} \, \text{farad/m}), \tag{A5}$$

$$\frac{1}{\sqrt{\mu_0 \epsilon_0}} = c = 3 \times 10^8 \, \text{m/sec}.$$

Of the four equations (A1) - (A4), only two are independent. From the divergence of both sides of (A1), if $\nabla \cdot \mathbf{B} = 0$ at any one instant, $\nabla \cdot \mathbf{B} = 0$ at all times. This gives (A3). From the divergence of (A2), the equation of continuity

$$\nabla \cdot \mathbf{j} + \frac{\partial \rho}{\partial t} = 0, \tag{A6}$$

leads to (A4).

On introducing the 4-potential $A_\mu = (A, \frac{i}{c}\phi)$

$$\mathbf{B} = \nabla \times \mathbf{A}, \tag{A7}$$

$$\mathbf{E} = -\nabla \phi - \frac{\partial \mathbf{A}}{\partial t}, \tag{A8}$$

and the Lorentz relation

$$\nabla \cdot \mathbf{A} + \frac{1}{c^2}\frac{\partial \phi}{\partial t} = 0, \tag{A9}$$

one gets from (A1) - (A4)

$$\partial_\mu \partial_\mu \mathbf{A} = -\mu_0 \mathbf{j}, \tag{A10}$$

$$\partial_\mu \partial_\mu \phi = -\frac{1}{\epsilon_0}\rho. \tag{A11}$$

If one makes the gauge transformation

$$\mathbf{A}' = \mathbf{A} + \nabla\chi, \qquad \phi' = \phi - \frac{\partial\chi}{\partial t}, \qquad \chi = \text{a scalar function}, \tag{A12}$$

with the condition

$$\partial_\mu \partial_\mu \chi = 0, \tag{A13}$$

the $\mathbf{B}$, $\mathbf{E}$ fields, the equation (A10), (A11) and the Lorentz relation (A9) are all unchanged.

In tensor notations, the above equations take the following forms. Define

$$x_\mu = (x_1, x_2, x_3, x_4 = ict), \tag{A14}$$

$$j_\mu = (\rho v_1, \rho v_2, \rho v_3, ic\rho), \tag{A15}$$

$$F_{\mu\nu} = \partial_\mu A_\nu - \partial_\nu A_\mu = -F_{\nu\mu}, \tag{A16}$$

then

$$(A1) + (A3): \qquad \partial_k F_{\ell m} + \partial_\ell F_{mk} + \partial_m F_{k\ell} = 0, \tag{A17}$$

$$(A2) + (A4): \qquad \partial_\nu F_{\rho\nu} = \mu_0 j_\rho, \tag{A18}$$

$$(A9): \qquad \partial_\nu A_\nu = 0, \tag{A19}$$

$$(A10) + (A11): \qquad \partial_\mu \partial_\mu A_\nu = -\mu_0 j_\nu, \tag{A20}$$

$$(A12): \qquad A'_\nu = A_\nu + \partial_\nu \chi. \tag{A21}$$

If a 4-vector f_μ is defined by

$$f_\mu = F_{\mu\nu} j_\nu, \qquad\qquad (A22)$$

then

$$f_k = \rho(E_k + [\mathbf{v} \times \mathbf{B}]_k), \qquad k = 1, 2, 3,$$

$$f_4 = \frac{i}{c}\mathbf{E} \cdot \mathbf{j} = \frac{i}{c}\rho \mathbf{E} \cdot \mathbf{v}, \qquad\qquad (A23)$$

and

$$(A21) + (A18): \qquad f_\rho = \frac{1}{\mu_0} F_{\rho\nu} \partial_\lambda F_{\nu\lambda}. \qquad\qquad (A24)$$

In empty space ($\rho = 0$, $\mathbf{j} = 0$), it is possible to choose a function Λ to make A_4 in (A21) zero, i.e.,

$$\phi = -icA_4 = 0. \qquad\qquad (A25)$$

With this choice, the Lorentz relation (A19) becomes

$$\nabla \cdot \mathbf{A} = 0. \qquad\qquad (A26)$$

This relation is not Lorentz invariant, but in any Lorentz frame, it is always possible to choose a gauge (Λ) such that $\phi = 0$.[31]

For $\rho = 0$, one may choose $\nabla \cdot \mathbf{A} = 0$, so that of the three components A_1, A_2, A_3, only two are linearly independent.

On expressing $A_j(x)$, $x = x_\mu$, in Fourier integral representation, one has

$$A_j(x) = \frac{1}{(2\pi)^{3/2}} \int a_j(k) e^{ik \cdot x} \, d^3k, \qquad\qquad (A27)$$

$$k_\nu = (k_j, k_4) = (\mathbf{k}, k_4),$$

$$\hbar k_4 = \frac{iE}{c}, \qquad\qquad E = \hbar\omega,$$

$$k_\nu^2 = k_\nu k_\nu = \mathbf{k}^2 - \frac{w^2}{c^2} = 0. \qquad\qquad (A28)$$

---

[31] *Only when $\rho = 0$ may we make $\phi = 0$; for $\rho \neq 0$, making $\phi = 0$ contradicts (A11).*

This last relation expresses the fact that the quantum (photon) of electromagnetic field has zero rest mass.[32]

From (A26), (A27), one has

$$\nabla \cdot \mathbf{A} = \partial_j A_j = \frac{1}{(2\pi)^{3/2}} \int k_j a_j(k) e^{ik_\nu x_\nu} d^3k = 0, \qquad (A29)$$

so that

$$k_j a_j = (k \cdot a) = 0, \qquad (A29a)$$

i.e., the vector potential $\mathbf{A}$ is perpendicular to the direction of propagation k of the electromagnetic wave.

From (A8), it is seen that for monochromatic electromagnetic wave, $\mathbf{E}$ is parallel to $\mathbf{A}$. Hence in vacuum ($\rho = 0$), electromagnetic waves are *transverse* waves. But this transverse property is not Lorentz invariant, coming as it does from $\nabla \cdot \mathbf{A} = 0$ which is not Lorentz invariant.

The solutions of equations (A11), (A10) are, on account of their time-reversal invariance, the "retarded" and the "advanced" potentials[33]

$$\left. \begin{matrix} \phi_{ret}(\mathbf{r},t) \\ \phi_{adv}(\mathbf{r},t) \end{matrix} \right\} = \frac{1}{4\pi\epsilon_0} \int d\mathbf{r}' \frac{\rho(\mathbf{r}',t')}{|\mathbf{r}-\mathbf{r}'|}, \qquad t' = \begin{cases} t - \frac{|\mathbf{r}-\mathbf{r}'|}{c} \\ t + \frac{|\mathbf{r}-\mathbf{r}'|}{c} \end{cases}, \qquad (A30)$$

$$\left. \begin{matrix} \mathbf{A}_{ret}(\mathbf{r},t) \\ \mathbf{A}_{adv}(\mathbf{r},t) \end{matrix} \right\} = \frac{\mu_0}{4\pi} \int d\mathbf{r}' \frac{\mathbf{j}(\mathbf{r}',t')}{|\mathbf{r}-\mathbf{r}'|}, \qquad t' = \begin{cases} t - \frac{|\mathbf{r}-\mathbf{r}'|}{c} \\ t + \frac{|\mathbf{r}-\mathbf{r}'|}{c} \end{cases}. \qquad (A31)$$

---

[32] *That the photon has zero rest mass is also reflected in the field equations (A10, 11) or (A17, 18) containing only $\partial_\nu A_\nu$, but not the A_ν themselves. The Klein-Gordon and the Dirac equations both contain the mass m_0 of the particle, when the equations are regarded as field equations of the Ψ.*

[33] *The general solutions of (A11), (A10) are obtained by adding to (A30), (A31) arbitrary solutions of the respective homogeneous equations*

$$\partial_\mu \partial_\mu \phi(r,t) = 0, \qquad \partial_\mu \partial_\mu A(r,t) = 0.$$

In the Appendix of Chapter **8**, we shall see that these integrals may be expressed in terms of Green's functions which are Lorentz invariant and which appear in the quantization conditions of the A_ν fields.

Exercises: *Chapter 6*

1. (a) Assume $\mathcal{L}(x) = \mathcal{L}(u^i(x), \partial_\mu u^i(x), \partial_\mu \partial_\nu u^i(x))$. Derive the generalized Euler-Lagrange equation, i.e. the analogue of Eq. (3).

   (b) Consider a vector field $B_\mu(x)$,

$$\mathcal{L}(x) = -\frac{1}{4} B_{\mu\nu} B_{\mu\nu} - \frac{1}{2} m^2 B_\mu B_\mu + \varepsilon(\partial_\lambda B_{\mu\nu}) B_\lambda B_{\mu\nu}$$

   with $B_{\mu\nu} \equiv \partial_\mu B_\nu - \partial_\nu B_\mu$ and ε some constant. Use (a) to derive field equations.

   (c) Add a total derivative to $\mathcal{L}(x)$:
   $\delta\mathcal{L}(x) = -\frac{\varepsilon}{2} \partial_\lambda \{ B_{\mu\nu} B_\lambda B_{\mu\nu} \}$.
   Then $\mathcal{L}'(x) = \mathcal{L}(x) + \delta\mathcal{L}(x)$ no longer depends on $\partial_\mu \partial_\nu u^i(x)$. Use Eq. (3) to derive the field equations. Discuss whether these equations are equivalent to those in (b).

   (d) Now impose the constraint $\partial_\mu B_\mu = 0$ and discuss its role.

2. *(Examples on Noether Theorem)*

   (a) As the first exercise, consider a rotation about the z-axis:

$$x_1' = x_1 - \varepsilon x_2,$$
$$x_2' = x_2 + \varepsilon x_1,$$
$$x_3' = x_3,$$

   with ε an infinitesimal. Obtain the expression for the conserved quantity in the case of the Dirac electron field.

   (b) Consider the gauge transformation of the first kind:

$$\phi_i(x) \rightarrow e^{i\alpha} \phi_i(x).$$

   Obtain the resultant conserved quantity, respectively, for a complex scalar field, the electromagnetic field, and the Dirac electron field.

Chapter 7.

Many-Body Systems

In Chapter 6, Sect. 1, we describe the underlying idea of the theory of quantized field, namely, of regarding a field as a system of nondenumerably infinitely many degrees of freedoms, which upon quantization represents a system of infinitely many "quanta", or "particles". Before treating the quantization of fields we need a theory of the symmetry property of a system of identical, or indistinguishable particles, and the appropriate representation of its states.

By indistinguishable particles, we mean that under any permutation P of the particles, the operator of any physical quantity Q, such as the Hamiltonian or the electric moment, of the system and their expectation values $< Q >$ are invariant,

$$PQ = QP,$$

$$< Q >=< \Psi \mid P^{-1}QP \mid \Psi >=< Q >_P . \qquad (1)$$

A question that arises is the effect of permutations of the indistinguishable particles on the state (or, the wave function of the state) of a system.

§.7.1. Permutations:

(1) Permutation group

Consider a system of n indistinguishable particles. There are $n!$ permutations P of these n particles. The simplest permutation operation is the transposition of a pair of particles j and k . As an example, the transposition of particle B, A in a system of 4 particles A, B, C, D may be represented by the scheme[1]

$$T_{AB} = \begin{pmatrix} A & B & C & D \\ B & A & C & D \end{pmatrix}. \qquad (2)$$

A permutation consisting of two transpositions T_{AB} and T_{AC} in succession is

---

[1] *The permutations (transpositions) are read vertically downward.*

$$T_{AC}T_{AB} = \begin{pmatrix} A & B & C & D \\ B & C & A & D \end{pmatrix} = P_{ABC}, \tag{3}$$

which may be denoted by P_{ABC} in which A, B, C is cyclically permuted. But the transpositions T_{AB} and T_{AC} taken in the reversed order lead to a different permutation

$$T_{AB}T_{AC} = \begin{pmatrix} A & B & C & D \\ C & A & B & D \end{pmatrix} = P_{ACB}. \tag{4}$$

It is seen that

$$P_{ABC}P_{ACB} = P_{ACB}P_{ABC} = 1. \tag{5}$$

i.e., P_{ABC} and P_{ACB} are the inverse of each other, and this can be generalized and expressed by saying that to every permutation P, there is an inverse P^{-1},

$$P^{-1}P = PP^{-1} = 1. \tag{5a}$$

It is readily seen that the $n!$ permutations P_j have the following properties
(i) Two successive permutations P_i, P_j give rise to another P_k,

$$P_iP_j = P_k. \tag{6}$$

(ii) The P_i has the associative property,

$$P_i(P_jP_k) = (P_iP_j)P_k. \tag{7}$$

(iii) Every P_i has an inverse satisfying Eq. (5a).

(iv) There is the identity permutation which leaves the system unchanged.

These properties define the P_i as a *group* - called the permutation group, or, the symmetric group.[2]

In the example of four particles above, it is seen that $T_{AB}, T_{AC}T_{AB}$ are elements of the group (of 4! elements) and from Eqs. (3) and (4), it is seen that

$$P_iP_j \neq P_jP_i. \tag{8}$$

---

[2] *The name "symmetric group" must not be confused with symmetry group.*

The group is non-commutative, and is said to be non-abelian.

Any permutation P_j can be regarded as made up of a number ν_j of transpositions of pairs. We shall call the P_j's "even" and "odd" by an ϵ_j such that

$$\epsilon_j = \begin{cases} 1, & \text{for} \quad \nu_j = \text{even integer;} \\ -1, & \text{for} \quad \nu_j = \text{odd integer.} \end{cases} \tag{9}$$

It is clear that P_j and P_j^{-1} have the same evenness or oddness, and that the ϵ of the product P_k of two permutations $P_i P_j$ is

$$\epsilon_k = \epsilon_i \epsilon_j. \tag{9a}$$

(2) Permutation operator

Consider a state $\mid a >$ of a system of n particles. Its representative in the coordinate representation is the wave function

$$\psi(x_1, x_2, ..., x_n).$$

Upon a permutation P of the particles, this becomes

$$P\psi(x_1, x_2, ..., x_n) = \psi(x_{\alpha_1}, x_{\alpha_2}, ..., x_{\alpha_n}).$$

Let us introduce an operator U_p

$$\psi(x_\alpha) = U_P \psi(x). \tag{10}$$

The U_{P_j} obviously have the group properties (i) - (iv) above, for example,

$$U_{P_i} U_{P_j} = U_{P_k}, \qquad U_{P^{-1}} = U_P^{-1}. \tag{11}$$

U_P is also unitary,[3]

---

[3] *This follows from the invariance of the integral of any two functions* $\psi_1(x)$, $\psi_2(x)$,

$$\begin{aligned} (\psi_1(x), \psi_2(x)) &= (\psi_1(x_\alpha), \psi_2(x_\alpha)) \\ &= (U_P \psi_1(x), U_P \psi_2(x)) \\ &= (U_P^\dagger U_P \psi_1(x), \psi_2(x)), \qquad etc. \end{aligned}$$

$$U_P^{\dagger} = U_P^{-1}. \tag{12}$$

The operator U_P commutes with any observable (hermitian operator Q),[4]

$$QU_P - U_PQ = 0 \tag{12a}$$

Let U_T be an operator for the transposition Eq. (2),

$$U_T\psi(x..., x_j, .., x_k, ..) = \psi(..., x_k, .., x_j, ..). \tag{13}$$

It is obvious

$$U_T U_T = 1$$

so that the eigenvalues of U_T are ± 1 , and corresponding to these eigenvalues ± 1 , the function ψ is $\left\{ \begin{array}{c} symmetric \\ antisymmetric \end{array} \right\}$,

$$U_T\psi_s = \psi_s$$

$$U_T\psi_a = -\psi_a. \tag{14}$$

If T^{-1} and T are defined as Eq. (5a), then

$$U_T = U_{T^{-1}} = (U_T)^{-1}. \tag{15}$$

From Eqs. (9), (11), and (15), it is seen that

$$\begin{aligned} U_P\psi_s &= \psi_s, \\ U_P\psi_a &= \epsilon_P\psi_a, \end{aligned} \tag{16}$$

The question is, in Nature, what is the symmetry property of a system of indistinguishable particles.

---

[4] *This follows from Eq. (1),*

$$\begin{aligned} (\psi(x), Q\psi(x)) &= (U_P\psi(x), QU_P\psi(x)) \\ &= (\psi(x), U_P^{\dagger}QU_P\psi(x)), \qquad etc. \end{aligned}$$

166

The answer is the empirical law of nature:

For particles with half-odd-integral spin, called fermions, such as electrons, protons, neutrons, muons, neutrinos, the wave function is always the antisymmetric ψ_a kind in Eq. (14).

For particles with integral spins, such as photons, pions, α-particles, and particles built up from even number of half-odd-integral spins, called bosons, the wave function is always of the ψ_s kind.

§.7.2. Symmetric and antisymmetric wave functions for fermions and bosons

Consider a system of n indistinguishable, noninteracting particles. Let K stand for the complete set of quantum numbers of a single particle.[5]

For bosons, the symmetric wave function $\Psi^{(s)}$ is

$$\Psi^{(s)}_{K,K,..,K_n} = \frac{1}{\sqrt{n!}} \sum_P U_P \psi_K(x_1) \psi_K(x_2)...\psi_{K_n}(x_n) \tag{17}$$

where the sum is over all $n!$ permutations of $x_1, x_2, ..., x_n$, and U_P is the permutation operator Eq. (10).

For fermions, the wave function is antisymmetric, and for non-interacting particles can be put in a determinant form

$$\Psi^{(a)}_{K,K,..,K_n} = \frac{1}{\sqrt{n!}} \begin{vmatrix} \psi_K(x_1) & \psi_K(x_2) & ... & \psi_K(x_n) \\ .. & .. & & .. \\ .. & .. & & .. \\ .. & .. & & .. \\ \psi_{K_n}(x_1) & \psi_{K_n}(x_2) & ... & \psi_{K_n}(x_n) \end{vmatrix}$$

$$= \frac{1}{\sqrt{n!}} \sum_P \epsilon_P U_P \psi_K(x_1) \psi_K(x_2)...\psi_{K_n}(x_n) \tag{18}$$

where the summation is over the $n!$ permutations of $x_1, ..., x_n$, and $\epsilon_P = \pm 1$ according as the permutation is $\left\{ \begin{matrix} even \\ odd \end{matrix} \right\}$ number of transpositions of pairs of x_i and x_j .

---

[5] *This means the totality of quantum numbers for the common eigenstates of a complete set of commuting observables for a single particle, in the sense of Ch. 5, Sect. 4(1), Vol. I ("Quantum Mechanics"). For example, for an electron in a central field, K stands for (n, ℓ, m_ℓ, m_s), or (n, ℓ, j, m) .*

From Eq. (18), it follows that

$$\Psi^{(a)}_{K_1,K_2,\ldots K_n} = 0, \qquad \text{when two sets} \quad K_j = K_i. \tag{19}$$

This is the Pauli exclusion principle.

For convenience, we introduce the following hermitian projection operators[6]

---

[6] *The hermitian character is seen as follows:*

$$S^\dagger = \frac{1}{n!} \sum_P U_P^\dagger = \frac{1}{n!} \sum U_{P^{-1}} = S, \tag{20a}$$

$$A^\dagger = \frac{1}{n!} \sum_P \epsilon_P U_P^\dagger = \frac{1}{n!} \sum_P \epsilon_{P^{-1}} U_{P^{-1}} = A, \tag{21a}$$

$$C^\dagger = 1 - S^\dagger - A^\dagger = 1 - S - A = C. \tag{22a}$$

That they are projection operators is seen as follows

$$U_{P_i} S = \frac{1}{n!} \sum U_{P_i} U_P = \frac{1}{n!} \sum U_{P_j} = S, \tag{20b}$$

$$U_{P_i} A = \frac{1}{n!} \sum U_{P_i} \epsilon_P U_P = \frac{1}{n!} \epsilon_{P_i} \sum \epsilon_{P_i} \epsilon_P U_{P_i} U_P$$
$$= \frac{1}{n!} \epsilon_{P_i} \sum \epsilon_{P_j} U_{P_j} = \epsilon_{P_i} A \quad [Eqs. \ (9a), \ (11)] \tag{21b}$$

From Eqs. (20b) and (21b), one obtains

$$S U_{P_i} = S = U_{P_i} S, \tag{23a}$$

$$A U_{P_i} = \epsilon_{P_i} A = U_{P_i} A, \tag{23b}$$

$$C U_{P_i} = U_{P_i} C, \tag{23c}$$

and

$$S^2 = \frac{1}{n!} \sum U_P S = S, \qquad [Eq. \ (23a)],$$

$$A^2 = \frac{1}{n!} \sum \epsilon_P^2 A = A. \tag{24}$$

$$S = \frac{1}{n!} \sum_P U_P, \tag{20}$$

$$A = \frac{1}{n!} \sum_P \epsilon_P U_P, \tag{21}$$

$$C = 1 - S - A, \tag{22}$$

where the sum is over all $n!$ permutations (of the symmetric group).
From Eqs. (23a) and (23b),

$$\begin{aligned} U_P(S\Psi) &= S\Psi, \\ U_P(A\Psi) &= \epsilon_P(A\Psi). \end{aligned} \tag{25}$$

Comparison with Eq. (16) shows that

$$\begin{aligned} S\Psi &\quad \text{is a symmetric function} \\ A\Psi &\quad \text{is an antisymmetric function} \end{aligned} \tag{26}$$

and reference to Eq. (14) shows that

$$\begin{aligned} U_T(S\Psi) &= (S\Psi), \\ U_T(A\Psi) &= -(A\Psi). \end{aligned} \tag{27}$$

From the orthogonality of $\Psi^{(s)}$ and $\Psi^{(a)}$ function, one has

$$(A\Psi, S\Psi) = 0,$$

$$A^\dagger S = AS = 0, \qquad S^\dagger A = SA = 0, \tag{28}$$

and from Eqs. (24) and (28), one obtains

$$C^2 = C, \tag{29}$$

so the C is also a projection operator.

From Eqs. (20), (21), and (22), one sees that an arbitrary ψ may be projected into three parts

$$\Psi = S\Psi + A\Psi + C\Psi. \tag{30}$$

Consider

$$S + A = \frac{1}{n!} \sum_P (1 + \epsilon_P) U_P.$$

Let U_P be the permutation operator corresponding to permutations made up of even number of transpositions so that $\epsilon_P = 1$. Then by Eqs. (23a) and (23b),

$$U_P (S + A) = \frac{2}{n!} \sum_P U_P = S + A,$$
$$U_P \{(S + A)\Psi\} = \{(S + A)\Psi\}, \tag{31}$$

i.e., $S + A$ picks out those states which do not change sign upon even number of transpositions T_{ij}. The rest, C, are states that do not have this symmetry.

As an example, for two particles, n = 2 , and

$$U_P = U_T;$$
$$S = \frac{1}{2}(1 + U_T), \qquad A = \frac{1}{2}(1 - U_T);$$
$$C = 0, \tag{32}$$

so that an arbitrary state can always be written

$$\Psi = S\Psi + A\Psi$$
$$= \Psi^{(s)} + \Psi^{(a)}. \tag{33}$$

For a system of n particles, the wave function in Eq. (10) is the representative in the coordinate representation of a ket $| K_1, ..., K_n >$

$$\Psi_{K, K, .., K_n}(x_1, ..., x_n) = < x_1, ..., x_n | K_1, K_2, ..., K_n > . \tag{34}$$

Now we introduce another representation

$$| n_1, n_2, n_3, ... >, \tag{35}$$

where n_j is the number of particles which have the set of quantum number K_j, and the total number of particles is

$$\sum_j n_j = n. \tag{36}$$

The representation (35) is called the "occupation number representation", and its representatives in the coordinate representation are

$$\Psi_{n_1,n_2,...}(x) = < x_1, x_2, ..., x_n \mid n_1, n_2, ... > . \tag{37}$$

By means of the projection operators S and A and Eq. (26), we can obtain the wave function for bosons and fermions

$$\Psi^B_{n_1,n_2,...}(x_1, ..., x_n) = c_n < x_1, ..., x_n \mid S \mid n_1, n_2, ... >, \tag{38}$$

$$\Psi^F_{n_1,n_2,...}(x_1, ..., x_n) = c_n < x_1, ..., x_n \mid A \mid n_1, n_2, ... >, \tag{39}$$

where c_n is the normalization constant.

If all the $n_1, n_2, ... = 1$, these two wave functions are just $\Psi^{(s)}, \Psi^{(a)}$, respectively, of Eqs. (17) and (18),

$$< x \mid S \mid 1, 1, 1, ... > = \frac{1}{\sqrt{n!}} \Psi^{(s)}(x) \tag{40}$$

$$< x \mid A \mid 1, 1, 1, ... > = \frac{1}{\sqrt{n!}} \Psi^{(a)}(x) \tag{41}$$

In $\mid n_1, n_2, ... >$ in Eq. (35), there may be infinite numbers of sets of quantum number K_j , so that there are infinitely many states $\mid n_1, n_2, ... >$.[7] We assume that the $\mid n_1, n_2, ... >$ are orthogonal and form a complete set,

$$< n_1, n_2, ... \mid n'_1, n'_2, ... > = \delta_{n_1,n'_1} \delta_{n_2,n'_2} ..., \tag{42}$$

$$\sum_{(n)} \mid n_1, n_2, ... > < n_1, n_2, ... \mid = 1, \tag{43}$$

where the summation is over all $n_1, n_2, n_3, ...$ subject to Eq. (36).

---

[7] *For example, for a particle in a central field, $K = (n, \ell, m, m_s)$, and $n = 1, 2, 3, ...$ infinite. Since the total number of particles n is fixed, there are many $n'_j s$ which are zero.*

We shall now calculate the c_n in Eqs. (38) and (39). We remember that in the case of $\Psi^B_{n_1,n_2,...}(x)$, the $n_1, n_2, n_3, ...$ do not have to be 1, and the normalization constant $c_n = n!$ in Eq. (40) is not for the general case. Let

$$\Psi_{K_1,K_2,...}(x_1, x_2, x_3, ...) \tag{44}$$

be the general wave function in which the particles are in various (single-particle) states K_1, K_2 , no longer restricted to $n_j = 1$ for all j, and let Ψ^B of Eq. (38) be

$$\begin{aligned}\Psi^B(x) &= c_n S \Psi_{K_1 K_2...}(x) \\ &= c_n \frac{1}{n!} \sum_P U_P \Psi_{K_1 K_2...}(x_1, x_2, ...).\end{aligned} \tag{45}$$

The permutation operator U_P acts on the $x_1, x_2, x_3, ...$, and the sum is over all $n!$ permutations. An equivalent view is to leave the $x_1, x_2, ...$ alone and permute the $K_1, K_2, ...$ by the inverse permutation U_{P-1}, i.e.,

$$U_P \Psi_{K_1,K_2,...}(x_1, x_2, ...) = \Psi_{U_{P-1}(K_1,K_2,...)}(x_1, x_2, ...) \tag{46}$$

As n_j particles have the set of quantum numbers K_j, K_j appears n_j times in $\Psi_{K_1 K_2...}(x_1,x_2,...)$ so that $n_1! n_2! ... = \Pi_j n_j!$ permutations have no effect on $\Psi_{K_1 K_2...}(x_1, x_2, ...)$.

From Eq. (45),

$$\begin{aligned}1 = (\Psi^B, \Psi^B) &= |c_n|^2 (S\Psi_{K_1 K_2...}, S\Psi_{K_1 K_2...}) \\ &= |c_n|^2 (\Psi_{K_1 K_2...}, S\Psi_{K_1 K_2...}) \quad \text{by Eq. (20a)} \\ &= |c_n|^2 \frac{1}{n!} \sum (\Psi_{K_1 K_2...}, \Psi_{U_{P-1}(K_1 K_2...)}) \quad \text{by Eq. (46)}\end{aligned}$$

one obtains, for bosons,

$$|c_n|^2 = \frac{n!}{\Pi_j n_j!}, \tag{47}$$

where

$$\sum_{j=1} n_j = n, \qquad \text{by Eq. (36).}$$

For fermions, one has, in place of Eq. (45),

$$\Psi^F(x) = c_n A \Psi_{K_1 K_2 \ldots}(x). \tag{48}$$

But now

$$n_j = 1, \quad \text{or} \quad 0,$$

and

$$|\, c_n \,|^2 = n!, \tag{49}$$

again with

$$\sum_{n=1} n_j = n, \quad \text{by Eq. (36)}.$$

§.7.3. Fock representation: creation and annihilation operators.

As exposed in Chapter 6, Sect. 1, we approach the problem of the quantum theory of fields by regarding a classical field as a dynamical system of nondenumerably infinite number of degrees of freedom and applying the quantization condition to the field variables. In brief, a quantized field is represented by a system of nondenumerably infinite number of particles. The most convenient mathematical method is to represent a field by such a system of harmonic oscillators. The quantum mechanics of harmonic oscillators is well known, and the Fock representation is particularly suitable for application to the problem of quantized fields. This Fock representation has been introuced in Appendix B, Ch. 3, Vol. I. We shall summarize the results there for one oscillator, and extend them to a system of oscillators.

(1) Harmonic oscillator

Without repeating the details, we shall simply collect the results of Appendix B, Ch. 3, Vol. I (*"Quantum Mechanics"*), below,

$$H = \hbar\omega \frac{1}{2}(\xi^2 - \frac{d^2}{d\xi^2}), \tag{B1}$$

$$b^\dagger = \frac{1}{\sqrt{2}}(\xi - \frac{d}{d\xi}), \qquad b = \frac{1}{\sqrt{2}}(\xi + \frac{d}{d\xi}). \tag{B2}$$

$$bb^\dagger - b^\dagger b = 1, \tag{B4}$$

$$b^\dagger b \,|\, \lambda > = \lambda \,|\, \lambda >, \tag{B7}$$

$$b^\dagger b(b \mid \lambda >) = (\lambda - 1)(b \mid \lambda >), \qquad (B9)$$

$$b^\dagger b(b^n \mid \lambda >) = (\lambda - n)(b^n \mid \lambda >). \qquad (B11)$$

Concept of the lowest state:

$$b \mid \lambda_{min} >= 0, \qquad (B12)$$

i.e.,

$$b^\dagger b \mid \lambda_{min} >= 0. \qquad (B13)$$

From Eq. (B7), one has

$$\lambda_{min} = 0.$$

We introduce the notation $\mid 0 >$ for the lowest state $\mid \lambda_{min} >$

$$\mid 0 >\equiv \mid \lambda_{min} > . \qquad (50)$$

$$bb^\dagger \mid \lambda >= (\lambda + 1) \mid \lambda > . \qquad (B17)$$

$$b^\dagger b(b^\dagger \mid \lambda >) = (\lambda + 1)(b^\dagger \mid \lambda >). \qquad (B10)$$

$$H = (b^\dagger b + \frac{1}{2})\hbar\omega, \qquad (B3)$$

$$H \mid \lambda >= (\lambda + \frac{1}{2}) \mid \lambda > \hbar\omega, \qquad (B3,7)$$

$$H(b^\dagger \mid \lambda >) = (\lambda + \frac{3}{2})(b^\dagger \mid \lambda >)\hbar\omega \qquad (B3) + (B10).$$

This last relations shows that $b^\dagger \mid \lambda >$ is the eigenstate of H for the eigenvalue $\lambda + 1$, i.e., $b^\dagger \mid \lambda >$ changes $\mid \lambda >$ into $\mid \lambda + 1 >$. When normalized,

$$b^\dagger \mid \lambda >= \sqrt{\lambda + 1} \mid \lambda + 1 > . \qquad (B18)$$

Similarly

$$b \mid \lambda >= \sqrt{\lambda} \mid \lambda - 1 > . \qquad (B23)$$

The corresponding eigenvalues and eigenstates of H are:

$$\frac{H}{\hbar\omega} = \frac{1}{2}, \quad \frac{3}{2}, \quad \frac{5}{2}, \quad ...(n + \frac{1}{2})... \tag{B14}$$

$$| \lambda > = | 0 >, \quad b^\dagger | 0 >, \quad \frac{1}{\sqrt{2}}(b^\dagger)^2 | 0 >, ... \frac{1}{\sqrt{n!}}(b^\dagger)^n | 0 > ... \tag{$B-19$}$$

The operators $b^\dagger, b^-$ are adjoint to each other, but are not self-adjoint (i.e., not hermitian). Therefore they are not "observables" as the $p.q$. Now the operator $b^\dagger$ does not commute with any operator $A(q, p)$, i.e., $b^\dagger$ by itself constitutes "a complete set of commutable operators" in the sense of Ch. 5, Sec. 4, Vol. I. Hence one may take the $b^\dagger$ representation, and the set of kets (B-19) above constitutes a complete set, in terms of which an arbitrary ket $| P >$ can be expressed

$$| P > = \sum_{k=0} a_k (b^\dagger)^k | 0 >, \tag{51}$$

i.e., $| P >$ is expressed in terms of the eigenkets of H.

In the $b^\dagger$ - representation , $b^\dagger$ is a multiplicative factor[8] but b has to be obtained from (B4) above.

By repeated application of (B-4), one obtains

$$b(b^\dagger)^n - (b^\dagger)^n b = n(b^\dagger)^{n-1}. \tag{52}$$

If one writes

$$f(b^\dagger) = \sum_{k=0} a_k (b^\dagger)^k, \tag{53}$$

then from Eq. (51),

$$b | P > = b f(b^\dagger) | 0 >, \tag{54}$$

and from Eq. (52),

$$(bf - fb) | 0 > = \sum_{n=1} a_n n (b^\dagger)^{n-1} | 0 > . \tag{55}$$

---

[8] *Just as in the Schrödinger, or coordinate, representation, q or x is a multiplicative number, while p is determined by the commutation relation.*

From (B12) and (50), $b \mid 0 >= 0$, hence

$$
\begin{aligned}
bf(b^\dagger) \mid 0 > &= \sum_{n=1} a_n n (b^\dagger)^{n-1} \mid 0 > \\
&= \frac{\partial f(b^\dagger)}{\partial b^\dagger} \mid 0 > .
\end{aligned}
\tag{56a}
$$

This gives the operator equation, for $\mid 0 >$,

$$
bf(b^\dagger) = \frac{\partial f(b^\dagger)}{\partial b^\dagger}
\tag{56b}
$$

which is consistent with the relation (B-4).

Assume $\mid 0 >$ to have been normalized,

$$
< 0 \mid 0 >= 1.
$$

From Eq. (56),

$$
\begin{aligned}
< 0 \mid b^n (b^\dagger)^n \mid 0 > &=< 0 \mid b^{n-1} b (b^\dagger)^n \mid 0 > \\
&= n < 0 \mid b^{n-1} (b^\dagger)^{n-1} \mid 0 > \\
&= n!
\end{aligned}
\tag{57}
$$

By repeated application of (B18), (B23), one finds

$$
< 0 \mid b^m (b^\dagger)^n \mid 0 >= 0 \qquad \text{for} \qquad m \neq n.
\tag{58}
$$

Let

$$
\mid Q >= \sum_{n=0} c_n (b^+)^n \mid 0 > .
\tag{59a}
$$

Then

$$
< Q \mid =< 0 \mid \sum_{n=0} c_n^* (b)^n
\tag{59b}
$$

and from Eq. (51),

$$
\begin{aligned}
< Q \mid P > &= \sum_{n,m} c_n^* a_m < 0 \mid (b)^n (b^+)^m \mid 0 > \\
&= \sum_n c_n^* a_n n! \qquad \text{by Eqs. (57)-(58)}
\end{aligned}
\tag{60}
$$

and

$$
1 =< P \mid P >= \sum_n \mid a_n \mid^2 n!
\tag{61}
$$

This leads to the interpretation of $|a_n|^2 n!$ as the probability that $|P>$ is in the n th state of the harmonic oscillator, i.e., the probability of the eigenvalue $(n + \frac{1}{2})\hbar\omega$.

(2) A system of harmonic oscillators

The results of the preceding subsection can readily be extended to a system of independent oscillators so that the Hamiltonian H is

$$H = \sum_k H_K,$$

$$H_K = \frac{1}{2}(\xi_k^2 - \frac{d^2}{d\xi_k^2})\hbar\omega_k, \tag{62}$$

$$H_j H_k - H_k H_j = 0,$$

$$b_k^\dagger = \frac{1}{\sqrt{2}}(\xi_k - \frac{d}{d\xi_k}), \qquad b_k = \frac{1}{\sqrt{2}}(\xi_k + \frac{d}{d\xi_k}),$$

$$b_k b_j^\dagger - b_j^\dagger b_k = \delta_{kj}, \qquad \text{for all} \quad k, j, \tag{63}$$

$$b_k \mid 0 >= 0, \qquad \text{for all} \quad k,$$

and as operators on $(b_k^\dagger) \mid 0 >$,

$$b_k = \frac{\partial}{\partial b_k^\dagger}. \tag{64}$$

The $\mid 0 >$ here is the lowest state of all H_k , and

$$< 0 \mid 0 >= 1,$$

and similarly to (B19),

$$(b_1^\dagger)^{n_1} (b_2^\dagger)^{n_2} (b_3^\dagger)^{n_3} \ldots \mid 0 > \tag{65}$$

represents the state in which the jth oscillator is in the n_j^{th} state, so that the eigenvalue corresponding to Eq. (65) is

$$\sum_j (n_j + \frac{1}{2})\hbar\omega_j. \tag{66}$$

If

$$|P> = \sum_{n_1,n_2,..} a_{n_1,n_2,..} (b_1^\dagger)^{n_1} (b_2^\dagger)^{n_2} ... |0>,$$

then

$$<P|P> = \sum_{n_1,n_2,..} |a_{n_1 n_2..}|^2 n_1! n_2! n_3! ...$$

i.e.,

$$|a_{n_1 n_2...}|^2 n_1! n_2! ... \tag{67}$$

is the probability that $|P>$ is a state in which the jth oscillator is in the $n_j{}^{th}$ state, the kth oscillator in the $n_k{}^{th}$ state, etc.

When the number of oscillators N is infinite, the system has an infinite number of degrees of freedom. This system is to represent a "field".

(3) Creation and annihilation operators

We have introduced the occupation-number representation in which a state is represented by the number n_j of particles in the single-particle state $|k_j>$, where k_j stands for the totality of quantum numbers specifying the one-particle state,

$$|n_1, n_2, n_3, ... > . \tag{68}$$

The coordinate representative of this ket is,

$$\Psi_{n_1 n_2...}(x_1, x_2, ... x_n) = <x_1, x_2, ... x_n \mid n_1, n_2, n_3, ... > . \tag{69}$$

(i) System of bosons

The state of a system is changed by a change in the occupation numbers n_j. The annihilation and creation operator, a and $a^\dagger$ respectively, are defined by

$$a_j \mid n_1, n_2, ..., n_j, ... > = \sqrt{n_j} \mid n_1, n_2, ..., n_j - 1, ... >, \tag{70}$$

$$a_j^\dagger \mid n_1, n_2, ..., n_j, ... > = \sqrt{n_j + 1} \mid n_1, n_2, ..., n_j + 1, ... >, \tag{71}$$

a_j decreasing the number of n_j in state $|k_j>$ by 1, $a_j^\dagger$ increasing n_j by 1.

In the x-representative $\Psi_{n_1 n_2...}(x)$, we have

$$(a_j^\dagger \Psi_{n_1 n_2...,n_j-1,...}, \Psi_{n_1 n_2...,n_j...})$$
$$= \sqrt{n_j}(\Psi_{n_1 n_2...n_j...}, \Psi_{n_1 n_2...n_j...})$$
$$= \sqrt{n_j},$$

$$(\Psi_{n_1 n_2...n_j-1...}, a_j \Psi_{n_1 n_2...n_j...})$$
$$= \sqrt{n_j}(\Psi_{n_1 n_2...n_j-1...}, \Psi_{n_1 n_2...n_j-1...})$$
$$= \sqrt{n_j}.$$

From these two equations, it is seen that

$$a_j \quad \text{and} \quad a_j^\dagger \qquad \text{are adjoint of each other.} \tag{72}$$

From the definitions (70) and (71), one obtains the following operator relations:

$$[a_i, a_j]_- = 0, \qquad [a_i^\dagger, a_j^\dagger]_- = 0, \tag{73a}$$

$$[a_i, a_j^\dagger]_- = \delta_{ij}, \tag{73b}$$

where

$$[A, B]_- = AB - BA. \tag{74}$$

The vacuum state of a system of harmonic oscillators is

$$| \, 0 >=| \, 0, 0, 0, ..., 0, ... > . \tag{75}$$

We introduce the "occupation number operator" N_j

$$N_j = a_j^\dagger a_j. \tag{76}$$

From Eq. (72), one has

$$(\Psi_{n_1 n_2...}, a_j^\dagger a_j \Psi_{n_1 n_2...}) = (a_j \Psi_{n_1 n_2...}, a_j \Psi_{n_1 n_2...}),$$

and

$$(a_j^\dagger a_j \Psi_{n_1 n_2...}, \Psi_{n_1 n_2...}) = (a_j \Psi_{n_1 n_2...}, a_j \Psi_{n_1 n_2...}).$$

Hence

$$(a_j^\dagger a_j)^\dagger = (a_j^\dagger a_j),$$

i.e.,

$$N_j = (a_j^\dagger a_j) \quad \text{is hermitian.} \tag{77}$$

From Eq. (76), one has

$$
\begin{aligned}
N_j \mid n_1, n_2, ..., n_j, ... > &= a_j^\dagger \sqrt{n_j} \mid n_1, n_2, ..., n_j - 1, ... > \\
&= n_j \mid n_1, n_2, ..., n_j, ... > .
\end{aligned}
\tag{78}
$$

i.e., the eigenvalues of N_j are n_j .

The $N_j, a_j^\dagger, a_j$ have the following relations

$$
\begin{aligned}
N_j a_j^\dagger \mid ..., n_j, ... > &= (n_j + 1) a_j^\dagger \mid ..., n_j, ... >, \\
a_j^\dagger N_j \mid ..., n_j, ... > &= n_j a_j^\dagger \mid ..., n_j, ... >, \\
N_j a_j \mid ..., n_j, ... > &= (n_j - 1) a_j \mid ..., n_j, ... >, \\
a_j N_j \mid ..., n_j, ... > &= n_j a_j \mid ..., n_j, ... > .
\end{aligned}
\tag{79}
$$

i.e., the operator relations

$$[N_i, a_j^\dagger]_- = a_j^\dagger \delta_{ij}, \tag{80a}$$

$$[N_i, a_j]_- = -a_j \delta_{ij}. \tag{80b}$$

By repeated application of the above relations, one obtains

$$N_j (a_j^\dagger)^k \mid ...n_j... >= (n_j + k)(a_j^\dagger)^k \mid ...n_j... >, \tag{81}$$

$$N_j (a_j)^k \mid ...n_j... >= (n_j - k)(a_j)^k \mid ...n_j... >, \tag{82}$$

where k is an arbitrary integer $< n_j$. For $k = n_j$, the last relation gives

$$
\begin{aligned}
N_j (a_j)^{n_j} \mid ...n_j... > &= N_j \mid ..., 0, ... > \\
&= 0
\end{aligned}
$$

If

$$N_j \mid ..., 0, ... >= 0 \quad \text{for all j,}$$

then

$$| 0, 0, ..., 0, ... > \equiv | 0 > \quad \text{is the vacuum state.} \tag{83}$$

From Eq. (71), one has

$$| n_1, n_2, ..., n_j, ... > = \prod \frac{1}{\sqrt{n_j!}} (a_j^\dagger)^{n_j} | 0 > . \tag{84}$$

If all oscillators are in the same state $| K >$ state, then

$$| n_1, n_2, ... > \equiv | n >$$
$$= \frac{1}{\sqrt{n!}} (a^\dagger)^n | 0 > . \tag{84a}$$

From Eqs. (70) and (71), one finds the non-vanishing matrix elements of $a_j^\dagger, a_j$

$$< n_j + 1 | a_j^\dagger | n_j > = < n_j | a_j | n_j + 1 >$$
$$= \sqrt{n_j + 1}. \tag{85}$$

(ii) System of fermions

For fermions, the Pauli exclusion principle postulates that the occupation number n_j be only either 0 or 1, and that the state function be antisymmetric with respect to the transposition of any two indistinguishable particles.

These two requirements can be expressed by the following form for the state ket

$$| k_1, k_2, ..., k_i, ..., k_j, ... > = a_1^\dagger a_2^\dagger ... a_i^\dagger ... a_j^\dagger ... | 0 >$$

so that upon the transposition of i and j ,

$$| k_1, k_2, ..., k_j, ... k_i, ... > = a_1^\dagger a_2^\dagger ... a_j^\dagger ... a_i^\dagger ... | 0 >$$
$$= - | k_1, k_2, ..., k_i, ..., k_j, ... > . \tag{86}$$

This condition, together with $n_j = 0$ or 1, leads to the following relations

$$(a_i)^2 | \Psi > = (a_i^\dagger)^2 | \Psi > = 0,$$
$$a_i a_j | \Psi > = -a_j a_i | \Psi >,$$
$$a_i^\dagger a_j^\dagger | \Psi > = -a_j^\dagger a_i^\dagger | \Psi >,$$
$$a_i a_j^\dagger | \Psi > = -a_j^\dagger a_i | \Psi >, \qquad i \neq j,$$
$$(a_i a_i^\dagger + a_i^\dagger a_i) | \Psi > = | \Psi > .$$

On defining the anticommutation operator (of Jordan and Wigner, 1927)

$$[A, B]_+ = AB + BA, \tag{87}$$

the above conditions can be expressed in the form

$$[a_i, a_j]_+ = 0, \qquad\qquad [a_i^\dagger, a_j^\dagger]_+ = 0, \tag{88a}$$

$$[a_i, a_j^\dagger]_+ = \delta_{ij}. \tag{88b}$$

For fermions, the relations (Eqs. 70, 71) are no longer valid. Instead, one now has

$$a_j \mid n_1, n_2, ..., n_j, ... > = (-1)^{f_j} n_j \mid n_1, n_2, ..., n_j - 1, ... > \tag{89}$$

$$a_j^\dagger \mid n_1, n_2, ..., n_j, ... > = (-1)^{f_j} (1 - n_j) \mid n_1, n_2, ..., n_j + 1, ... > \tag{90}$$

where

$$f_j = \sum_{i=1}^{j-1} n_i. \tag{91}$$

When in Eq. (89) $n_j = 0$ initially, $a_j \mid n_1, ..., n_j, ... > = 0$ does not exist and this is expressed by the vanishing of the righthand side. Similarly, when in Eq. (90) $n_j = 1$ initially, $a_j^\dagger \mid ..., n_j = 1, ... >$ does not exist.

References

Fock, V. A., Zeits. f. Physik **75**, 622 (1932).

Jordan, P. and Klein, O., *ibid.* **45**, 751 (1927), for systems obeying Bose-Einstein statistics.

Jordan, P. and Winger, E., *ibid.* **47**, 631 (1928), for systems obeying Fermi-Dirac statistics.

E. Mishkin, *Lecture Notes on Relativistic Quantum Mehanics* (1964), Poly. Tech. Brooklyn.

Chapter 8.
Quantization of Free Fields

We shall quantize the scalar field Φ of the Klein-Gordon equation, the vector electromagnetic field A_μ and the spinor field Ψ of the Dirac equation, all uncoupled to any source of charges or currents.

§.8.1. Klein-Gordon, real (pseudo) scalar field $\phi(x_\mu)$

The solutions of the Klein-Gordon equation Eq. (67), Ch. 6, are

$$e^{ik_\nu x_\nu} = e^{i(\mathbf{k}\cdot\mathbf{r} - k_0 x_0)}, \tag{1}$$

which form a complete set of functions.[1] We expand $\phi(x_\nu)$ in the form

$$\phi(x_\nu) = \frac{1}{(2\pi)^3} \int d^4k\,\delta(k_\nu^2 + \mu^2)\,C a_\mathbf{k} e^{ik_\nu x_\nu}, \tag{3}$$

where $\delta(k_\nu^2 + \mu^2)$ is given in (A2) of the appendix at the end of this chapter, C is a constant and $a_\mathbf{k}$ is an operator yet to be determined. As $\phi(x_\nu)$ is a real field, $\phi(x_\nu)$ is a hermitian operator. Changing k_ν into $-k_\nu$ changes $a_\mathbf{k}$ into $a_{-\mathbf{k}}$. It will be seen that $a_{-\mathbf{k}}$ is the adjoint of $a_\mathbf{k}$ (see Eq. (14) below). $\phi(x_\nu)$ is the Fourier transform of $\delta(k_\nu^2 + \mu^2)C a_\mathbf{k}$.

---

[1] *For notations, see the appendix at the end of this chapter,*

$$
\begin{aligned}
&x_4 = ix_0 = it, \\
&k_4 = ik_0, \quad k_0 = E_\mathbf{k} = \omega_\mathbf{k} = \sqrt{\mathbf{k}^2 + \mu^2}, \quad \mu = m_0, \\
&k_\nu^2 = k_\nu k_\nu = \mathbf{k}^2 - k_0^2 = -\mu^2, \\
&\mathbf{k}^2 + \mu^2 = k_0^2.
\end{aligned} \tag{2}
$$

Or, in ordinary units,

$$
\begin{aligned}
&x_4 = ix_0 = ict, \\
&k_4 = ik_0, \quad k_0 = \frac{E_\mathbf{k}}{\hbar c} = \frac{\omega_\mathbf{k}}{c} = \sqrt{\mathbf{k}^2 + \mu^2}, \quad \mu = m_0 c.
\end{aligned} \tag{A1a)-(A1c}
$$

On using the step function $\vartheta(x_0)$ in (A11) and the last relation of (A12), one can write

$$\phi(x_\lambda) = \frac{1}{(2\pi)^3} \int d^4k [\vartheta(k_0) + \vartheta(-k_0)] \delta(k_\nu^2 + \mu^2) Ca_{\mathbf{k}} e^{ik_\nu x_\nu}$$

$$= \phi_+(x_\nu) + \phi_-(x_\nu), \tag{4}$$

where

$$\phi_+(x_\lambda) = \frac{1}{(2\pi)^3} \int d^4k\, \vartheta(k_0) \delta(k_\nu^2 + \mu^2) Ca_{\mathbf{k}} a^{ik_\nu x_\nu}$$

$$\phi_-(x_\lambda) = \frac{1}{(2\pi)^3} \int d^4k\, \vartheta(k_0) \delta(k_\nu^2 + \mu^2) Ca_{-\mathbf{k}} e^{-ik_\nu x_\nu} \tag{5}$$

From (A2), (A11) and $k_0 = \omega_{\mathbf{k}}$ in Eq. (2), it is seen that[2]

$$\begin{aligned}
\vartheta(k_0)\delta(k_\nu^2 + \mu^2) &= \vartheta(k_0)\delta(k_0^2 - \omega_{\mathbf{k}}^2) \\
&= \vartheta(k_0)\frac{1}{2\omega_{\mathbf{k}}}[\delta(k_0 - \omega_{\mathbf{k}}) + \delta(k_0 + \omega_{\mathbf{k}})] \\
&= \frac{1}{2\omega_{\mathbf{k}}}\delta(k_0 - \omega_{\mathbf{k}}),
\end{aligned} \tag{6}$$

so that[3]

$$\begin{aligned}
\phi_+(x_\lambda) &= \frac{1}{2(2\pi)^3} \int d^3k \frac{1}{\omega_{\mathbf{k}}} Ca_{\mathbf{k}} e^{i(\mathbf{k}\cdot\mathbf{r} - \omega_{\mathbf{k}}t)}, \\
\phi_-(x_\lambda) &= \frac{1}{2(2\pi)^3} \int d^3k \frac{1}{\omega_{\mathbf{k}}} Ca_{-\mathbf{k}} e^{-i(\mathbf{k}\cdot\mathbf{r} - \omega_{\mathbf{k}}t)}.
\end{aligned} \tag{7}$$

From Eq. (69) of Ch. 6, the conjugate to $\phi(x_\lambda)$ is

$$\Pi(x_\lambda) = \frac{\partial}{\partial t}\phi(x_\lambda).$$

One writes

$$\Pi(x_\lambda) = \Pi_+(x_\lambda) + \Pi_-(x_\lambda), \tag{8}$$

$$\Pi_+(x_\lambda) = -\frac{i}{2(2\pi)^3} \int d^3k\, Ca_{\mathbf{k}} e^{i(\mathbf{k}\cdot\mathbf{r} - \omega_{\mathbf{k}}t)},$$

---

[2]

$$\vartheta(k_0)\delta(k_\nu^2 + \mu^2) = \frac{c}{2\omega_{\mathbf{k}}}\delta(k_0 - \frac{\omega_{\mathbf{k}}}{c}). \tag{6'}$$

[3] *In ordinary units, Eqs. (7), (9) have a factor c on the right.*

$$\Pi_-(x_\lambda) = \frac{i}{2(2\pi)^3} \int d^3k\, C a_{-\mathbf{k}} e^{-i(\mathbf{k}\cdot\mathbf{r}-\omega_\mathbf{k} t)}. \tag{9}$$

We shall take a (large cubic) volume of the field $V = L^3$ and assume periodic conditions

$$k_i = \frac{2\pi}{L} n_i, \quad i = x, y, z, \quad n_i = \pm \text{ integers},$$

which quantize k_i. We choose[4]

$$C = \sqrt{2\omega_\mathbf{k} L^3}. \tag{10}$$

For large L, the integrals over k may be replaced by sums over discrete k,

$$\frac{1}{(2\pi)^3} \int d^3k \quad \rightarrow \quad \frac{1}{L^3} \sum_\mathbf{k}$$

and[5]

$$\phi(x_\lambda) = \sqrt{\frac{1}{2L^3}} \sum \frac{1}{\sqrt{\omega_\mathbf{k}}} \{ a_\mathbf{k} e^{i(\mathbf{k}\cdot\mathbf{r}-\omega_\mathbf{k} t)} + a_{-\mathbf{k}} e^{-i(\mathbf{k}\cdot\mathbf{r}-\omega_\mathbf{k} t)} \},$$
$$\Pi(x_\lambda) = i\sqrt{\frac{1}{2L^3}} \sum \sqrt{\omega_\mathbf{k}} \{ -a_\mathbf{k} e^{i(\mathbf{k}\cdot\mathbf{r}-\omega_\mathbf{k} t)} + a_{-\mathbf{k}} e^{-i(\mathbf{k}\cdot\mathbf{r}-\omega_\mathbf{k} t)} \}. \tag{11}$$

These are the general solutions of the Klein-Gordon equation.

Quantization of the field may be accomplished by application of Dirac correspondence principle, which states that, to an elementary Poisson bracket $\{a(t), b(t)\} = c$ (which c some c-number) in a classical system, there corresponds an elementary commutator $[a(t), b(t)] = ic$ in the corresponding quantized system. In the present case, we find, for $t' - t = 0$,

$$[\phi(\mathbf{r}),\ \phi(\mathbf{r}')] = 0, \tag{12a}$$

$$[\phi(\mathbf{r}),\ \Pi(\mathbf{r}')] = i\delta(\mathbf{r} - \mathbf{r}'), \tag{12b}$$

$$[\Pi(\mathbf{r}),\ \Pi(\mathbf{r}')] = 0. \tag{12c}$$

---

[4]
$$C = \frac{1}{c}\sqrt{2\hbar\omega_\mathbf{k} L^3}. \tag{10'}$$

[5] Eqs. (11) have a factor $\sqrt{\hbar}$ on the right.

It is to be noted that these last three relations are not Lorentz invariant since they hold only for $t = t'$. That $\delta(\mathbf{r}-\mathbf{r}')$ appears in Eq. (12b) has a simple reason. Two points spatially separated cannot communicate by any instantaneous signal so that $\phi(\mathbf{r}), \Pi(\mathbf{r}')$ are independent and can both be measured with arbitrary accuracy.

It is straightforward to demonstrate that Eqs. (12a)-(12c) hold if and only if the following relations are assumed:

$$[a_{\mathbf{k}}, a_{\mathbf{k}'}] = [a_{-\mathbf{k}}, a_{-\mathbf{k}'}] = 0,$$
$$[a_{\mathbf{k}}, a_{-\mathbf{k}'}] = \delta_{\mathbf{k}\mathbf{k}'}. \tag{13}$$

Let us calculate the commutator

$$[\phi(\mathbf{x}_\lambda), \phi(\mathbf{x}'_\lambda)] = \phi(\mathbf{x}_\lambda)\phi(\mathbf{x}'_\lambda) - \phi(\mathbf{x}'_\lambda)\phi(\mathbf{x}_\lambda).$$

Let us write for brevity

$$R \equiv \mathbf{k} \cdot \mathbf{r} - \omega_{\mathbf{k}} t = k_\mu x_\mu, \qquad R' \equiv \mathbf{k}' \cdot \mathbf{r}' - \omega_{\mathbf{k}'} t = k'_\mu x'_\mu. \tag{14}$$

Then[6]

$$[\phi(\mathbf{x}_\lambda), \phi(\mathbf{x}'_\lambda)] = \frac{1}{2L^3} \sum_{k,k'} \frac{1}{\sqrt{\omega_{\mathbf{k}}\omega_{\mathbf{k}'}}} \{[a_{\mathbf{k}}, a_{\mathbf{k}'}]e^{i(R+R')} + [a_{-\mathbf{k}}, a_{-\mathbf{k}'}]e^{-i(R+R')}$$
$$+ [a_{\mathbf{k}}, a_{-\mathbf{k}'}]e^{i(R-R')} - [a_{\mathbf{k}'}, a_{-\mathbf{k}}]e^{-i(R-R')}\}. \tag{15}$$

Using Eq. (13), we find

$$[\phi(\mathbf{x}_\lambda), \phi(\mathbf{x}'_\lambda)] = \frac{1}{2L^3} \sum \frac{1}{\omega_{\mathbf{k}}} \{e^{i(R-R')} - e^{-i(R-R')}\}$$
$$= \frac{1}{2(2\pi)^3} \int \frac{d^3k}{\omega_{\mathbf{k}}} \{e^{i(k_\nu y_\nu)} - e^{-i(k_\nu y_\nu)}\} \tag{16}$$
$$= -\frac{1}{(2\pi)^3} \int d^3k \frac{\sin k_\nu y_\nu}{k_4},$$

where

$$y_\nu = x_\nu - x'_\nu, \qquad k_4 = i\omega_{\mathbf{k}}.$$

---

[6] *In ordinary units, Eq. (15) has $\hbar$ on the right; Eqs. (17) and (18) have $\frac{\hbar}{c}$ as a factor on the right; Eq. (19) $\frac{\hbar}{c^2}$ on the right.*

Comparison with the $\triangle(x)$ function of (A-6b) now leads to

$$[\phi(x_\lambda), \phi(x'_\lambda)] = \frac{1}{i}\triangle(x_\lambda - x'_\lambda), \tag{17}$$

which is a Lorentz invariant commutation relation.
Similarly,

$$
\begin{aligned}
[\phi(x_\lambda), \Pi(x'_\lambda)] &= \frac{i}{(2\pi)^3}\int d^3k \cos k_\nu y_\nu \\
&= -\frac{1}{i}\frac{\partial}{\partial t}\triangle(x_\lambda - x'_\lambda),
\end{aligned}
\tag{18}
$$

$$
\begin{aligned}
[\Pi(x_\lambda), \Pi(x'_\lambda)] &= -\frac{1}{(2\pi)^3}\int d^3k k_4 \sin k_\nu y_\nu \\
&= -i\frac{\partial^2}{\partial t^2}\triangle(x_\lambda - x'_\lambda).
\end{aligned}
\tag{19}
$$

Let us come back to the relations Eq. (13). They are identical with the commutation relations for the creation and annihilation operators $a_k^\dagger$ and a_k if we identify them as follows:

$$a_k = a_k, \qquad a_{-k} = a_k^\dagger. \tag{20}$$

To fully bring out the meaning of a_k and a_{-k} in Eq. (11), let us take the Heisenberg equations of motion[7]

$$\frac{d\phi}{dt} = -i[\phi, H], \qquad \frac{d\Pi}{dt} = -i[\Pi, H] \tag{21}$$

where

$$[A, B] = AB - BA.$$

From Eq. (11), one obtains

$$\sum_k \sqrt{\omega_k}[a_k e^{iR} - a_{-k}e^{-iR}] = \sum_k \frac{1}{\sqrt{\omega_k}}[(a_k e^{iR} + a_{-k}e^{-iR}), H], \tag{22}$$

where for brevity as in Eq. (12), $R \equiv \mathbf{k} \cdot \mathbf{r} - \omega_k t$.

---

[7] *In ordinary units, Eqs. (21), (22), (25), and (26) have a factor $\hbar$ on the left; (23), (27) a factor $\hbar$ on the right.*

From Eq. (21), one obtains

$$[a_{\mathbf{k}}, H] = \omega_{\mathbf{k}} a_{\mathbf{k}},$$
$$[a_{-\mathbf{k}}, H] = -\omega_{\mathbf{k}} a_{-\mathbf{k}}. \tag{23}$$

The momentum $P_j, j = 1, 2, 3$, is, from Eq. (74) of Ch. 6,

$$P_j = \int_{V_3} \frac{1}{2} \{\Pi(x_\lambda) \partial_j \phi(x_\lambda) + (\partial_j \phi(x_\lambda)) \Pi(x_\lambda)\} d^3 x. \tag{24}$$

By means of the commutation relations Eq. (13), it can be shown that for any function $F(\phi, \pi)$ of $\phi(x_\lambda)$ and $\pi(x_\lambda)$, the following relation holds

$$\nabla F = -i[\mathbf{P}, F]. \tag{25}$$

In particular, for $F = \phi(x_\lambda)$ of Eq. (11), this relation gives

$$\sum \mathbf{k}(a_{\mathbf{k}} e^{iR} - a_{-\mathbf{k}} e^{-iR}) = -\sum_{\mathbf{k}} [\mathbf{P}, (a_{\mathbf{k}} e^{iR} + a_{-\mathbf{k}} e^{-iR})]. \tag{26}$$

From this, one gets, in a way similar to Eq. (23),

$$[a_{\mathbf{k}}, \mathbf{P}] = \mathbf{k} a_{\mathbf{k}},$$
$$[a_{-\mathbf{k}}, \mathbf{P}] = -\mathbf{k} a_{-\mathbf{k}}. \tag{27}$$

Let $E_0, | E_0 >, \mathbf{P}_0, | \mathbf{P}_0 >$ be the eigenvalue and eigenstate of H and $\mathbf{P}$

$$(H - E_0) | E_0 >= 0, \quad (\mathbf{P} - \mathbf{P}_0) | \mathbf{P}_0 >= 0. \tag{28}$$

From Eq. (27) and Eq. (28), one obtains[8]

$$H(a_{\mathbf{k}} | E_0 >) = (E_0 - \omega_{\mathbf{k}})(a_{\mathbf{k}} | E_0 >),$$
$$\mathbf{P}(a_{\mathbf{k}} | \mathbf{P}_0 >) = (\mathbf{P}_0 - \mathbf{k})(a_{\mathbf{k}} | \mathbf{P}_0 >),$$
$$H(a_{-\mathbf{k}} | E_0 >) = (E_0 + \omega_{\mathbf{k}})(a_{-\mathbf{k}} | E_0 >),$$
$$\mathbf{P}(a_{-\mathbf{k}} | \mathbf{P}_0 >) = (\mathbf{P}_0 + \mathbf{k})(a_{-\mathbf{k}} | \mathbf{P}_0 >). \tag{29}$$

These show that $a_{\mathbf{k}}$ lowers the eigenvalue of H by $\omega_{\mathbf{k}}$ and the eigenvalue of $\mathbf{P}$ by k, while $a_{-\mathbf{k}}$ raises that of H by $\omega_{\mathbf{k}}$ and that of $\mathbf{P}$ by k. This completes the identification Eq. (20) of $a_{-\mathbf{k}}$ with the creation operator $a_{\mathbf{k}}^{\dagger}$, and $a_{\mathbf{k}}$ with the

---

[8] *In ordinary units, $\omega_{\mathbf{k}}, k$ on the right in (29) stand for $\hbar\omega_{\mathbf{k}}, \hbar k$ respectively.*

annihilation operator a_k of Eqs. (73a, b), Eq. (11) then brings out the particle aspect of the field $\phi(x_\lambda)$.

If one uses the occupation number representation of Eq. (34) in Ch. 7, then

$$| \, n_1, n_2, n_3, ... >$$

is a state in which n_i oscillators have the momentum

$$\mathbf{k}_i$$

and energy

$$E_{k_i} = \sqrt{\mathbf{k}_i^2 + m_0^2}, \quad i = 1, 2, 3, ...$$

From Eqs. (70) and (71) of Ch. 7, one has

$$
\begin{aligned}
a_{k_i} \, | \, n_1, n_2, ..., n_i, ... > \, &= \sqrt{n_i} \, | \, n_1, n_2, ..., n_i - 1, ... >, \\
a_{k_i}^\dagger \, | \, n_1, n_2, ..., n_i, ... > \, &= \sqrt{n_i + 1} \, | \, n_1, n_2, ..., n_i + 1, ... >,
\end{aligned}
\tag{30}
$$

with the occupation-number operator N_k of Eq. (76) in Ch. 7

$$N_k = a_k^\dagger a_k, \tag{31}$$

one has, for the energy and momentum operator,[9]

$$
\begin{aligned}
H &= \sum_k E_k N_k = \sum_k E_k a_k^\dagger a_k, \\
\mathbf{P} &= \sum_k \mathbf{P}_k N_k = \sum_k \mathbf{k} a_k^\dagger a_k.
\end{aligned}
\tag{32}
$$

and

$$
\begin{aligned}
H \, | \, n_1, n_2, ... > \, &= (\sum_k n_k E_k) \, | \, n_1, n_2, ... >, \\
\mathbf{P} \, | \, n_1, n_2, ... > \, &= (\sum_k n_k \mathbf{k}) \, | \, n_1, n_2, ... > .
\end{aligned}
\tag{33}
$$

From Eq. (24), one finds that, for $E_2 - E_1 > 0$,

$$(E_2 - E_1 - \omega_k) < E_2 \, | \, a_k^\dagger \, | \, E_1 >= 0,$$

---

[9] *In ordinary units, k in Eqs. (32) and (33), ω_k in Eqs. (34a) and (34b) stand for $\hbar k, \hbar \omega_k$ respectively.*

so that

$$< E_2 \mid a_{\mathbf{k}}^{\dagger} \mid E_1 >= 0 \quad \text{unless} \quad \omega_{\mathbf{k}} = E_2 - E_1. \tag{34a}$$

Similarly, for $E_1 - E_2 > 0$,

$$< E_2 \mid a_{\mathbf{k}} \mid E_1 >= 0 \quad \text{unless} \quad \omega_{\mathbf{k}} = E_1 - E_2. \tag{34b}$$

The real, scalar Klein-Gordon field $\phi(x_\lambda)$, when quantized, may be applied to the neutral pion π^0 which is experimentally known to have zero spin, and is a pseudo scalar (i.e., $\phi(x_\lambda)$ changes sign upon space inversion $\mathbf{r} \to -\mathbf{r}$).

To close this section, it is useful to note that the momentum operator given by Eq. (32) differs from that in Eq. (24) by a c-number. We write

$$P_j = \int_{V_3} : \frac{1}{2} \{ \Pi(x_\lambda) \partial_j \phi(x_\lambda) + (\partial_j \phi(x_\lambda)) \Pi(x_\lambda) \} : d^3x, \tag{35}$$

where the double colon " : : " indicates "normal ordering" of the operators, meaning that creation operators always precede (i.e. are to the left of) all annihilation operators. In general, we have, for an arbitrary operator $\hat{Q}$,

$$< 0 \mid : \hat{Q} : \mid 0 >= 0. \tag{36}$$

Analogously, by normal ordering the quantized Noether's current, we obtain the energy operator as given by Eq. (32).

§.8.2. Klein-Gordon complex, scalar field

The quantization of the real scalar fields $\psi_1(x_\lambda), \psi_2(x_\lambda)$ of Eq. (78) in Ch. 6 follows the same procedure as in Eqs. (10)-(19) of Ch. 7. Thus for the hermitian operators ψ_1, ψ_2 , analogously to Eq. (11), we now have, on writing

$$a(\mathbf{k}), \, a^{\dagger}(\mathbf{k}) \quad \text{for} \quad a_{\mathbf{k}}, \, a_{\mathbf{k}}^{\dagger}, \tag{37}$$

$$\psi_j(x_\lambda) = \sqrt{\frac{1}{2L^3}} \sum_{\mathbf{k}} \frac{1}{\sqrt{\omega_{\mathbf{k}}}} \{ a_j(\mathbf{k}) e^{iR} + a_j^{\dagger}(\mathbf{k}) e^{-iR} \}, \tag{38}$$

$$j = 1, 2, \quad R \equiv k_\nu x_\nu = \mathbf{k} \cdot \mathbf{r} - \omega_{\mathbf{k}} t, \quad R' \equiv \mathbf{k}' \cdot \mathbf{r}' - \omega_{\mathbf{k}'} t'.$$

$$[a_j(\mathbf{k}), \, a_j(\mathbf{k}')] = [a_j^{\dagger}(\mathbf{k}), \, a_j^{\dagger}(\mathbf{k}')] = 0,$$
$$[a_j(\mathbf{k}), a_j^{\dagger}(\mathbf{k}')] = \delta_{\mathbf{k}\mathbf{k}'}. \tag{39}$$

Now we introduce

$$a_{\mathbf{k}} = \frac{1}{\sqrt{2}}(a_1(\mathbf{k}) + ia_2(\mathbf{k})),$$

$$a_{\mathbf{k}}^{\dagger} = \frac{1}{\sqrt{2}}(a_1^{\dagger}(\mathbf{k}) - ia_2^{\dagger}(\mathbf{k})),$$

$$b_{\mathbf{k}} = \frac{1}{\sqrt{2}}(a_1(\mathbf{k}) - ia_2(\mathbf{k})),$$ (40)

$$b_{\mathbf{k}}^{\dagger} = \frac{1}{\sqrt{2}}(a_1^{\dagger}(\mathbf{k}) + ia_2^{\dagger}(\mathbf{k})),$$

and by analogy with Eq. (20),

$$a_{-\mathbf{k}} = b_{\mathbf{k}}^{\dagger}, \qquad b_{-\mathbf{k}} = a_{\mathbf{k}}^{\dagger}. \tag{41}$$

Then from Eqs. (39) and (40), one obtains

$$[a_{\mathbf{k}}, b_{\mathbf{k}'}] = [a_{\mathbf{k}}^{\dagger}, b_{\mathbf{k}'}^{\dagger}] = [a_{\mathbf{k}}, b_{\mathbf{k}'}^{\dagger}] = [b_{\mathbf{k}}, a_{\mathbf{k}'}^{\dagger}] = 0,$$

$$[a_{\mathbf{k}}, a_{\mathbf{k}'}^{\dagger}] = [b_{\mathbf{k}}, b_{\mathbf{k}'}^{\dagger}] = \delta_{\mathbf{k}\mathbf{k}'}. \tag{42}$$

In terms of the $a_{\mathbf{k}}$, $b_{\mathbf{k}}$, $a_{\mathbf{k}}^{+}$, $b_{\mathbf{k}}^{+}$, one obtains for the non-hermitian operators $\Psi(x_\lambda)$, $\Psi^{+}(x_\lambda)$, $\Pi(x_\lambda)$, $\Pi^{+}(x_\lambda)$ corresponding to Ψ, Ψ^*, Π, Π^* in Eqs. (76) and (79) of Ch. 6, the expressions[10]

$$\Psi(x_\lambda) = \sqrt{\frac{1}{2L^3}} \sum_{\mathbf{k}} \frac{1}{\sqrt{\omega_{\mathbf{k}}}} \{a_{\mathbf{k}} e^{iR} + b_{\mathbf{k}}^{\dagger} e^{-iR}\},$$

$$\Psi^{\dagger}(x_\lambda) = \sqrt{\frac{1}{2L^3}} \sum_{\mathbf{k}} \frac{1}{\sqrt{\omega_{\mathbf{k}}}} \{b_{\mathbf{k}} e^{iR} + a_{\mathbf{k}}^{\dagger} e^{-iR}\},$$

$$\Pi(x_\lambda) = i\sqrt{\frac{1}{2L^3}} \sum_{\mathbf{k}} \sqrt{\omega_{\mathbf{k}}} \{-b_{\mathbf{k}} e^{iR} + a_{\mathbf{k}}^{\dagger} e^{-iR}\},$$ (43)

$$\Pi^{\dagger}(x_\lambda) = i\sqrt{\frac{1}{2L^3}} \sum_{\mathbf{k}} \sqrt{\omega_{\mathbf{k}}} \{-a_{\mathbf{k}} e^{iR} + b_{\mathbf{k}}^{+} e^{-iR}\}.$$

Using the commutation relation (17) for the two real field $\psi_1(x_\lambda), \psi_2(x_\lambda)$, one can show that

$$[\Psi(x_\lambda), \Psi^{\dagger}(x_\lambda')] = \frac{1}{i}\triangle(x_\lambda - x_\lambda') \tag{44}$$

---

[10] *In ordinary units, a factor $\sqrt{\hbar}$ on the right in Eq. (43); $\frac{\hbar}{c}$ on the right in (44); $\hbar$ on the right in (46c); $\frac{1}{\hbar}$ on the right in Q, in (47).*

192

which is Lorentz invariant, and

$$[\Psi(x_\lambda),\ \Psi(x'_\lambda)] = [\Psi^\dagger(x_\lambda),\ \Psi^\dagger(x'_\lambda)] = 0. \tag{45}$$

Corresponding to Eqs. (12a), (12b), and (12c) for a space-like separation $\mathbf{r} - \mathbf{r}'$, (i.e., $t - t' = 0$) , one obtains

$$[\Psi(\mathbf{r},t),\ \Psi^\dagger(\mathbf{r}',t)] = [\Pi(\mathbf{r},t),\ \Pi^+(\mathbf{r}',t)] = 0, \tag{46a}$$

$$[\Psi(\mathbf{r},t),\ \Psi^\dagger(\mathbf{r}',t)] = [\Psi^\dagger(x_\lambda),\ \Pi(x_\lambda)] = 0, \tag{46b}$$

$$[\Psi(\mathbf{r},t),\ \Pi(\mathbf{r}',t)] = -[\Psi^\dagger(\mathbf{r},t),\ \Pi^\dagger(\mathbf{r}',t)] = -\frac{1}{i}\delta(\mathbf{r} - \mathbf{r}'). \tag{46c}$$

The same remark following Eq. (12b) applies here to Eq. (46c).

That the $a_\mathbf{k}, b_\mathbf{k}, a_\mathbf{k}^\dagger, b_\mathbf{k}^\dagger$ have the meaning of annihilation and creation operators can be seen by following through the same procedure as in Eqs. (23)-(34).

From Eq. (89) of Ch. 6, we have the total electric charge of the field in a volume V_3

$$Q = \int_{V_3} \rho d^3x = -ie \int_{V_3} (\Psi\dot\Psi^* - \Psi^*\dot\Psi)d^3x,$$

upon symmetrizing,

$$Q = -\frac{ie}{2}\int_{V_3}(\Psi\Pi + \Pi\Psi - \Psi^\dagger\Pi^\dagger - \Pi^\dagger\Psi^\dagger)d^3x, \tag{47}$$

and on using Eq. (41)

$$Q = e\sum_\mathbf{k}(a_\mathbf{k}^\dagger a_\mathbf{k} - b_\mathbf{k}^\dagger b_\mathbf{k}). \tag{47a}$$

This suggests the interpretation that $a_\mathbf{k}^\dagger, a_\mathbf{k}$ are the creation and annihilation operator respectively of positively charged particle of the quantized field, $b_\mathbf{k}^\dagger, b_\mathbf{k}$ are similarly for the negatively charge particle of the quantized field.

Analogously to Eq. (31), we define the occupation number operators for + and - charged particles

$$N_{k(+)} = a_\mathbf{k}^\dagger a_\mathbf{k}, \quad N_{k(-)} = b_\mathbf{k}^\dagger b_\mathbf{k},$$

and

$$N_+ = \sum_k N_{k(+)}, \quad N_- = \sum_k N_{k(-)}. \tag{48}$$

In terms of these operators,

$$Q = e(N_+ - N_-). \tag{47b}$$

From Eqs. (40) and (45a), we obtain

$$\begin{aligned} &[a_\mathbf{k}, Q] = ea_\mathbf{k}, \qquad &&[a_\mathbf{k}^\dagger, Q] = -ea_\mathbf{k}^\dagger, \\ &[b_\mathbf{k}, Q] = -eb_\mathbf{k}, \qquad &&[b_\mathbf{k}^\dagger, Q] = eb_\mathbf{k}^\dagger. \end{aligned} \tag{49}$$

The Heisenberg equations of motion of Ψ and Ψ^+ are[11]

$$\frac{d}{dt}\Psi = -i[\Psi, H], \qquad \frac{d}{dt}\Psi^+ = -i[\Psi^\dagger, H], \tag{50}$$

and analogously to Eq. (25),

$$\nabla\Psi = -i[\mathbf{P}, \Psi], \qquad \nabla\Psi^\dagger = -i[\mathbf{P}, \Psi^\dagger]. \tag{51}$$

From these equations, one obtains[12]

$$\begin{aligned} &[a_\mathbf{k}, H] = E_\mathbf{k} a_\mathbf{k}, \qquad &&[a_\mathbf{k}^\dagger, H] = -E_\mathbf{k} a_\mathbf{k}^\dagger, \\ &[b_\mathbf{k}, H] = E_\mathbf{k} b_\mathbf{k}, \qquad &&[b_\mathbf{k}^\dagger, H] = -E_\mathbf{k} b_\mathbf{k}^\dagger, \\ &[a_\mathbf{k}, \mathbf{P}] = \mathbf{k} a_\mathbf{k}, \qquad &&[b_\mathbf{k}^\dagger, \mathbf{P}] = -\mathbf{k} b_\mathbf{k}^\dagger, \\ &[b_\mathbf{k}, \mathbf{P}] = \mathbf{k} b_\mathbf{k}, \qquad &&[b_\mathbf{k}^\dagger, \mathbf{P}] = -\mathbf{k} b_\mathbf{k}^\dagger. \end{aligned} \tag{52}$$

Let $|0>$ be the vacuum state

$$a_\mathbf{k} |0> = b_\mathbf{k} |0> = 0,$$

$$H |0> = 0, \quad \mathbf{P} |0> = 0. \tag{53}$$

---

[11] In ordinary units, a factor $\frac{1}{\hbar}$ on the right in (50), (51).

[12] In ordinary units, k on the right in (52), (54) stands for $\hbar k$.

Then one gets

$$H = \sum_{\mathbf{k}} E_{\mathbf{k}}(N_{k(+)} + N_{k(-)})$$
$$= \sum_{\mathbf{k}} E_{\mathbf{k}}(a_{\mathbf{k}}^{\dagger} a_{\mathbf{k}} + b_{\mathbf{k}}^{\dagger} b_{\mathbf{k}}), \qquad (54)$$
$$\mathbf{P} = \sum_{\mathbf{k}} \mathbf{k}(a_{\mathbf{k}}^{\dagger} a_{\mathbf{k}} + b_{\mathbf{k}}^{\dagger} b_{\mathbf{k}}).$$

The charged Klein-Gordon fields Ψ, Ψ^*, or ψ_1, ψ_2 can be taken to apply to the Π^+ and Π^- pions which have experimintally been found to have spin zero.

Again, it is useful to note that the operators Q, H, and P [Eqs. (47b) and (54)] are related to the quantized conserved Noether's currents [cf. §6.3.] by normal ordering. In other words, Noether's theorem gives rise to conserved dynamical invariants which, upon quantization and normal ordering, describe the corresponding quantities in the quantized theory.

§.8.3. Electromagnetic fields

The general procedure of the quantization of fields has been illustrated by the examples of the scalar fields of the Klein-Gordon equation in the two preceding sections. For the electromagnetic field A_μ of the Maxwell equations, the general method is the same, but there are a few special considerations.[13]

(i) The rest mass m_0 of the quantized field particles (photons) is zero.

(ii) In the Lorentz gauge, the condition Eq. (94) of Ch. 6.

$$\partial_\mu A_\mu = 0, \qquad (55)$$

imposes one relation among the A_μ so that only three of the four A_μ's are independent.

Furthermore, if one chooses, in any Lorentz frame, a gauge in which $A_4 = 0$, one has Eq. (100) of Ch. 6,

$$div\, A = \partial_{\mathbf{k}} A_{\mathbf{k}} = 0 \qquad (56)$$

so that only two of the A_μ are independent.

---

[13] See, e.g. I. J. R. Aitchison, An Informal Introduction to Gauge Field Theories (Cambridge University Press, London, 1982).

(iii) Since the (physical) fields **E** and **B** are invariant under the gauge transformation (97), Ch. 6, it is necessary to see the relation of this to the definition of the operators and the physical state of an electromagnetic field.

(1) Without the Lorentz condition

For reasons that will become clear later in the section, we need to start *without* the Lorentz condition (55) so that the A_μ are independent. We use the following notations

$$A(A_1, A_2, A_3, A_4 = iA_0), \tag{57}$$

A_1, A_2, A_3, A_0 are hermitian operators, so that A_4 is anti-hermitian, i.e.,

$$\begin{aligned} A_j^\dagger(x) &= A_j(x), & j &= 1,2,3, \\ A_0^\dagger(x) &= A_0(x), & A_4^\dagger(x) &= -A_4(x). \end{aligned} \tag{58}$$

The energy-momentum relation is

$$k^2 = \mathbf{k}^2 - k_0^2 = 0, \tag{59a}$$

where[14]

$$E_\mathbf{k} = |\,\mathbf{k}\,| = \omega_\mathbf{k} = k_0. \tag{59b}$$

The solutions of Eq. (101) of Ch. 6,

$$\partial_\nu \partial_\nu A_\mu(x) = 0, \tag{60}$$

namely

$$e^{ikx}, \quad kx = k_\nu x_\nu, \tag{61}$$

form a complete set, so that one may expand $A_\mu(x)$ as in Eq. (3)

$$A_\mu(x) = \frac{1}{(2\pi)^3} \int d^4k\, \delta(k_\nu^2) C a_\mu(\mathbf{k}) e^{ikx}, \tag{62}$$

14

$$E_\mathbf{k} = c\hbar\,|\,\mathbf{k}\,| = \hbar\omega_\mathbf{k} = c\hbar k_0. \tag{59b'}$$

196

which, on account of the $\delta(k_\nu^2)$ function, is really a 3-dimensional integral. We choose, as in Eq. (10),[15]

$$C = \sqrt{2\omega_{\mathbf{k}} L^3}. \tag{63}$$

On account of the hermitian properties Eq. (58), we have

$$a_j(\mathbf{k}) = a_j^\dagger(-\mathbf{k}), \qquad a_0(\mathbf{k}) = a_0^\dagger(-\mathbf{k}). \tag{64}$$

From Eq. (6), we have[16]

$$\delta(k^2) = \frac{1}{2\omega_{\mathbf{k}}} \big(\delta(k_0 + \omega_{\mathbf{k}}) + \delta(k_0 - \omega_{\mathbf{k}}) \big). \tag{65}$$

Hence

$$A_\mu(x) = \frac{1}{2(2\pi)^3} \int d^3k \frac{1}{\omega_{\mathbf{k}}} C(a_\mu(\mathbf{k}) e^{iR} + a_\mu^\dagger(\mathbf{k}) e^{-iR}) \tag{66}$$

$$R = \mathbf{k} \cdot \mathbf{r} - \omega_{\mathbf{k}} t. \tag{67}$$

Again, as in Eq. (11), we quantize $\mathbf{k}$ by taking the field in a large volume L^3 and assuming periodic conditions

$$k_j = \frac{2\pi}{L} n_j, \qquad n_j = \pm \text{integers}, \tag{68}$$

and replacing

$$\frac{1}{(2\pi)^3} \int d^3k \quad \text{by} \quad \frac{1}{L^3} \sum_{\mathbf{k}} \tag{69}$$

so that Eq. (6) becomes[17]

$$\mathbf{A}(x) = \sqrt{\frac{1}{2L^3}} \sum_{\mathbf{k}} \frac{1}{\sqrt{\omega_{\mathbf{k}}}} \{ a(\mathbf{k}) e^{iR} + a^\dagger(\mathbf{k}) e^{-iR} \} \tag{70a}$$

---

[15]

$$C = \frac{1}{2} \sqrt{2\omega_{\mathbf{k}} L^3}. \tag{63'}$$

[16] *In ordinary units, $\omega_{\mathbf{k}}$ stands for $\frac{\omega_{\mathbf{k}}}{c}$ in (65), (66).*

[17] A factor $\sqrt{\hbar}$ on the right in (70a), (71a); a factor $\frac{\hbar}{c}$ on the right in (74), (74a).

$$\equiv \mathbf{A}_+(x) + \mathbf{A}_-(x), \tag{70b}$$

$$A_0(x) = \sqrt{\frac{1}{2L^3}} \sum_{\mathbf{k}} \frac{1}{\sqrt{\omega_{\mathbf{k}}}} \{a_0(\mathbf{k})e^{iR} + a_0^\dagger(\mathbf{k})e^{-iR}\} \tag{71a}$$

$$= A_{0+}(x) + A_{0-}(x). \tag{71b}$$

From Eqs. (105) and (102) of Ch. 6, we have

$$\dot{A}_j(x) = \Pi_j(x) + i\partial_j A_4(x),$$
$$\dot{A}_4(x) = \Pi_4(x) - i\partial_j A_j(x), \tag{72a}$$

$$\Pi_j(x) = -i(-\partial_4 A_j + \partial_j A_4),$$
$$\Pi_4(x) = i(\partial_4 A_4 + \partial_j A_j). \tag{72b}$$

The $a_j(\mathbf{k}), a_j^\dagger(\mathbf{k}), a_0(\mathbf{k}), a_0^\dagger(\mathbf{k}), j = 1, 2, 3$, operators satisfy the relations

$$\begin{aligned}
[a_i(\mathbf{k}), a_j(\mathbf{k}')] &= [a_i^\dagger(\mathbf{k}), a_j^\dagger(\mathbf{k}')] = 0, \qquad i, j = 1, 2, 3, \\
[a_0(\mathbf{k}), a_0(\mathbf{k}')] &= [a_0^\dagger(\mathbf{k}), a_0^\dagger(\mathbf{k}')] = 0, \\
[a_i(\mathbf{k}), a_j^\dagger(\mathbf{k}')] &= \delta_{ij}\delta_{\mathbf{k}\mathbf{k}'}, \\
[a_0(\mathbf{k}), a_0^\dagger(\mathbf{k}')] &= -\delta_{\mathbf{k}\mathbf{k}'},
\end{aligned} \tag{73}$$

all other commutators vanishing.[18]

From Eqs. (70a), (71a), and (73), calculations give[19]

$$[A_\mu(x), A_\nu(x')] = -\frac{1}{(2\pi)^3} \int d^3k \frac{\sin k_\lambda (x - x')_\lambda}{k_4} \delta_{\mu\nu} \tag{74}$$

$$= \frac{1}{i} D(x - x')\delta_{\mu\nu}, \tag{74a}$$

where $D(x - x')$ is the invariant function in (A-16a). The above commutator relation is Lorentz invariant.

---

[18] *The negative sign in the last of the relations in Eq. (73) comes from the anti-hermiticity of A_4, namely $A_4^\dagger = -A_4$, in Eq. (58). On account of this negative sign, one may regard $a_0^\dagger$, a_0 as the annihilation and creation operators respectively.*

[19] *In ordinary units, a factor $\frac{\hbar}{c}$ on the right in (74), (74a).*

For $t = t'$, the above relation reduces to

$$[A_\mu(\mathbf{r}, t), A_\nu(\mathbf{r}', t)] = 0, \quad \text{by } (A21). \tag{74b}$$

Similarly[20]

$$
\begin{aligned}
[A_j(x), \Pi_\ell(x')] &= \frac{i}{(2\pi)^3} \int d^3k \cos(k_\nu(x - x')_\nu) \\
&= i\partial_t D(x - x')\delta_{j\ell}, \quad j, \ell = 1, 2, 3,
\end{aligned} \tag{75}
$$

$$
\begin{aligned}
[A_j(x), \Pi_4(x')] &= [A_4(x), \Pi_j(x')] \\
&= -\partial_j D(x - x'),
\end{aligned} \tag{76}
$$

$$[A_4(x), \Pi_4(x')] = i\partial_t D(x - x'), \tag{77}$$

$$[A_j(\mathbf{r}, t), \Pi_\ell(\mathbf{r}', t)] = i\delta(\mathbf{r} - \mathbf{r}')\delta_{j\ell}, \tag{75a}$$

$$[A_4(\mathbf{r}, t), \Pi_4(\mathbf{r}', t)] = i\delta(\mathbf{r} - \mathbf{r}'). \tag{77a}$$

All the above results have been obtained without making use of the Lorentz condition Eq. (55),

$$\partial_\mu A_\mu = 0.$$

(2) Polarization of photons

From[21]

$$
\begin{aligned}
\mathbf{B} &= \nabla \times \mathbf{A}, &\quad \mathbf{E} &= -(\partial_0 \mathbf{A} + \nabla A_0), \\
\partial_0 &= \frac{\partial}{\partial x_0} = \frac{\partial}{\partial t},
\end{aligned} \tag{78}
$$

and

$$\mathbf{B}(x) \equiv \mathbf{B}_+(x) + \mathbf{B}_-(x)$$

---

[20] *In ordinary units, a factor $\hbar\epsilon_0$ on the right of (76), (75a), (77a); a factor $\frac{\hbar\epsilon_0}{c}$ on the right of (75), (77).*

[21]

$$
\begin{aligned}
\mathbf{E} &= -c(\partial_0 \mathbf{A} + \nabla A_0), \\
\partial_0 &= \frac{\partial}{\partial x_0} = \frac{1}{c}\frac{\partial}{\partial t}.
\end{aligned} \tag{78'}
$$

$$\mathbf{E}(x) \equiv \mathbf{E}_+(x) + \mathbf{E}_-(x), \tag{79}$$

one obtains from Eq. (70a)[22]

$$\mathbf{B}_+(x) = i\sqrt{\frac{1}{2L^3}} \sum_{\mathbf{k}} \frac{1}{\sqrt{\omega_{\mathbf{k}}}} [\mathbf{k} \times \mathbf{a}(\mathbf{k})] e^{iR},$$

$$\mathbf{B}_-(x) = -i\sqrt{\frac{1}{2L^3}} \sum_{\mathbf{k}} \frac{1}{\sqrt{\omega_{\mathbf{k}}}} [\mathbf{k} \times \mathbf{a}^\dagger(\mathbf{k})] e^{-iR}, \tag{80}$$

$$\mathbf{E}_+(x) = i\sqrt{\frac{1}{2L^3}} \sum_{\mathbf{k}} \frac{1}{\sqrt{\omega_{\mathbf{k}}}} \{|\mathbf{k}| \mathbf{a}(\mathbf{k}) - \mathbf{k}\, \mathbf{a}_0(\mathbf{k})\} e^{iR},$$

$$\mathbf{E}_-(x) = -i\sqrt{\frac{1}{2L^3}} \sum_{\mathbf{k}} \frac{1}{\sqrt{\omega_{\mathbf{k}}}} \{|\mathbf{k}| \mathbf{a}^\dagger(\mathbf{k}) - \mathbf{k}\, \mathbf{a}_0^\dagger(\mathbf{k})\} e^{-iR}. \tag{81}$$

Owing to the unusual sign for the last equation of (73), we must introduce the physical vacuum state $| 0 >$, in a careful manner.

From Eqs. (80) and (81), it is seen that $| 0 >$ may be defined by

$$\mathbf{B}_+(x) \,|\, 0 >= 0, \tag{82}$$

$$\mathbf{E}_+(x) \,|\, 0 >= 0, \tag{83}$$

i.e., for any $\mathbf{k}$, one must have

$$[\mathbf{k} \times \mathbf{a}(\mathbf{k})] \,|\, 0 >= 0, \tag{84}$$

$$(|\mathbf{k}| \mathbf{a}(\mathbf{k}) - \mathbf{k}\, \mathbf{a}_0(\mathbf{k})) \,|\, 0 >= 0. \tag{85}$$

Let us choose a rectangular coordinate system with the x_3-axis along $\mathbf{k}$, and $a_1(\mathbf{k}), a_2(\mathbf{k})$ are transverse to $\mathbf{k}$, whereas $a_3(\mathbf{k})$ is "longitudinal" along $\mathbf{k}$. In this coordinate system, $k_1 = k_2 = 0$, and Eq. (84) gives for the transverse parts:

$$a_1(\mathbf{k}) \,|\, 0 >= 0, \qquad a_2(\mathbf{k}) \,|\, 0 >= 0 \tag{86}$$

and Eq. (85) gives for the longitudinal part

---

[22] *In ordinary units, a factor $c\sqrt{\hbar}$ on the right in (80); a factor $c^2\sqrt{\hbar}$ on the right in (81).*

$$(a_3(\mathbf{k}) - a_0(\mathbf{k})) \mid 0 >= 0. \tag{87}$$

From Eq. (73), one finds

$$[a_3(\mathbf{k}) - a_0(\mathbf{k}), \ a_3^\dagger(\mathbf{k}) - a_0^\dagger(\mathbf{k})] = 0. \tag{88}$$

From Eqs. (87) and (88), it is seen that[23]

$$(a_3^\dagger(\mathbf{k}) - a_0^\dagger(\mathbf{k})) \mid 0 > = 0, \tag{89}$$
$$(a_3^\dagger(\mathbf{k}) - a_0^\dagger(\mathbf{k}))^n \mid 0 > = 0, \qquad n = 2, 3, ... \tag{90}$$

We introduce

$$a_1^\dagger(\mathbf{k}) \mid 0 > = \mid 1, 0 >,$$
$$a_2^\dagger(\mathbf{k}) \mid 0 > = \mid 0, 1 >, \tag{91}$$

which represent states, respectively, with one photon with momentum $\mathbf{k}$ along x_1 and x_2 axis. One says that the photon is polarized along the x_1 or x_2 axis. The notation

$$\mid n, m > = \frac{1}{\sqrt{n!m!}}((a_1^\dagger(\mathbf{k}))^n (a_2^\dagger(\mathbf{k}))^m \mid 0 >, \tag{92}$$

represents a state with n photons polarized along the x_1 - axis and m photons polarized along the x_2 - axis.
From

$$[a_1^\dagger, \ a_3 - a_0] = 0, \qquad\qquad [a_2^\dagger, \ a_3 - a_0] = 0,$$

it follows that

$$[(a_1^\dagger)^n, \ a_3 - a_0] = [(a_2^\dagger)^m, \ a_3 - a_0] = 0,$$

and

$$[(a_1^\dagger)^n (a_2^\dagger)^m, a_3 - a_0] = 0,$$

---

[23] *Eq. (88) yields*

$$< 0 \mid (a_3 - a_0)(a_3^\dagger - a_0^\dagger) \mid 0 > = < 0 \mid (a_3^\dagger - a_0^\dagger)(a_3 - a_0) \mid 0 > = 0,$$

so that Eq. (89) follows as a consequence.

and

$$(a_1^\dagger)^n (a_2^\dagger)^m (a_3 - a_0) = (a_3 - a_0)(a_1^\dagger)^n (a_2^\dagger)^m,$$

and

$$(a_1^\dagger)^n (a_2^\dagger)^m (a_3 - a_0) \mid 0 >= (a_3 - a_0)(a_1^\dagger)^n (a_2^\dagger)^m \mid 0 > .$$

By Eqs. (87) and (92), it is seen that

$$(a_3(\mathbf{k}) - a_0(\mathbf{k})) \mid n, m >= 0, \tag{93}$$

for arbitrary state $\mid n, m >$. From the definition of $A_{\mu+}(x)$ in Eqs. (70) and (71), it is seen that Eq. (93) is equivalent to

$$\partial_\mu A_{\mu+}(x) \mid n, m >= 0, \tag{94}$$

for any state $\mid n, m >$. This is a condition on the operators $A_{\mu+}$ of the quantized electromagnetic field. It is called the "subsidiary condition", and it takes the place of the Lorentz condition (55)

$$\partial_\mu A_\mu = 0, \tag{55}$$

in the classical theory.[24]

From Eqs. (50) and (51),

$$\partial_t A_\mu(x) = -i[A_\mu(x), H],$$
$$\nabla A_\mu(x) = -i[\mathbf{P}, A_\mu(x)], \tag{95}$$

and Eqs. (70) and (71) for $A_\mu(x)$, one obtains, as Eqs. (23) and (27),

$$[\mathbf{a}(\mathbf{k}), H] = E_\mathbf{k}\mathbf{a}(\mathbf{k}), \qquad [\mathbf{a}^\dagger(\mathbf{k}), H] = -E_\mathbf{k}\mathbf{a}^\dagger(\mathbf{k}),$$
$$[a_j(\mathbf{k}), \mathbf{P}] = \mathbf{k}a_j(\mathbf{k}), \qquad [a_j^\dagger(\mathbf{k}), \mathbf{P}] = -\mathbf{k}a_j^\dagger(\mathbf{k}). \tag{96}$$

The vacuum state $\mid 0 >$ satisfies

---

[24] *Following the step leading to Eq. (89), we see that*

$$(a_3^\dagger(\mathbf{k}) - a_0^\dagger(\mathbf{k})) \mid n, m >= 0. \tag{93a}$$

The contributions due to time-like photons are always cancelled by those due to longitudinal photons.

202

$$H \mid 0 >= 0, \qquad \mathbf{P} \mid 0 >= 0. \tag{97}$$

Let the occupation numbers for states with polarization along x_1 and x_2 axis be N_1 and N_2 .

$$N_1(\mathbf{k}) = a_1^\dagger(\mathbf{k})a_1(\mathbf{k}), \qquad N_2(\mathbf{k}) = a_2^\dagger(\mathbf{k})a_2(\mathbf{k}). \tag{98}$$

Then, analogously to Eq. (32), one has

$$H = \sum_{\mathbf{k}} E_{\mathbf{k}}(N_1(\mathbf{k}) + N_2(\mathbf{k})),$$
$$\mathbf{P} = \sum_{\mathbf{k}} \mathbf{k}(N_1(\mathbf{k}) + N_2(\mathbf{k})). \tag{99}$$

Only a_1, $a_1^\dagger$, a_2, $a_2^\dagger$, the "transverse" operators, appear in H and P; there are no "longitudinal" photons.

As the time-like and longitudinal photons are included, H and P contain terms proportional to

$$N_3(\mathbf{k}) - N_0(\mathbf{k}) = a_3^\dagger(\mathbf{k})a_3(\mathbf{k}) - a_0^\dagger(\mathbf{k})a_0(\mathbf{k}).$$

The expectation value is, from Eq. (93),

$$< n,m \mid [a_3^\dagger a_3 - a_0^\dagger a_0] \mid n,m >$$
$$= < n,m \mid [a_3^\dagger - a_0^\dagger]a_3 \mid n,m > \tag{100}$$
$$=0,$$

indicating that the expectation value involves only transverse photons. In other words, the physically allowed states may contain certain number of time-like and longitudinal photons but effects due to these two types of photons cancel each other. A gauge transformation changes the admixture of these "unphysical" photons such that cancellation always holds.

(3) Gauge invariance

The condition for the Lorentz condition $\partial_\mu A_\mu = 0$ for the gauge transformation

$$A_\mu \rightarrow A_\mu + \partial_\mu\chi(x), \tag{101a}$$

is

$$\partial_\nu \partial_\nu \chi(x) = 0, \tag{101b}$$

where $\chi(x)$ is a real, scalar function $\chi(x_\mu)$, i.e., $\chi(x)$ is a hermitian operator. Let the solution $\chi(x)$ of $\partial_\nu \partial_\nu \chi(x) = 0$ be Fourier analyzed as the $A_\mu(x)$ in Eq. (66), and , with the relation (20)

$$\mathbf{a}(-\mathbf{k}) = \mathbf{a}^\dagger(\mathbf{k}),$$

and the hermiticity of χ , similar to Eqs. (7) and (9) ,

$$\chi(-\mathbf{k}) = \chi^\dagger(\mathbf{k}), \tag{102}$$

one obtains the following transformation relations of the $a(\mathbf{k})$, $a^\dagger(\mathbf{k})$, $a_0(\mathbf{k})$, $a_0^\dagger(\mathbf{k})$ operators

$$\begin{aligned}
\mathbf{a}(\mathbf{k}) &\rightarrow \mathbf{a}(\mathbf{k}) + i\mathbf{k}\chi(\mathbf{k}), \\
\mathbf{a}^\dagger(\mathbf{k}) &\rightarrow \mathbf{a}^\dagger(\mathbf{k}) - i\mathbf{k}\chi^\dagger(\mathbf{k}), \\
a_0(\mathbf{k}) &\rightarrow a_0(\mathbf{k}) + i \mid \mathbf{k} \mid \chi(\mathbf{k}), \\
a_0^\dagger(\mathbf{k}) &\rightarrow a_0^\dagger(\mathbf{k}) - i \mid \mathbf{k} \mid \chi^\dagger(\mathbf{k}).
\end{aligned} \tag{103}$$

From these relations and $k_1 = k_2 = 0$ (by the choice of the coordinate system in Eq. (86), it is seen that

$$\begin{aligned}
a_1(\mathbf{k}) &\rightarrow a_1(\mathbf{k}), & a_2(\mathbf{k}) &\rightarrow a_2(\mathbf{k}), \\
a_1^\dagger(\mathbf{k}) &\rightarrow a_1^\dagger(\mathbf{k}), & a_2^\dagger(\mathbf{k}) &\rightarrow a_2^\dagger(\mathbf{k}),
\end{aligned} \tag{104}$$

i.e., the "transverse" operators are invariant under gauge transformation Eq. (101), and, since $k_3 = \mid \mathbf{k} \mid$,

$$a_3(\mathbf{k}) - a_0(\mathbf{k}) \text{ is also invariant.} \tag{105}$$

This, on account of Eq. (87)

$$(a_3(\mathbf{k}) - a_0(\mathbf{k})) \mid 0 >= 0, \tag{106}$$

means that a gauge transformation Eq. (101) does not affect the vacuum state and energy and momentum of the field Eq. (99).

In a formal way, the above theory of the quantized free electromagnetic field is successful, by the creation and annihilation operators, in bringing out the photon (particle) aspect

$$E_{\mathbf{k}} = \omega_{\mathbf{k}}, \qquad \mathbf{p_k} = \mathbf{k} \qquad (107)$$

which are the Einstein relations.

There is a troublesome point in the theory, namely, each degree of freedom has a zero point energy $\frac{1}{2}\omega_{\mathbf{k}}$, and since a field has an infinite number of degrees of freedom, the zero point energy of the field is also infinite.

§.8.4. Dirac electron-positron field[25]

As mentioned in Chapter 3, Sect. 4, in the Majorana representation the 4-component Dirac wave function can be separated into two pairs of mutually complex-conjugated functions.

$$\begin{pmatrix} \Psi^{(1)} \\ \Psi^{(2)} \end{pmatrix} \qquad \text{and} \qquad \begin{pmatrix} \Psi^{(3)} = \Psi^{*(1)} \\ \Psi^{(4)} = \Psi^{*(2)} \end{pmatrix} \qquad (108)$$

For a free electron field, the solutions are plans waves

$$\begin{aligned} \Psi^{(s)}(x) &= u_s e^{ikx}, & s &= 1,2, \\ \Psi^{(t)}(x) &= u_t e^{-ikx}, & t &= 3,4, \end{aligned} \qquad (109)$$

$$\Psi^{(3)} = \Psi^{*(1)}, \qquad \Psi^{(4)} = \Psi^{*(2)}. \qquad (110)$$

The u_s, u_t and their normalization have been given in Eqs. (39, 40a, b, c, d), Ch. 3.

On expanding $\Psi(x)$, $\Psi^*(x)$ in the complete set of plane waves Eqs. (88), (89), Ch. 3, one has

$$\begin{aligned} \Psi(x) &= \frac{1}{\sqrt{L^3}} \sum_{\mathbf{k}} \{ \sum_s a_s u_s e^{iR} + \sum_t b_t u_t e^{-iR} \}, \\ \Psi^*(x) &= \frac{1}{\sqrt{L^3}} \sum_{\mathbf{k}} \{ \sum_s a_s^* u_s^* e^{-iR} + \sum_t b_t^* u_t^* e^{iR} \}, \end{aligned} \qquad (111)$$

where a_s, a_s^*, a_t, a_t^* are all complex numbers (i.e., not operators), and the summation of s is from 1 to 2; of t from 3 to 4, and

$$R \equiv k_\mu x_\mu = \mathbf{k} \cdot \mathbf{r} - \omega_{\mathbf{k}} t. \qquad (112)$$

---

[25] See Exercise 8.1. on quantization of the Dirac particle in the Pauli-Dirac repsentation.

The energy-momentum relation is

$$k_\mu^2 = k_\mu k_\mu = \mathbf{k}^2 - k_0^2 = -m_0^2$$
$$\pm k_0 = E_\mathbf{k} = \omega_\mathbf{k} \tag{113}$$
$$E_\mathbf{k} = \sqrt{\mathbf{k}^2 + m_0^2},$$

and

$$u_s^\dagger u_s = 1, \qquad u_t^\dagger u_t = 1. \tag{114}$$

Note that the indices $s = 1, 2$, $t = 3, 4$ do *not* refer to the spinor components, but to the 4-component u . Thus $u_s, s = 1$, is a column 4-component spinor, $u_s^\dagger$ a row 4-component spinor, etc. In Majorana's representation (110), $u_1, u_2, u_3 = u_1^*$, $u_4 = u_2^*$ are four column 4-component spinors.

To quantize the field Ψ, Ψ^* , the first step is to replace the functions Ψ, Ψ^* by non-hermitian operators $\Psi(x), \Psi^\dagger(x)$, and the a_s, a_s^*, b_t, b_t^* by annihilation and creation operators a_s, $a_s^\dagger$, b_t, $b_t^\dagger$

$$\Psi(x) = \frac{1}{\sqrt{L^3}} \sum_\mathbf{k} \sum_s \{a_s(\mathbf{k}) u_s(\mathbf{k}) e^{iR} + b_s^\dagger(\mathbf{k}) u_s^*(\mathbf{k}) e^{-iR}\}$$
$$\Psi^\dagger(x) = \frac{1}{\sqrt{L^3}} \sum_\mathbf{k} \sum_s \{a_s^\dagger(\mathbf{k}) u_s^\dagger(\mathbf{k}) e^{-iR} + b_s(\mathbf{k}) \tilde{u}_s(\mathbf{k}) e^{iR}\}. \tag{115}$$

where u_s, u_s^* are column matrices, $u_s^\dagger \equiv \tilde{u}_s^*$, $\tilde{u}_s$ are row matrices. The summation over s is from 1 to 2.

The meaning of Eq. (115) is as follows:

$a_s(\mathbf{k})$, $a_s^\dagger(\mathbf{k})$ are the annihilation and creation operators respectively of electron in the state of charge $-e$, momentum $\mathbf{p}$ and energy $E_\mathbf{k}$.

$b_s(\mathbf{k})$, $b_s^\dagger(\mathbf{k})$ are the annihilation and creation operators respectively of positron $(+e, \mathbf{p}, E_\mathbf{k}$ state).

These operators obey the anti-commutation relations,

$$\{a_s(\mathbf{k}), a_{s'}(\mathbf{k}')\}_+ = \{b_s(\mathbf{k}), b_{s'}(\mathbf{k}')\}_+ = 0,$$
$$\{a_s^\dagger(\mathbf{k}), a_{s'}^\dagger(\mathbf{k}')\}_+ = \{b_s^\dagger(\mathbf{k}), b_{s'}^\dagger(\mathbf{k}')\}_+ = 0,$$
$$\{a_s(\mathbf{k}), b_{s'}^\dagger(\mathbf{k}')\}_+ = \{a_s^\dagger(\mathbf{k}), b_{s'}(\mathbf{k}')\}_+ = 0, \tag{116}$$
$$\{a_s(\mathbf{k}), b_{s'}(\mathbf{k}')\}_+ = \{a_s^\dagger(\mathbf{k}), b_{s'}^\dagger(\mathbf{k}')\}_+ = 0,$$
$$\{a_s(\mathbf{k}), a_{s'}^\dagger(\mathbf{k}')\}_+ = \{b_s(\mathbf{k}), b_{s'}^\dagger(\mathbf{k}')\}_+ = \delta_{ss'} \delta_{\mathbf{k}\mathbf{k}'},$$

where

$$\{A, B\}_+ = AB + BA. \tag{117}$$

We introduce the electron and positron occupation number $N_{s-}(\mathbf{k}), N_{s+}(\mathbf{k})$

$$N_{s-}(\mathbf{k}) = a_s^\dagger(\mathbf{k})a_s(\mathbf{k}), \qquad N_{s+}(\mathbf{k}) = b_s^\dagger(\mathbf{k})b_s(\mathbf{k}) \tag{118}$$

The vacuum $| 0 >$ is such that

$$a_s(\mathbf{k}) \mid 0 >= 0, \qquad b_s(\mathbf{k}) \mid 0 >= 0, \tag{119}$$

and therefore

$$N_{s-}(\mathbf{k}) \mid 0 >= 0, \qquad N_{s+}(\mathbf{k}) \mid 0 >= 0. \tag{120}$$

From Eq. (116), one obtains the commutation relations

$$\begin{aligned}
[N_{s\pm}(\mathbf{k}), N_{s'\mp}(\mathbf{k}')]_- &= 0, \\
[N_{s-}(\mathbf{k}), a_{s'}(\mathbf{k}')]_- &= -\delta_{ss'}\delta_{\mathbf{k}\mathbf{k}'}a_s(\mathbf{k}), \\
[N_{s+}(\mathbf{k}), b_{s'}(\mathbf{k}')]_- &= -\delta_{ss'}\delta_{\mathbf{k}\mathbf{k}'}b_s(\mathbf{k}), \\
[N_{s-}(\mathbf{k}), a_{s'}^\dagger(\mathbf{k}')]_- &= \delta_{ss'}\delta_{\mathbf{k}\mathbf{k}'}a_s^\dagger(\mathbf{k}), \\
[N_{s+}(\mathbf{k}), b_{s'}^\dagger(\mathbf{k}')]_- &= \delta_{ss'}\delta_{\mathbf{k}\mathbf{k}'}b_s^\dagger(\mathbf{k}).
\end{aligned} \tag{121}$$

If in the equation (115) for $\Psi(x)$, $\Psi^+(x)$ one makes the interchange

$$\begin{aligned}
a_s(\mathbf{k}) &\longleftrightarrow b_s(\mathbf{k}), & a_s^\dagger(\mathbf{k}) &\longleftrightarrow b_s^\dagger(\mathbf{k}), \\
u_s(\mathbf{k}) &\longleftrightarrow u_s^\dagger(\mathbf{k}), & (i.e.,\ u_s^*(\mathbf{k}) &\longleftrightarrow \tilde{u}_s(\mathbf{k})),
\end{aligned} \tag{122}$$

then in Eq. (115),

$$\left\{ \begin{array}{c} \Psi(x) \\ \Psi^\dagger(x) \end{array} \right\} \text{ goes into } \left\{ \begin{array}{c} \Psi^\dagger(x) \\ \Psi(x) \end{array} \right\}, \tag{123}$$

i.e., the electron goes into the positron and vice versa. This is the charge conjugation transformation. In a free electron field (i.e., no electromagnetic interaction), the quantized field theory is symmetrical in the electron-positron interchange.

From Chapter 6, Eqs. (115)-(118) for the Hamiltonian, the 4-current and the total electric charge, one has, on using the $\Psi, \Psi^\dagger$ of Eq. (115), and the anticommutation relations (116),

$$H = \sum_{\mathbf{k}} \sum_{s=1}^{2} E_s [a_s^\dagger(\mathbf{k}) a_s(\mathbf{k}) + b_s^\dagger(\mathbf{k}) b_s(\mathbf{k})], \qquad (124a)$$

$$Q = -e \sum_{\mathbf{k}} \sum_{s=1}^{2} [a_s^\dagger(\mathbf{k}) a_s(\mathbf{k}) - b_s^\dagger(\mathbf{k}) b_s(\mathbf{k})], \qquad (125a)$$

and if $| \, 0 >$ is the vacuum state of Eqs. (119) and (120) ,

$$H = \sum_{\mathbf{k}} \sum_{s=1}^{2} E_{\mathbf{k}} (N_{s-}(\mathbf{k}) + N_{s+}(\mathbf{k})), \qquad (124b)$$

$$Q = -e \sum_{\mathbf{k}} \sum_{s=1}^{2} (N_{s-}(\mathbf{k}) - N_{s+}(\mathbf{k})). \qquad (125b)$$

From this and the preceding three sections on the Klein-Gordon, electromagnetic field, one summarizes the following result:

For complex conjugate field (Klein-Gordon complex $\Psi(x)$, $\Psi^*(x)$, Dirac $\Psi(x)$, $\Psi^*(x)$), the non-hermitian operators $\Psi(x)$, $\Psi^\dagger(x)$ correspond, respectively, to particle and anti-particle (π^+ and π^- for the Klein-Gordon field; the electron-positron for the Dirac field).

For real field (Klein-Gordon real, scalar field $\phi(x)$ and the electromagnetic field $A_\mu(x)$) , the operators $\Psi(x), \Psi^\dagger(x)$ are hermitian, and the antiparticle of the field is the particle itself (π^0 of the Klein-Gordon field and the photon of the electromagnetic field).

§.8.5. Dirac's theory of emission and absorption of radiation

In the preceding sections of the present chapter, we have treated a few examples of the quantization of 'pure' fields, for example, electromagnetic fields in regions of space free of electric charges and currents.

We now proceed to consider the following physical situation: an atom with one electron, in an electromagnetic field.

In the Schrödinger theory, the system is described by the equation[26]

---

[26] *The interaction* $-\frac{e}{mc}(\mathbf{A} \cdot \mathbf{p})$ *with the electromagnetic field represented by a vector potential* $\mathbf{A}$ *comes from Eq. (16) of Chapter 5. For this section, we shall not impose the* $\hbar = c = 1$ *unit system.*

$$i\hbar\frac{\partial\Psi}{\partial t} = (H_0 + H_1)\Psi, \tag{126}$$

where

$$H_0 = -\frac{\hbar}{2m}\nabla^2 + V(r) + H_f$$

$$H_1 = -\frac{e}{mc}(\mathbf{A}\cdot\mathbf{p}). \tag{127}$$

the first two terms in H_0 are the kinetic and potential energy of the electron in the atom; H_f is the Hamiltonian of the free electromagnetic field. H_1 is the interaction between the electron and the field.

From Eq. (32), one has

$$H_f = \sum_{k\lambda}\hbar\omega_{\mathbf{k}}a_\lambda^\dagger(\mathbf{k})a_\lambda(\mathbf{k}), \tag{128}$$

where λ is to denote the state of polarization of the photon. The vector potential A is, from Eq. (70a)

$$\mathbf{A}(x_\mu) = \sum_{k,\lambda}c\sqrt{\frac{2\pi h}{L^3\omega_{\mathbf{k}}}}\mathbf{S}_\lambda(\mathbf{k})[a_\lambda(\mathbf{k})e^{iR} + a_\lambda^\dagger(\mathbf{k})e^{-iR}], \tag{129}$$

$$R = \mathbf{k}\cdot\mathbf{r} - \omega_{\mathbf{k}}t,$$

where $\mathbf{S}_\lambda(\mathbf{k})$ is a unit vector with $\lambda = 1,2$ for the two directions of polarizations as discussed in Eq. (70a).[27]

For the system H_0, the solution of the Schrödinger is

$$\Phi_j = v_j(r)\psi_j(x_\mu), \tag{130}$$

where

$$(-\frac{\hbar}{2m}\nabla^2 + V(r) - \epsilon_j)v_j(r) = 0, \tag{131}$$

and $\psi_j(x_\mu)$ is the state function of the electromagnetic field which in the occupation number representation is

$$a_\lambda^\dagger(\mathbf{k})\psi_{...n_{k\lambda}...} = \sqrt{n_{k\lambda} + 1}\psi_{...n_{k\lambda}+1...},$$
$$a_\lambda(\mathbf{k})\psi_{...n_{k\lambda}...} = \sqrt{n_{k\lambda}}\psi_{...n_{k\lambda}-1...}. \tag{132}$$

---

[27] *This expression for $A(x)$ differs from that in Eq. (70a) by a factor $\sqrt{4\pi}$ for a change to the e.s.u system.*

$n_{k\lambda}$ being the number of photons of momentum $\hbar\mathbf{k}$ and polarization λ $(\lambda = 1, 2)$.

The interaction H_1 is

$$H_1 = -\frac{e}{m}\sqrt{\frac{2\pi h}{L^3}} \sum_{k\lambda} \frac{1}{\sqrt{\omega_k}}(\mathbf{S}_\lambda \cdot \mathbf{p})\{a_\lambda(\mathbf{k})e^{iR} + a_\lambda^\dagger(\mathbf{k})e^{-iR}\}. \tag{133}$$

On account of this interaction, the Ψ in Eq. (126) is not a stationary state but is time dependent. Let the Ψ be expanded in the complete set of $v_j\psi_j$,

$$\Psi = \sum_j c_j(t)v_j\psi_j \, \exp(-\frac{i}{\hbar}\epsilon_j t). \tag{134}$$

From Eq. (126), one obtains

$$\frac{dc_\ell}{dt} = -\frac{i}{\hbar}\sum_j (\Phi_\ell, H_1\Phi_j)c_j(t) \, \exp(i\omega_{\ell j} t), \tag{135}$$

$$\hbar\omega_{\ell j} = \epsilon_\ell - \epsilon_j. \tag{136}$$

To integrate the equation for c_ℓ , let the initial state be an atom in a state v_u and the field with $n_{k\lambda}$ photons in state $\mathbf{k}, \lambda$,

$$\Phi = v_u\psi_{...n_{k\lambda}...}, \tag{137}$$

so that[28]

$$c_u(t=0) = 1, \qquad c_j(t=0) = 0, \qquad j \neq u. \tag{138}$$

Then from Eqs. (133) and (132),

$$(\Phi_\ell, H_1\Phi_u) = \frac{e}{m}\sqrt{\frac{2\pi\hbar}{L^3}} \sum_{k\lambda} \sqrt{\frac{n_k + 1}{\omega_k}}(v_\ell, (\mathbf{S}_\lambda \cdot \mathbf{p})e^{-i\mathbf{k}\cdot\mathbf{r}}v_u)e^{i\omega_k t}. \tag{139}$$

The probability that the system - atom + radiation field - has gone to state v_ℓ in time t is

---

[28] *The perturbation theory method is that of Chapter 7, Sect. 2(1), Vol. I ("Quantum Mechanics"). See (VII-32) - (VII-36), etc., in Vol. I.*

$$|c_\ell(t)|^2$$

provided t is not too long and $|c_\ell(t)|^2$ is still very small compared with 1.

$$|c_\ell(t)|^2 = \frac{1}{\hbar L^3}\left(\frac{2\pi e}{m}\right)^2 \sum_{k\lambda} \frac{n_k+1}{\omega_k} |(v_\ell, (\mathbf{S}_\lambda \cdot \mathbf{p})e^{-i\mathbf{k}\cdot\mathbf{r}}v_n)|^2 \frac{\sin^2 \xi}{\xi^2}\left(\frac{t^2}{2\pi}\right), \quad (140)$$

where

$$\xi = \frac{\omega_{\ell u} + \omega_k}{2}t.$$

Consider the situation when the initial state u is such that the atom is in an excited state v_u , and v_ℓ is a lower state,

$$\epsilon_u - \epsilon_\ell = \hbar\omega_{u\ell} > 0, \quad (141)$$

so that

$$\xi = \left(\frac{\omega_k - \omega_{u\ell}}{2}\right)t. \quad (142)$$

We are then seeking the probability that an atom in an excited state v_u makes a transition to a lower state v_ℓ increasing a photon $\hbar\omega_k$ in the field.

The function $\frac{\sin^2 \xi}{\xi^2}$ has a very sharp maximum at $\xi = 0$, and

$$\frac{1}{\pi}\int_{-\infty}^{\infty} \frac{\sin^2 \xi}{\xi^2}d\xi = 1.$$

We shall replace

$$\frac{1}{\pi}\frac{\sin^2 \xi}{\xi^2} \quad \text{by} \quad \delta(\xi) = \frac{2}{t}\delta(\omega_k - \omega_{u\ell}). \quad (143)$$

To evaluate the matrix element in Eq. (140), we make the dipole approximation (as in (VII-47)-(VII-50), Vol. I, for $kr = \frac{2\pi r}{\lambda} \ll 1$, i.e., for wavelengths long compared with atomic size) and obtain (by using (VII-50), Vol. I.)

$$\mathbf{S}_\lambda \cdot (v_\ell, \mathbf{p}e^{-i\mathbf{k}\cdot\mathbf{r}}v_u) \simeq \mathbf{S}_\lambda \cdot (v_\ell, \mathbf{p}v_u)$$
$$= im\omega_{\ell u}(\mathbf{S}_\lambda \cdot \mathbf{r}_{\ell u}), \quad (144)$$

where

$$\mathbf{r}_{\ell u} = \int v_\ell^* \mathbf{r} v_u d^3 x. \quad (145)$$

The polarization factor $S_\lambda \cdot r$ can be evaluated by choosing the coordinate system of Eq. (86), namely, with the x_3 -axis along k.

The S_1, S_2 are unit vectors of polarization transverse to k. We shall now take S_1 to be in the plane of k and r , and S_2 normal to this plane. If ϑ is the angle between k and r , then

$$(S_1 \cdot r_{\ell u}) = \sin \vartheta r_{\ell u}, \qquad (S_2 \cdot r_{\ell u}) = 0. \qquad (146)$$

The summation over the discrete k in Eq. (140) is again replaced, for a large volume L^3, by an integral

$$\sum_k \rightarrow (\frac{L}{2\pi})^3 \int d^3 k$$

$$= \frac{1}{c^3}(\frac{L}{2\pi})^3 \int d\Omega \omega_k^2 d\omega_k,$$

where $d\Omega$ is the element of solid angle for $d^3 k$. Thus

$$| c_\ell(t) |^2 = t\frac{e^2}{2\pi\hbar c} \int d\Omega \int d\omega_k (n_k + 1)\omega_k^3 \sin^2 \theta |< \ell | r | u >|^2 \delta(\omega_k - \omega_{u\ell}).$$
$$(147)$$

(1) Spontaneous emission coefficient $A(u \rightarrow \ell)$

There are two terms in Eq. (147). Consider the term not containing n_k , i.e., independent of the presence or absence of photons n_k in the field. Upon integrating over ω_k and $\sin^2 \vartheta d \cos \vartheta d\varphi$ (directions of photon k), one obtains for the probability per unit time $\frac{1}{t} | c_\ell(t) |^2$,

$$P = \frac{1}{t} | c_\ell(t) |^2 = \frac{4\omega_{u\ell}^3}{3\hbar c^3} |< \ell | er | u >|^2 . \qquad (148)$$

This transition probability, being independent of the $n_k (E_k = \hbar\omega_k)$ of the electromagnetic field, can be identified with Einstein's spontaneous emission coefficient $A(u \rightarrow \ell)$. In fact, the expression in Eq. (148) agrees with the expression (VII-54), Vol. I, for $A(u \rightarrow \ell)$,

$$A(u \rightarrow \ell) = \frac{64\pi^4 \nu^3}{3\hbar c^3} |< \ell | er | u >|^2 \qquad (149),$$

which is there obtained indirectly from the Einstein relation

$$A(u \to \ell) = \frac{8\pi h\nu^3}{c^3} B(u \leftarrow \ell) \tag{150}$$

and the absorption coefficient $B(u \leftarrow \ell)$ is obtained from the perturbation theory.

(2) Induced emission and absorption coefficient, $B(u \to \ell)$ and $B(u \leftarrow \ell)$

Consider the term containing n_k in the integral in Eq. (147). This term, being proportional to the number n_k of quanta of energy $\hbar\omega_k$, can be identified with the induced emission process of Einstein. Let $\rho(\nu)d\nu$ be the energy density of the radiation field having frequendy between ν and $\nu + d\nu$. The n_k in Eq. (147) or (140), is

$$n_k = \frac{L^3 \rho(\nu_k) d\nu_k}{2h\omega_k}, \tag{151}$$

the factor 2 in the denominator coming from the fact that in an unpolarized field of radiation, half the energy is associated with each direction of polarization. Hence from Eqs. (140), (141), and (146), one obtains

$$\begin{aligned} P &= \frac{1}{t} \mid c_\ell(t) \mid^2 = \frac{2\pi}{3\hbar^2} \mid< \ell \mid er \mid u >\mid^2 \rho(\nu) \\ &\equiv B(u \to \ell)\rho(\nu), \end{aligned} \tag{152}$$

which gives indeed the Einstein $B(u \to \ell)$ coefficient for induced emission (or, negative absorption) as obtained in (VII-53), Vol. I, and in Chapter 1, Sect. 6, Vol. I.

For absorption, we take the second equation in Eq. (132), i.e., $a_\lambda(\mathbf{k})\psi$. This leads to n_k replacing $n_k + 1$ in Eqs. (132), (138), and (147). All the steps of calculation above are the same. The result is, for the absorption of one photon $\hbar\omega_k = \hbar\omega_{u\ell}$ by the atom, the probability per second is

$$\begin{aligned} P(u \leftarrow \ell) &= \frac{2}{3\hbar^2} \mid< \ell \mid er \mid u >\mid^2 \rho(\nu) \\ &\equiv B(u \leftarrow \ell)\rho(\nu), \end{aligned} \tag{153}$$

so that

$$B(u \leftarrow \ell) = B(u \to \ell),$$

as obtained by Einstein from the equilibrium radiation energy distribution (1917).[29]

The above theory, of treating the emission and absorption of radiation by the method of quantized field, is due to Dirac (1927). This is the beginning of the theory of quantized field, which is extended immediately by P. Jordan, O. Klein, E. Wigner, W. Pauli in 1928.

---

[29] *A Einstein, Physik Zeit.* **18**, *121 (1917); Mitt. d. Phys. Ges. Zurich,* **18** *(1916); Verhandlungen der Deutschen Physik Gesells* **18**, *318 (1916).*

Appendix

Green's Functions $\triangle$ and D

We shall obtain here the Green's functions of the equations

$$\partial_\nu \partial_\nu A_\mu(x) = 0, \qquad x = (x_1, x_2, x_3, x_4),$$

$$(\partial_\nu \partial_\nu - \mu^2)\phi(x) = 0, \qquad x_4 = ix_0 = ict,$$

which are the classical equations, respectively, of the 4-potentials of electromagnetic field and the scalar field of the Klein-Gordon equation. These equations are in the relativistic form, and their Green's functions are Lorentz invariant. The Green's functions appear in the commutation relations when the fields are quantized.

We shall summarize our notations:

$$x = (x_1, x_2, x_3, x_4) = (\mathbf{r}, ict = ix_0),$$

$$p = (p_1, p_2, p_3, p_4) = (\mathbf{p}, p_4 = \frac{iE}{c} \equiv ip_0), \qquad (A1a)$$

$$p^2 = p_\nu^2 = \mathbf{p}^2 + p_4^2 = \mathbf{p}^2 - \frac{E^2}{c^2} = -m_0^2 c^2,$$

$$\mathbf{k} = \frac{1}{\hbar}\mathbf{p}, \qquad k_4 = ik_0, \qquad (A1b)$$

$$\mu = \frac{m_0 c}{\hbar}, \quad (\frac{1}{\mu} = \frac{1}{2\pi}\lambda_c, \quad \lambda_c = \text{Compton wave length})$$

Hence

$$k^2 = k_\nu^2 = \mathbf{k}^2 + k_4^2$$
$$= \mathbf{k}^2 - k_0^2 = -\mu^2$$
$$E_\mathbf{k} = c\sqrt{\mathbf{p}^2 + m_0^2 c^2} \qquad (A1c)$$
$$= \hbar c\sqrt{\mathbf{k}^2 + \mu^2}$$

In the following, we shall use the "natural units" in which

$$\hbar = c = 1.$$

Thus

$$k_\nu^2 + \mu^2 = \mathbf{k}^2 + \mu^2 - k_0^2$$
$$= E_\mathbf{k}^2 - k_0^2 \qquad (A1d)$$

$k_\nu^2 + \mu^2 = 0$ is a Lorentz invariant. Thus the Dirac δ function $\delta(k_\nu^2 + \mu^2)$ is a Lorentz invariant operator. It is expressible in the form[30]

$$\delta(k^2 + \mu^2) = \frac{1}{2E_{\mathbf{k}}}(\delta(k_0 + E_{\mathbf{k}}) + \delta(k_0 - E_{\mathbf{k}})) \qquad (A2)$$

We define a step function $\epsilon(k_0)$ of the sign of k_0

$$E(k_0) = \begin{cases} 1, & k_0 > 0 \\ -1, & k_0 < 0 \end{cases} \qquad (A3)$$

$k_4 \, (ik_0)$ is the time-component of the 4-vector k . Under proper Lorentz transformations inside the light cone

$$r^2 - c^2 t^2 < 0,$$

the sign of t, and of k_0 , does not change. Hence $\epsilon(k_0)$ is a Lorentz invariant function.

From (A-2) and (A-3), we have, since $E_{\mathbf{k}}$ is always positive,

$$\epsilon(k_0)\delta(k_\nu^2 + \mu^2) = \frac{1}{2E_{\mathbf{k}}}\{\delta(k_0 - E_{\mathbf{k}}) - \delta(k_0 + E_{\mathbf{k}})\}. \qquad (A4)$$

---

[30] *From*

$$\delta(ax) = \frac{\delta(x)}{|a|}$$

one sees

$$\int_{-\infty}^{\infty} f(x)\delta[(x - a)(x - b)]dx = \frac{1}{|a - b|}[f(a) + f(b)]$$

Also

$$\int_{-\infty}^{\infty} f(x)\frac{\delta(x - a) + \delta(x - b)}{|a - b|}dx = \frac{1}{|a - b|}[f(a) + f(b)].$$

Hence (A-2).

(1) $\triangle(x)$ function

Let us define the Lorentz invariant $\triangle(x)$ function[31]

$$\triangle(x) = \frac{i}{(2\pi)^3} \int d^4 k e^{ik_\nu x_\nu} \epsilon(k_0) \delta(k_\nu^2 + \mu^2), \qquad (A5)$$

where

$$d^4 k = d^3 k dk_0, \qquad (ik_0 = k_4), \qquad (A5a)$$

i.e., $\triangle(x)$ is the Fourier transform of $\epsilon(k_0)\delta(k_\nu^2 + \mu^2)$. On account of the $\delta(k_\nu^2 + \mu^2)$ function, the integral is really over a 3-dimensional surface of the 4-dimensional k space. On using (A-4), we have $(x_0 = ct, \quad x_4 = ix_0)$

$$\triangle(x) = \frac{i}{(2\pi)^3} \int d^3 k e^{ik \cdot r} \frac{1}{2E_k} \int_{-\infty}^{\infty} dk_0 e^{-k_0 x_0} \{\delta(k_0 - E_k) - \delta(k_0 + E_k)\} \quad (A6)$$

$$= \frac{1}{(2\pi)^3} \int d^3 k e^{ik \cdot r} \frac{\sin x_0 \sqrt{k^2 + \mu^2}}{\sqrt{k^2 + \mu^2}} \qquad (A6a)$$

$$= -\frac{i}{(2\pi)^3} \int d^3 k \frac{\sin k_\nu x_\nu}{k_4}. \qquad (A6b)$$

The $\triangle(x)$ above defined has the following properties:

(i)
$$\triangle(x)_{x_0=0} = 0, \quad \text{from (A-6a)}. \qquad (A8)$$

(ii)
$$\frac{\partial}{\partial t}\triangle(x)\mid_{t=0} = \frac{1}{(2\pi)^3} \int d^3 k e^{ik \cdot r}$$
$$= \delta(r), \qquad (\delta(r) = \delta(x_1)\delta(x_2)\delta(x_3)). \qquad (A9)$$

(iii) $\triangle(x)$ satisfies the Klein-Gordon equation

---

[31] *The $\triangle(x)$ function can also be defined by the integral*

$$\triangle(x) = \frac{1}{(2\pi)^4} \int_C d^4 k \frac{e^{ik_\nu x_\nu}}{k_\nu^2 + \mu^2}. \qquad (A7)$$

where the contour C is a large circle about $k_0 = 0$ on the complex k_0 plane and two small circles, all in the counter clockwise sense, about the two poles $k_0 = +E_k$ of $k_\nu^2 + \mu^2 = (E_k - k_0)(E_k + k_0)$ in (A1d). The $\triangle(x)$ so defined is identical with (A6a).

$$(\partial_\nu \partial_\nu - m_0^2)\triangle(x) = 0, \qquad (A10)$$

as is seen from (A-5).

(iv) Let us introduce the step function

$$\vartheta(x_0) = \begin{cases} 1, & x_0 > 0; \\ 0, & x_0 < 0. \end{cases} \qquad (A11)$$

For the same reason as for $\epsilon(k_0)$ in (A-3) , $\vartheta(x_0)$ is an invariant function under proper Lorentz transformations. This $\vartheta(x_0)$ has the following properties:

$$\begin{aligned} \epsilon(x_0) &= 2\vartheta(x_0) - 1, \\ \epsilon^2(x_0) &= 1, \\ \vartheta^2(x_0) &= \vartheta(x_0), \\ \vartheta(x_0)\vartheta(-x_0) &= 0, \\ \vartheta(x_0) \pm \vartheta(-x_0) &= \begin{cases} 1 \\ \epsilon(x_0). \end{cases} \end{aligned} \qquad (A12)$$

In view of the last relation, $\triangle(x)$ in (A5) can be written

$$\triangle(x) = \triangle_+(x) - \triangle_-(x), \qquad (A13)$$

where

$$\begin{aligned} \triangle_+(x) &= -\frac{i}{(2\pi)^3} \int d^4 k e^{ik_\nu x_\nu} \vartheta(k_0)\delta(k_\nu^2 + \mu^2) \\ &= \frac{1}{(2\pi)^4} \int_{C_+} d^4 k \frac{e^{ik_\nu x_\nu}}{k_\nu^2 + \mu^2}, \qquad d^4 k = dk dk_0, \end{aligned} \qquad (A13a)$$

$$\triangle_- = \frac{1}{(2\pi)^4} \int_{C_-} d^4 k \frac{e^{ik_\nu x_\nu}}{k_\nu^2 + \mu^2}, \qquad d^4 k = dk dk_0, \qquad (A13b)$$

where C_+ is a closed curve, in the counter clockwise sense, on the complex k_0-plane about the pole $k_0 = E_{\mathbf{k}}$, and C_- is a closed curve, in the clockwise sense, about the pole $k_0 = -E_{\mathbf{k}}$.

(2) D(x) function

For fields with $m_0 = 0$ $(\mu = \frac{m_0 c}{\hbar} = 0)$, (A1d) is

$$k_\nu^2 = \mathbf{k}^2 + k_4^2 = \mathbf{k}^2 - k_0^2 = 0, \qquad (A14a)$$

$$E_{\mathbf{k}} = |\,\mathbf{k}\,|. \qquad (A14b)$$

and (A4) is now

$$\epsilon(k_0)\delta(k_\nu^2) = \frac{1}{2\,|\,\mathbf{k}\,|}\{\delta(k_0 - |\,\mathbf{k}\,|) - \delta(k_0 + |\,\mathbf{k}\,|)\} \qquad (A15)$$

and $D(x)$ is similarly defined as the (negative) Fourier transform of $\epsilon(k_0)\delta(k^2)$, i.e., (from (A6), (A6a)),

$$D(x) = -\frac{i}{(2\pi)^3}\int d^4k\, e^{ik_\nu x_\nu}\epsilon(k_0)\delta(k_\nu^2) \qquad (A16)$$

$$= -\frac{1}{(2\pi)^3}\int d^3k\, e^{i\mathbf{k}\cdot\mathbf{r}}\frac{\sin x_0\,|\,\mathbf{k}\,|}{|\,\mathbf{k}\,|}. \qquad (A16a)$$

Similarly to (A-7), $D(x)$ can also be defined by the integral

$$D(x) = \frac{1}{(2\pi)^3}\int_C d^4k\,\frac{e^{ik_\nu x_\nu}}{k_\nu^2}, \qquad k_4 = i\,|\,\mathbf{k}\,| \qquad (A17)$$

where the contour is similar to that in (A7), the two poles here being similar to that in (A7), the poles being $k_0 = \pm\,|\,\mathbf{k}\,|$ (or $\pm E_{\mathbf{k}}$). From (A17), one obtains (A16a).

To evaluate $D(x)$ in (A16a), let us use spherical polar coordinates for $\mathbf{k}$, and denote $|\,\mathbf{k}\,|$ by K, so that

$$d^3k = K^2 dK\, d\cos\theta\, d\varphi,$$

and

$$\begin{aligned}
D(x) &= -\frac{1}{2\pi^2 r}\int_0^\infty \sin x_0 K \sin rK\, dK \\
&= -\frac{1}{4\pi r}\int_0^\infty \{\cos K(x_0 - r) - \cos K(x_0 + r)\}dK \\
&= -\frac{1}{4\pi r}\{\delta(x_0 - r) - \delta(x_0 + r)\} \qquad (A18a) \\
&= -\frac{1}{2\pi}\epsilon(x_0)\delta(x^2), \quad \text{using (A15)}. \qquad (A18)
\end{aligned}$$

As

$$\delta(x^2) = \delta(\mathbf{r}^2 - x_0^2) = \delta(\mathbf{r}^2 - t^2),$$

it is seen that $D(x) = 0$ except on the light cone $x^2 - t^2 = 0$.

Thus $D(x)$ has the following properties

(i)

$$D(x) = 0 \qquad \text{except on the light cone.} \tag{A19}$$

(ii) From (A16) and (A18),

$$D(x) \text{ is its own Fourier transform.} \tag{A20}$$

(iii)

$$D(x)\,|_{x_0=0} = 0, \qquad \text{from (A16a).} \tag{A21}$$

(iv)

$$\frac{\partial}{\partial t} D(x)\,|_{x_0=0} = -\delta(\mathbf{r}) \tag{A22}$$

(v) $D(x)$ satisfies the d'Alembertian equation

$$\partial_\nu \partial_\nu D(x) = 0, \qquad \text{from (A14a).} \tag{A23}$$

(vi) From (A18) and the last relation in (A12),

$$\begin{aligned}
D(x) &= -\frac{1}{2\pi}\epsilon(x_0)\delta(x^2) \\
&= \frac{1}{2\pi}\{\vartheta(-x_0) - \vartheta(x_0)\}\delta(x^2).
\end{aligned} \tag{A24}$$

We define the retarded and advanced $D(x)$

$$\begin{aligned}
D_{ret}(x) &\equiv \frac{1}{2\pi}\vartheta(x_0)\delta(x^2) = \frac{1}{4\pi r}\delta(x_0 + r), \\
D_{adv}(x) &\equiv \frac{1}{2\pi}\vartheta(-x_0)\delta(x^2) = \frac{1}{4\pi r}\delta(x_0 - r).
\end{aligned} \tag{A25}$$

$D_{adv}(x), D_{ret}(x)$ both satisfy the equation

$$\partial_\nu \partial_\nu D(x) = -\delta^4(x) = -\delta^3(\mathbf{r})\delta(x_0). \tag{A26}$$

References

Dirac, P. A. M., Proc. Roy. Soc. London **A114**, 243, 710 (1927). The general theory of quantization of fields as qualitatively outlined in Sect. 1, and the theory of emission and absorption of radiation (the Einstein $A_{(m \to n)}$ coefficient), is first given by Dirac.

Jordan, P. and Klein, O., Zeits. f. Physik **45**, 751 (1927), for systems satisfying Bose-Einstein statistics.

Jordan P. and Wigner, E., ibid. **47**, 631 (1928), for systems satisfying Fermi-Dirac statistics.

Heisenberg, W. and Pauli, W., ibid. **56**, 1 (1929), for canonical formalism of field quantization.

Pauli, W. and Weisskopf, V., Helvetica Phys. Acta **7**, 709 (1934), for the Klein-Gordon field.

Jordan, P. and Pauli, W., Zeits. f. Physik **47**, 151, (1928), for the $D(x_\nu)$ function.

Dirac, P. A. M., Proc. Cambridge Phil. Soc. **30**, 150 (1934), for the $\triangle(x_\nu)$ and $D(x_\nu)$ functions.

Heitler, W., *The Quantum Theory of Radiation*, 3rd ed. (Oxford Univ. Press, London, 1954).

Wentzel, G., *Quantum Theory of Fields* (Interscience Publ., New York, 1949).

Jauch, J. M. and Rohlich, F., *The Theory of Photons and Electrons* (Addison-Wesley, Mass., 1955).

Bjorken. J. D. and Drell, S. D., *Relativistic Quantum Field* (McGraw Hill, N. Y., 1967).

Itzykson, C. and Zuber, J. -B., *Quantum Field Theory* (McGraw-Hill, New York, 1980).

Lee, T. D., *Particle Physics and Introduction to Field Theory* (Harwood, Chur-London-New York, 1981).

Aitchison, I. J. R., *An Informal Introduction to Gauge Field Theories* (Cambridge University Press, London, 1982).

Cheng, T. P., and Li, L. -F., *Gauge Theory of Elementary Particle Physics* (Clarendon Press, Oxford, 1984).

Exercises: *Chapter 8*

1. (a) Show that Eq. (13) is equivalent to Eqs. (12a)-(12c).

(b) In the case of the electromagnetic field, write down the elementary equal-time commutators and show that your results are equivalent to Eqs. (73).

(c) The "normal-mode" expansion for the Dirac field may also be specified by

$$\psi(\mathbf{x},t) = \int \frac{d^3p}{(2\pi)^{\frac{3}{2}}} \sum_s \{a_s^{(+)}(\mathbf{p})u(\mathbf{p},s)e^{ip\cdot x} + a_s^{(-)*}(\mathbf{p})v(\mathbf{p},s)e^{-ip\cdot x}\},$$

$$(154)$$

where u- and v-spinors are already given in Ch. 2 (Exercise 1). The momentum conjugate to $\psi(x)$ is then specified by

$$\pi(\mathbf{x},t) = \frac{\partial \mathcal{L}}{\partial(\partial_0\psi)} = +i\bar{\psi}\gamma_4 = i\psi^\dagger(\mathbf{x},t). \tag{155}$$

Quantization of the Dirac field is accomplished by imposing the following equal-time anti-commutation relations:

$$\begin{aligned}
\{\psi_i(\mathbf{x},t),\ \psi_j(\mathbf{y},t)\} &= 0, \\
\{\psi_i^\dagger(\mathbf{x},t),\ \psi_j^\dagger(\mathbf{y},t)\} &= 0, \\
\{\psi_i(\mathbf{x},t),\ i\psi_j^\dagger(\mathbf{y},t)\} &= i\delta_{ij}\delta^3(\mathbf{x}-\mathbf{y}).
\end{aligned} \tag{156}$$

We introduce the following set of anti-commutation relations,

$$\begin{aligned}
\{a_s(+)(\mathbf{p}),\ a_s(+)*(\mathbf{p}')\} &= \delta_{ss'}\delta^3(\mathbf{p}-\mathbf{p}'), \\
\{a_s(-)(\mathbf{p}),\ a_s(-)*(\mathbf{p}')\} &= \delta_{ss'}\delta^3(\mathbf{p}-\mathbf{p}'),
\end{aligned} \tag{157}$$

all other elementary anticommutators $= 0$.

Demonstrate that Eqs. (156) and Eqs. (157) are completely equivalent to each other, showing that quantization of the fields may also be accomplished equally well in the Pauli-Dirac representation.

2. We define

$$T(\varphi_i(x)\varphi_j(y)) \equiv \varphi_i(x)\varphi_j(y), \quad \text{if} \quad x_0 > y_0;$$

$$\pm \varphi_j(y)\varphi_i(x), \quad \text{if} \quad y_0 > x_0. \tag{158}$$

(+ for bosons & − for fermions.)

The chronological pairing, or T-pairing, is defined to be the vacuum expectation value of $T(\varphi_i(x)\varphi_j(y))$.

(a) Use Eqs. (70) - (73) to show that, up to $\delta_{\mu\nu} \to \delta_{\mu\nu} - \lambda\frac{k_\mu k_\nu}{k^2}$,

$$\overbrace{A_\mu(x)A_\nu(y)} \equiv -i\delta_{\mu\nu}D_0^c(x-y)$$

$$= \frac{\delta_{\mu\nu}}{(2\pi)^4 i} \int d^4 k e^{-ik\cdot(x-y)} \frac{1}{k^2 - i\varepsilon} \tag{159}$$

(b) Use Eqs. (111) - (116) to show that

$$\overbrace{\psi(x)\bar\psi(y)} \equiv -iS^c(x-y)$$

$$= \frac{1}{(2\pi)^4 i} \int d^4 p e^{ip\cdot(x-y)} \frac{m - i\gamma\cdot p}{m^2 + p^2 - i\varepsilon} \tag{160}$$

Chapter 9.
Quantum Electrodynamics I: S-Matrix Elements

As already mentioned earlier (Ch. 0 and Ch. 6), the theory of quantized electromagnetic fields began with the work of Dirac in 1927, followed immediately by the work of Jordan and Wigner, and by Fermi in 1930. The quantum theory of a pure electron field presents no difficulties, nor does the quantized theory of a free electromagnetic field. For a coupled system of an electron interacting with an electromagnetic field, however, a relativistic quantum field theory, known as "quantum electrodynamics" or QED, presents serious deep-rooted difficulties arising from the persistent presence of infinities in the results of calculations of physical amplitudes. A major breakthrough came in the mid 1940's with the work of Tomonaga in Japan and Schwinger, Feynman, and Dyson in the United States. As shall be explained later in this and the next chapters, the theory succeeds in "subtracting" these infinities away in a systematic manner so that finite results are obtained, which have been found to be in remarkable agreement with the observed Lamb shifts and the "g anomaly".

Recent developments in particle physics have led to the general acceptance of the $SU(3)$ gauge theory of strong interactions among quarks and gluons, known as quantum chromodynamics or QCD, and the Glashow-Salam-Weinberg (GSW) $SU(2)_L \times U(1)$ theory of electroweak interactions as the standard model. The GSW theory not only contains QED as one of its important components, but uses QED as its prototype in the sense that both gauge principle and renormalizability are served as the guideline for constructing the theory. In a similar vein, QCD has been proposed, and tested to some extent, as the candidate theory of strong interactions. Despite the amazing successes of the standard model which we shall summarize at the end of this volume, we need to look for the deeper reasons why theories of this kind work so well, in which presence of infinities is a persistent feature. Till now, there seems not yet a satisfactory answer to this very question.

In this chapter, we introduce the calculational scheme for QED, leaving discussion of renormalization to the subsequent chapter. This will then pave the way for our presentation of the standard model in the last part of the book (Part III). A pedestrian, rather than formal, approach is adopted in our introduction to this important subject. Thus, treatments, such as through the path-integral formulation, which are technical in nature, will either be relegated to appendices, or omitted entirely. By giving up the opportunity of phrasing the theory perhaps in a more elegant manner, we wish to present the subject in terms of the concepts which we had become familiar from lessons in classical dynamics and ordinary quantum mechanics. The conventional pedestrian approach has both the benefit of learning to know how to make specific predictions from the theory and that of gaining strong intuition (and perhaps better insights) toward the subject.

§.9.1. The Evolution Operator and the S-Matrix

Consider a physical system described by the Hamiltonian,

$$H = H_0 + H', \tag{1}$$

where H_0 is the unperturbed Hamiltonian and H' is a small perturbation. It is assumed that the perturbation H' is turned on during the period $-T_0 < t < T_0$ with T_0 some large time. Consider the state $\mid t_0 >$ which is assumed to be an eigenstate of H at time t_0. The state will evolve with time:

$$\mid t >= U(t, t_0) \mid t_0 >, \tag{2}$$

where $U(t, t_0)$ is referred to as the "evolution operator". For the sake of simplicity, we suppress the known time-dependent phase factor $e^{-iE_0 t}$ related to H_0 so that $U(t, t_0)$ is determined entirely by H'.

It is clear that the evolution operator $U(t', t)$ satisfies the following properties:

i.

$$U(t_2, t_0) = U(t_2, t_1)U(t_1, t_0), \qquad \text{for } t_2 > t_1 > t_0. \tag{3a}$$

It arises because $\mid t_2 >= U(t_2, t_0) \mid t_0 >$ and $\mid t_2 >= U(t_2, t_1) \mid t_1 >= U(t_2, t_1)(U(t_1, t_0) \mid t_0 >)$.

ii.

$$U^{-1}(t_1, t_0) = U(t_0, t_1), \qquad \text{(Existence of the inverse operator).} \tag{3b}$$

iii.

$$U^\dagger(t_1,t_0)U(t_1,t_0) = 1, \qquad \text{(unitarity)}. \tag{3c}$$

It comes from $< t_1 \mid t_1 > = < t_0 \mid t_0 >$ and $\mid t_1 > = U(t_1,t_0) \mid t_0 >$.

iv.

$$U(t,t) = 1, \qquad \text{(Existence of the identity operator)}. \tag{3d}$$

We find

$$\begin{aligned}
i\frac{d}{dt} U(t,t_0) &\equiv i \lim_{\Delta t \to 0} \frac{U(t+\Delta t, t_0) - U(t,t_0)}{\Delta t} \\
&= i \lim_{\Delta t \to 0} \frac{U(t+\Delta t, t)U(t,t_0) - U(t,t_0)}{\Delta t} \\
&= i\{ \lim_{\Delta t \to 0} \frac{U(t+\Delta t, t) - 1}{\Delta t} \}U(t,t_0).
\end{aligned} \tag{4}$$

We define the interaction Hamiltonian:

$$H_{int}(t) \equiv i \lim_{\Delta t \to 0} \frac{U(t+\Delta t, t) - 1}{\Delta t}, \tag{5}$$

so that $H_{int}(t)$ charaterizes how the system evolves with time at t. It follows that

$$i\frac{d}{dt}U(t,t_0) = H_{int}(t)U(t,t_0). \tag{6}$$

Note that

$$i\frac{d}{dt}U(t,t_0) \mid t_0 > = H_{int}(t)U(t,t_0) \mid t_0 >,$$

or,

$$i\frac{d}{dt} \mid t > = H_{int}(t) \mid t >, \tag{7}$$

which suggests that $H_{int}(t) = H'$ with H' given by Eq. (1).

Eq. (6) can readily be solved:

$$\begin{aligned}
U(t,t_0) =& 1 - i \int_{t_0}^{t} dt_1 H_{int}(t_1)U(t_1,t_0) \\
=& 1 - i \int_{t_0}^{t} dt_1 H_{int}(t_1) \\
&+ (-i)^2 \int_{t_0}^{t} dt_1 \int_{t_0}^{t_1} dt_2 H_{int}(t_1)H_{int}(t_2) \\
&+ ....
\end{aligned} \tag{8}$$

Define the chronological or T-product:

$$T(A(t_1)B(t_2)) = \begin{cases} A(t_1)B(t_2) & \text{for } t_1 > t_2 \; ; \\ B(t_2)A(t_1) & \text{for } t_1 < t_2 \; . \end{cases} \qquad (9)$$

Here $A(t)$ and $B(t)$ are *even* functions in fermion operators; that is, the number of fermion operators which enter $A(t)$ or $B(t)$ is even. An extra minus sign should be introduced in Eq. (9) for $t_1 < t_2$ if $B(t_2)A(t_1)$ differs from $A(t_1)B(t_2)$ by *odd* permutations of fermion operators. Note that

$$\int_{t_0}^{t} dt_1 \int_{t_0}^{t_1} dt_2 H_{int}(t_1) H_{int}(t_2)$$
$$= \frac{1}{2} \int_{t_0}^{t} dt_1 \int_{t_0}^{t} dt_2 T(H_{int}(t_1) H_{int}(t_2)), \qquad (10a)$$

and, more generally,

$$\int_{t_0}^{t} dt_1 \int_{t_0}^{t_1} dt_2 ... \int_{t_0}^{t_{n-1}} dt_n H_{int}(t_1) H_{int}(t_2) ... H_{int}(t_n)$$
$$= \frac{1}{n!} \int_{t_0}^{t} dt_1 \int_{t_0}^{t} dt_2 ... \int_{t_0}^{t} dt_n T(H_{int}(t_1) H_{int}(t_2) ... H_{int}(t_n)). \qquad (10b)$$

Therefore, Eq. (8) becomes

$$U(t, t_0)$$
$$= T(\exp(-i) \int_{t_0}^{t} dt H_{int}(t)) \qquad (11)$$
$$\equiv 1 + \sum_{n=1}^{\infty} \frac{(-i)^n}{n!} \int_{t_0}^{t} dt_1 \int_{t_0}^{t} dt_2 ... \int_{t_0}^{t} dt_n T(H_{int}(t_1) H_{int}(t_2) ... H_{int}(t_n)).$$

The scattering operator (matrix), or S-matrix, is defined by

$$S = \lim_{t \to \infty} \lim_{t_0 \to -\infty} U(t, t_0), \qquad (12)$$

provided that the double limit exists in a certain operator sense. Eqs. (1)-(12) provide an intuitive and simple way to introduce the concept of the scattering matrix which is tied closely to the evolution operator. The major problem which we encounter in practice is that, for a system of interacting fields, the limiting procedures appearing in these equations such as Eqs. (12) and (5) can hardly be categorized in a mathematically rigorous manner.

Consider

$$H_{int}(t) = \int d^3x H_{int}(x,t)$$
$$\equiv -\int d^4x \mathcal{L}_{int}(x,t),$$

so that Eqs. (12) and (11) yield

$$S = T(\exp i \int d^4x \mathcal{L}_{int}(x)). \tag{14}$$

It is customary to introduce the transition operator, or T-matrix, as follows:

$$< f \mid S \mid i > = \delta_{fi} + i(2\pi)^4 \delta^4 (\sum_f p_f - \sum_i p_i) T_{fi}. \tag{15}$$

For the scattering problem, H_0 is often taken to be the free Hamiltonian so that the initial state (at $t < -T_0$) and the final state (at $t > T_0$) are plane waves. For instance, the initial and final states in Bhabha scattering, $e^+e^- \to e^+e^-$, are specified by

$$\mid i > = b_{s_1}^+(\mathbf{p}_1) a_{s_2}^+(\mathbf{p}_2) \mid 0 >, \tag{16a}$$

$$\mid f > = b_{s_1'}^+(\mathbf{p}_1') a_{s_2'}^+(\mathbf{p}_2') \mid 0 > . \tag{16b}$$

Here $a_s^+(\mathbf{p}) [b_s^+(\mathbf{p})]$ is the creation operator for an electron [positron] of spin s and three-momentum $\mathbf{p}$. The cross section, which is the transition probability per unit volume per unit time, is given by

$$\sigma = \frac{1}{f} \int \frac{d^3p_1'}{N_0} \int \frac{d^3p_2'}{N_0} \overline{\sum} |< f \mid S \mid i >|^2 \cdot \frac{1}{VT}, \tag{17}$$

where f is the flux factor $(= \mid v_1 - v_2 \mid$ in the case of two-particle scattering), N_0 is some phase-space normalization factor (a specific power of 2π in our normalization), and $\overline{\Sigma}$ denotes suitable summation and averaging over internal indices such as spins. V and T are the total volume and time, respectively, so that

$$| (2\pi)^4 \delta^4 (\sum p_f - \sum p_i) |^2$$

$$= | \int d^4x \exp ix \cdot (\sum p_f - \sum p_i) |^2$$

$$= \int d^4x \exp ix \cdot (\sum p_f - \sum p_i) \cdot (2\pi)^4 \delta^4 (\sum p_f - \sum p_i) \tag{18}$$

$$= VT(2\pi)^4 \delta^4 (\sum p_f - \sum p_i).$$

Accordingly, we find, for $f \neq i$,

$$d\sigma = \frac{1}{f} (\frac{d^3 p_1'}{N_0})(\frac{d^3 p_2'}{N_0})(2\pi)^4 \delta^4 (p_1' + p_2' - p_1 - p_2) \sum \overline{| T_{fi} |^2} . \tag{19}$$

The phase-space factor N_0 depends on the choice for the various normalization factors associated with wave functions and others. Later in this chapter, we shall set up Feynman rules for calculating T_{fi} in such a way that all powers of 2π are suitably taken into account by an appropriate choice of N_0 . In this way, we obtain

$$N_0 = (2\pi)^3. \tag{20}$$

On the flux factor, $f = 1$ for a high energy fixed target scattering experiment while $f = 2$ for a collider experiment.

Finally, it should be stressed that the formulae obtained in this section may be modified slightly in order to describe bound-state problems such as Lamb shifts or the "g anomaly". We shall turn to these problems later in Ch. 10.

§.9.2. S-Matrix Elements and Feynman Rules

Consider the interaction of the electron (positron) field $\psi(x)$ with the electromagnetic field $A_\mu(x)$:

$$\mathcal{L}_{int}(x) = -ie : \bar{\psi}(x)\gamma_\mu \psi(x) A_\mu(x) : \tag{21}$$

where $: \alpha\beta...\gamma :$ denotes the "normal ordering" of the operator product $\alpha\beta...\gamma$. Normal ordering is specified as follows:

i. If $\alpha_1, \alpha_2, ...\alpha_n$ are creation or annihilation operators, then

$$: \alpha_1 \alpha_2 ... \alpha_n := \eta_N \alpha_{i_1} \alpha_{i_2} ... \alpha_{i_n} \tag{22}$$

where $\{i_1, i_2, ..., i_n\}$ is a permutation of $1, 2, ..., n$ such that all creation operators must precede annihilation operators. The phase factor η_N is -1 if the number of permutations for fermions is odd and $+1$ otherwise.

ii. Let $\alpha_1^i, \alpha_2^i, ..., \alpha_n^i$ be creation and annihilation operators and c_i be c-numbers. Then

$$: \sum_i c_i \alpha_1^i \alpha_2^i ... \alpha_n^i := \sum_i c_i : \alpha_1^i \alpha_2^i ... \alpha_n^i : \qquad (23)$$

Accordingly, we have, for the vacuum state $\mid 0 >$,

$$< 0 \mid : \alpha\beta...\gamma : \mid 0 >= 0. \qquad (24)$$

It is clear that, for any given operator product $\alpha\beta...\gamma$, the normal ordering $: \alpha\beta...\gamma :$ is uniquely specified since any two creation (annihilation) operators either commute or anticommute.

There is a problem because of elevation of $\mathcal{L}_{int}(x)$ as a c-number in classical field theory (Lagrangian formalism) to a q-number as given by Eq. (21). Derivation of the Euler-Lagrange equation (the field equation) from a Lagrangian density $\mathcal{L}(x)$, or application of Noether's theorem from a given $\mathcal{L}(x)$ in generating physical observables (such as four-momentum and the angular-momentum tensor) involves variation of a functional with respect to a field function. Mathematical rigor is lost as field functions are treated as operators. Normal ordering just introduced makes the situation even worse, since a specific basis for defining creation and annihilation operators must be chosen as a reference point. Thus, the canonical formalism as stressed by T. D. Lee and many others (where field functions are treated as operators from the outset) or the path-integral formulation as initiated by R. P. Feynman provides a more coherent treatment of the problem than the naive Lagrangian formalism. Nevertheless, the situation seems less problematic when the $\mathcal{L}_{int}(x)$ of Eq. (21) is used only in connection with the S-matrix of Eq. (14).

We write

$$\psi(\mathbf{x}, t) = \int \frac{d^3p}{(2\pi)^{\frac{3}{2}}} \sum_s \{a_s^-(\mathbf{p})u(\mathbf{p}, s)e^{ip\cdot x} + b_s^+(\mathbf{p})v(\mathbf{p}, s)e^{-ip\cdot x}\}, \qquad (25)$$

which is just Eq. (109), Ch. 8. Using Eq. (114), Ch. 8, we have

$$\{a_s^-(\mathbf{p}), a_{s'}^+(\mathbf{p}')\} = \delta_{ss'}\delta^3(\mathbf{p} - \mathbf{p}'), \tag{26a}$$

$$\{b_s^-(\mathbf{p}), b_{s'}^+(\mathbf{p}')\} = \delta_{ss'}\delta^3(\mathbf{p} - \mathbf{p}'), \tag{26b}$$

$$\text{all other anticommutators} = 0. \tag{26c}$$

It is then straightforward to show that

$$\overline{\psi(x)\bar{\psi}(y)} = \frac{1}{(2\pi)^4 i} \int d^4p \, e^{ip\cdot(x-y)} \frac{m - i\gamma \cdot p}{m^2 + p^2 - i\varepsilon}, \tag{27}$$

where we have introduced the chronological pairing or T-pairing as follows:

$$\overline{\psi(x)\bar{\psi}(y)} \equiv\, <0 \mid T(\psi(x)\bar{\psi}(y)) \mid 0 > \equiv -iS^c(x - y). \tag{28}$$

Analogously, we write, for the electromagnetic field,

$$A_\mu(x) = \int \frac{d^3k}{(2\pi)^{\frac{3}{2}}} \frac{1}{\sqrt{2\omega}} \sum_{\lambda=1}^2 \{c_\lambda(\mathbf{k})\varepsilon_\mu^\lambda(\mathbf{k})e^{ik\cdot x} \\ + c_\lambda^+(\mathbf{k})\varepsilon_\mu^{\lambda*}(\mathbf{k})e^{-ik\cdot x}\}, \tag{29}$$

where $\varepsilon_\mu^\lambda(\mathbf{k})$ with $\lambda = 1, 2$ describe two possible transverse polarizations:

$$\varepsilon^\lambda \cdot \varepsilon^{\lambda\dagger} = \varepsilon^\lambda \cdot \varepsilon^{\lambda*} - \varepsilon_0^\lambda\varepsilon_0^{\lambda*} = 1, \tag{30a}$$

$$k \cdot \varepsilon^\lambda = \mathbf{k} \cdot \varepsilon^\lambda - k_0\varepsilon_0^\lambda = 0. \tag{30b}$$

(Cf. Eq. (127), Ch. 8.)
We have

$$[c_\lambda(\mathbf{k}), c_{\lambda'}^+(\mathbf{k}')] = \delta_{\lambda\lambda'}\delta^3(\mathbf{k} - \mathbf{k}'), \tag{31a}$$

$$\text{all other commutators} = 0 . \tag{31b}$$

Accordingly, we find

$$\overline{A_\mu(x)A_\nu(y)} \equiv -i\delta_{\mu\nu}D_0^c(x - y) \\ = \frac{\delta_{\mu\nu}}{(2\pi)^4 i} \int d^4k \, e^{-ik\cdot(x-y)} \frac{1}{k^2 - i\varepsilon}. \tag{32}$$

Now, consider Bhabha scattering,

$$e^+(p_1) + e^-(p_2) \rightarrow e^+(p_1') + e^-(p_2'). \qquad (33)$$

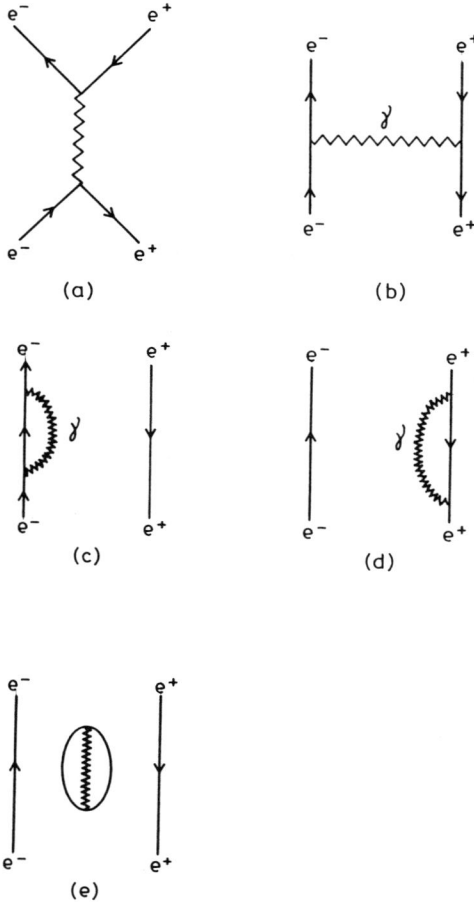

Fig. 1. Pictorial representation of the second-order S-matrix element for the Bhabha scattering.

The initial and final states are specified by Eqs. (16a) and (16b), respectively. We wish to evaluate the S-matrix element S_{fi}:

$$S_{fi} \equiv < f \mid S \mid i > = \sum_{n=0}^{\infty} S_{fi}^{(n)}, \qquad (34)$$

with

$$S_{fi}^{(n)} = < f \mid \frac{i^n}{n!} \int d^4 x_1 ... d^4 x_n T(\mathcal{L}_{int}(x_1)...\mathcal{L}_{int}(x_n)) \mid i > . \qquad (35)$$

It is straightforward to obtain

$$S_{fi}^{(0)} = \delta_{fi} \equiv \delta^3(\mathbf{p}_1 - \mathbf{p}_1')\delta_{s_1 s_1'} \delta^3(\mathbf{p}_2 - \mathbf{p}_2')\delta_{s_2 s_2'}, \qquad (36)$$

$$S_{fi}^{(1)} = 0. \qquad (37)$$

Eq. (37) follows from $< 0 \mid A_\mu(x) \mid 0 > = 0$. The leading nontrivial contribution $S_{fi}^{(2)}$ is given by

$$\begin{aligned} S_{fi}^{(2)} &= < f \mid \frac{i^2}{2!} \int d^4 x_1 d^4 x_2 T(\mathcal{L}_{int}(x_1)\mathcal{L}_{int}(x_2)) \mid i > \\ &= < f \mid \frac{i^2 \cdot (-ie)^2}{2} \int d^4 x_1 d^4 x_2 \\ & \quad T(: \bar{\psi}(x_1)\gamma_\mu \psi(x_1) A_\mu(x_1) :: \bar{\psi}(x_2)\gamma_\nu \psi(x_2) A_\nu(x_2) :) \mid i > \\ &= \frac{i^2 \cdot (-ie)^2}{2} \int d^4 x_1 d^4 x_2 \overline{A_\mu(x_1)A_\nu}(x_2) < 0 \mid a_{s_2'}(\mathbf{p}_2')b_{s_1'}(\mathbf{p}_1') \\ & \quad \cdot T(: \bar{\psi}(x_1)\gamma_\mu \psi(x_1) :: \bar{\psi}(x_2)\gamma_\nu \psi(x_2) :)b_{s_1}^+(\mathbf{p}_1)a_{s_2}^+(\mathbf{p}_2) \mid 0 > . \end{aligned}$$

$$(38)$$

The standard way to simplify the expression (38) is to pair off each creation operator by an annihilation operators until all creation and annihilation operators are paired off. We obtain

$$S_{fi}^{(2)} = \frac{i^2 \cdot (-ie)^2}{2} \int d^4x_1 d^4x_2 \overline{A_\mu(x_1) A_\nu(x_2)}$$

$$\cdot \{ \overline{ab} : \overline{\psi}(x_1)\gamma_\mu\psi(x_1) :: \overline{\psi}(x_2)\gamma_\nu\psi(x_2) : b^+ a^+$$

$$+ \left(\text{1st term with } x_1 \leftrightarrow x_2 \text{ and } \mu \leftrightarrow \nu\right)$$

$$+ ab : \overline{\psi}(x_1)\gamma_\mu\psi(x_1) :: \overline{\psi}(x_2)\gamma_\nu\psi(x_2) : b^+ a^+$$

$$+ \left(\text{3rd term with } x_1 \leftrightarrow x_2 \text{ and } \mu \leftrightarrow \nu\right)$$

$$+ ab : \overline{\psi}(x_1)\gamma_\mu\psi(x_1) :: \overline{\psi}(x_2)\gamma_\nu\psi(x_2) : b^+ a^+ \tag{39}$$

$$+ \left(\text{5th term with } x_1 \leftrightarrow x_2 \text{ and } \mu \leftrightarrow \nu\right)$$

$$+ ab : \overline{\psi}(x_1)\gamma_\mu\psi(x_1) :: \overline{\psi}(x_2)\gamma_\nu\psi(x_2) : b^+ a^+$$

$$+ \left(\text{7th term with } x_1 \leftrightarrow x_2 \text{ and } \mu \leftrightarrow \nu\right)$$

$$+ ab : \overline{\psi}(x_1)\gamma_\mu\psi(x_1) :: \overline{\psi}(x_2)\gamma_\nu\psi(x_2) : b^+ a^+ \}.$$

Here the 1st, 3rd, 5th, 7th, and 9th terms may be represented pictorially by Figs. 1(a) - 1(e), respectively.

The last three diagrams, Figs. 1(c) - 1(e), represent "renormalization" of the lowest-order graph, i.e., $S_{fi}^{(0)}$, and they do not contain explicitly an interaction between e^+ and e^-. Accordingly, we shall focus our attention on the first two diagrams in Fig. 1. Using Eqs. (25) and (26), we obtain

$$\overline{\psi(\mathbf{x},t)b_s^+(\mathbf{p})} \equiv <0 \mid T(\overline{\psi}(\mathbf{x},t)b_s^+(\mathbf{p};t=-\infty)) \mid 0>$$
$$= \frac{1}{(2\pi)^{\frac{3}{2}}} \bar{v}(\mathbf{p},s)e^{ip\cdot x}, \tag{40a}$$

$$\overline{\psi(x)a_s^+(\mathbf{p})} = \frac{1}{(2\pi)^{\frac{3}{2}}} u(\mathbf{p},s)e^{ip\cdot x}, \tag{40b}$$

$$\overline{a_s(\mathbf{p})\overline{\psi}(x)} = \frac{1}{(2\pi)^{\frac{3}{2}}} u(\mathbf{p},s)e^{-ip\cdot x}, \tag{40c}$$

$$b_s(\mathbf{p})\psi(x) = \frac{1}{(2\pi)^{\frac{3}{2}}}\, v(\mathbf{p},s)e^{-ip\cdot x}. \tag{40d}$$

Thus, the first four terms of Eq. (39) may be written in the following form:

$$S_{fi}^{\prime(2)} = (\frac{1}{(2\pi)^{\frac{3}{2}}})^4 (2\pi)^4 \delta^4(p_1' + p_2' - p_1 - p_2)i^2(-ie)^2$$

$$\cdot \{\bar{u}(p_1')\gamma_\mu v(p_2')\frac{1}{i}\frac{1}{(p_1 + p_2)^2 - i\varepsilon}\bar{v}(p_2)\gamma_\nu u(p_1) \tag{41}$$

$$- \bar{u}(p_1')\gamma_\mu u(p_1)\frac{1}{i}\frac{1}{(p_1' - p_2')^2 - i\varepsilon}\bar{v}(p_2)\gamma_\nu v(p_2')\},$$

where the minus sign related to the second term comes from an odd number of permutations among the fermion operators. It is clear that Eq. (41) may also be obtained via application of a set of Feynman rules in momentum space:

(a) Fermion propagator:

$$\Leftrightarrow \quad \frac{1}{i}\frac{m - i\gamma\cdot p}{m^2 + p^2 - i\varepsilon} \tag{42a}$$

(b) Photon propagator:

$$\Leftrightarrow \quad \frac{1}{i}\frac{\delta_{\mu\nu}}{k^2 - i\varepsilon} \tag{42b}$$

(c) Vertex:

$$\Leftrightarrow \quad i\cdot(-ie)\gamma_\mu \tag{42c}$$

(d) Summation over dummy discrete indices or integration over the internal momentum $[(2\pi)^{-4}\int d^4p]$ is always implied.

$$\tag{42d}$$

(e) External lines:

$u(\mathbf{p}, s)$	for incoming spinor,
$\bar{u}(\mathbf{p}, s)$	for outgoing spinor;
$\bar{v}(\mathbf{p}, s)$	for incoming antispinor,
$v(\mathbf{p}, s)$	for outgoing antispinor;
$\dfrac{\varepsilon_\mu^\lambda(\mathbf{k})}{\sqrt{2k_0}}$	for a photon in the initial state,
$\dfrac{\varepsilon_\mu^{\lambda*}(\mathbf{k})}{\sqrt{2k_0}}$	for a photon in the final state.

$$\tag{42e}$$

(f) There is an additional sign for each fermion loop, some counting factor for a given set of identical particle (just to ensure the proper normalization).

There is a non-essential factor $(-i)$ in going from S_{fi} to T_{fi} [cf. Eq. (15)]. Finally, factors of 2π are counted altogether at the end such that the cross section (or decay rate) is given by

$$d\sigma = \frac{1}{f}\left(\prod_i \frac{d^3 p_i'}{(2\pi)^3}\right)(2\pi)^4 \delta^4\left(\sum p_i' - \sum p_i\right)\overline{\sum} \mid T_{fi}\mid^2, \qquad (42f)$$

with f the flux factor, $\{p_i'\}$ the final momenta, and $\overline{\Sigma}$ denoting the appropriate summation and averaging over discrete indices.

(g) There is an additional minus sign if an odd number of permutations among the fermion operators of the same kind is required to separate out completely the designated T-pairings.

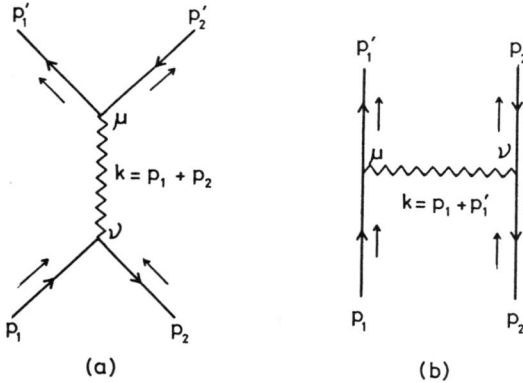

Fig. 2. Feynman diagrams for the Bhabha scattering.

Application of Feynman rules to the Bhabha scattering diagrams as illustrated in Figs. 2(a) and 2(b) yields

$$T_{fi} = (-i)\{\bar{u}(p_1')(e\gamma_\mu)v(p_2')\frac{1}{i}\frac{1}{(p_1 + p_2)^2 - i\varepsilon}\,\bar{v}(p_2)(e\gamma_\mu)u(p_1)$$
$$- \bar{u}(p_1')(e\gamma_\mu)u(p_1)\frac{1}{i}\frac{1}{(p_1' - p_1)^2 - i\varepsilon}\,\bar{v}(p_2)(e\gamma_\mu)v(p_2')\}, \qquad (43)$$

which may be obtained from Eqs. (41) and (15) *except* the known overall factor of (2π) [associated with Eqs. (40a)-(40d)]. The expression (43) is to be used in connection with Eq. (42f).

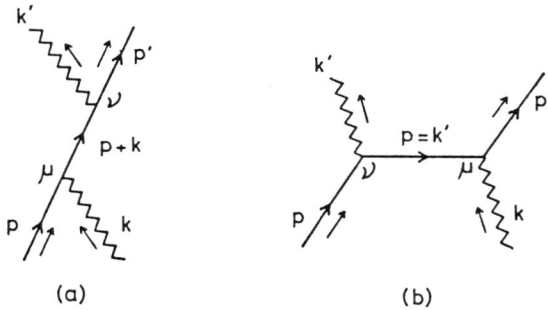

Fig. 3. Feynman diagrams for Compton scattering.

Analogously, application of Feynman rules to Compton scattering, as illustrated by Figs. 3(a) and 3(b), yields

$$
\begin{aligned}
T_{fi} = (-i)\{\bar{u}(p')(e\gamma_\nu)&\frac{\varepsilon_\nu^{\prime\lambda'*}(k')}{\sqrt{2k_0'}}\frac{1}{i}\frac{m - i\gamma\cdot(p+k)}{m^2 + (p+k)^2 - i\varepsilon}\\
&\cdot(e\gamma_\mu)\frac{\varepsilon_\mu^\lambda(k)}{\sqrt{2k_0}}u(p)\\
+ \bar{u}(p')(e\gamma_\mu)&\frac{\varepsilon_\mu^\lambda(k)}{\sqrt{2k_0}}\frac{1}{i}\frac{m - i\gamma\cdot(p-k')}{m^2 + (p-k')^2 - i\varepsilon}\\
&\cdot(e\gamma_\nu)\frac{\varepsilon_\nu^{\prime\lambda'*}(k')}{\sqrt{2k_0'}}u(p)\}.
\end{aligned}
\tag{44}
$$

This amplitude will be used later in §.9.3.

In closing this section, we wish to make remarks concerning the last three diagrams in Fig. 1. *First*, we note that the disconnected bubble of Fig. 1(e) appears as a term in S_0 :

$$
S_0 \equiv\, <0\mid S\mid 0> \equiv 1 + \underbrace{}_{} + ...
\tag{45}
$$

It is clear that, for any given diagram D, DS_0^i will also be a legitimate Feynman diagram where S_0^i is the i-th diagram in S_0 . These contributions are dropped since S_0 represents vacuum fluctuations that cannot be observed.

238

Second, an electron line which appears in S_{fi} always looks like

$$(46)$$

so that we may take into account Figs. 1(c) and 1(d) by interpreting the full electron line as the "physical" one (i.e., by using the observed electron mass and electric charge in the evaluation of a given diagram).

We shall return to the problem of renormalization in Ch. 10, where questions related to these diagrams will be addressed in some detail.

§.9.3. Calculation of Cross Sections

(1) $e^+e^- \to e^+e^-$ at High Energies

In the framework of QED, the transition amplitude for $e^+e^- \to e^+e^-$ is given by

$$T_{fi} = e^2 \{ \frac{1}{s} \bar{u}(p_1')\gamma_\mu v(p_2')\bar{v}(p_2)\gamma_\mu u(p_1)$$
$$- \frac{1}{t} \bar{u}(p_1')\gamma_\mu u(p_1)\bar{v}(p_2)\gamma_\mu v(p_2') \}, \tag{47}$$

where $s \equiv -(p_1 + p_2)^2$ and $t \equiv -(p_1' - p_1)^2$ [cf. Eq. (43)]. In an e^+e^- collider experiment, the cross section is determined via the formula,

$$d\sigma = \frac{1}{2} \frac{d^3 p_1'}{(2\pi)^3} \frac{d^3 p_2'}{(2\pi)^3} (2\pi)^4 \delta^4(p_1' + p_2' - p_1 - p_2) \overline{\sum} |T_{fi}|^2 . \tag{48}$$

We find, for $\sqrt{s} \gg m_e$,

$$\frac{d\sigma}{d\Omega} = \frac{1}{4} (\frac{\sqrt{s}}{2})^2 \frac{1}{(2\pi)^2} \overline{\sum} |T_{fi}|^2, \tag{49}$$

where Ω is the solid angle defined by $\mathbf{p'}_1$.

In an experiment in which no polarizations (spins) are detected, we have

$$\overline{\sum} |T_{fi}|^2 = \frac{1}{4} \sum_{\text{all spins}} e^4 \{ \frac{1}{s} \bar{u}(p_1')\gamma_\mu v(p_2')\bar{v}(p_2)\gamma_\mu u(p_1)$$
$$- \frac{1}{t} \bar{u}(p_1')\gamma_\mu u(p_1)\bar{v}(p_2)\gamma_\mu v(p_2') \}$$
$$\cdot \{ \frac{1}{s} \bar{u}(p_1')\gamma_\nu v(p_2')\bar{v}(p_2)\gamma_\nu u(p_1)$$
$$- \frac{1}{t} \bar{u}(p_1')\gamma_\nu u(p_1)\bar{v}(p_2)\gamma_\nu v(p_2') \}^*. \tag{50}$$

Note that

$$\sum_s u(p, s)\bar{u}(p, s) = \frac{-i\gamma \cdot p + m}{2E}, \tag{51a}$$

$$\sum_s v(p, s)\bar{v}(p, s) = \frac{-i\gamma \cdot p - m}{2E}. \tag{51b}$$

240

Thus, we find

$$
\begin{aligned}
\sum & \overline{|T_{fi}|}^2 \\
=& e^4 \frac{1}{4} \{ \frac{1}{s^2} Tr \frac{-i\gamma \cdot p_2' - m}{2E_2'} \gamma_\nu \frac{-i\gamma \cdot p_1' + m}{2E_1'} \gamma_\mu \\
& \cdot Tr \frac{-i\gamma \cdot p_2 - m}{2E_2} \gamma_\mu \frac{-i\gamma \cdot p_1 + m}{2E_1} \gamma_\nu \\
& - \frac{1}{st} [Tr \left(\frac{-i\gamma \cdot p_1' + m}{2E_1'} \gamma_\mu \frac{-i\gamma \cdot p_2' - m}{2E_2'} \gamma_\nu \frac{-i\gamma \cdot p_2 - m}{2E_2} \gamma_\mu \frac{-i\gamma \cdot p_1 + m}{2E_1} \gamma_\nu \right) \\
& + h.c.] \\
& + \frac{1}{t^2} Tr \frac{-i\gamma \cdot p_1' + m}{2E_1'} \gamma_\mu \frac{-i\gamma \cdot p_1 + m}{2E_1} \gamma_\nu \\
& \cdot Tr \frac{-i\gamma \cdot p_2' - m}{2E_2'} \gamma_\nu \frac{-i\gamma \cdot p_2 - m}{2E_2} \gamma_\mu \}.
\end{aligned}
\tag{52}
$$

Note that $\gamma_\mu \gamma_\nu + \gamma_\nu \gamma_\mu = 2\delta_{\mu\nu}$ and $\gamma_5 = \gamma_1 \gamma_2 \gamma_3 \gamma_4$ yield

$$
Tr \, \gamma_\mu \gamma_\nu = 4\delta_{\mu\nu},
\tag{53a}
$$

$$
Tr \, \gamma_\mu \gamma_\nu \gamma_\sigma \gamma_\rho = 4(\delta_{\mu\nu}\delta_{\sigma\rho} - \delta_{\mu\sigma}\delta_{\nu\rho} + \delta_{\mu\rho}\delta_{\nu\sigma}),
\tag{53b}
$$

$$
Tr \, \gamma_5 \gamma_\mu \gamma_\nu \gamma_\sigma \gamma_\rho = 4\varepsilon_{\mu\nu\sigma\rho}.
\tag{53c}
$$

For $\sqrt{s} \gg m_e$, Eq. (52) becomes

$$
\begin{aligned}
\sum \overline{|T_{fi}|}^2 =& e^4 \cdot \frac{1}{4} \cdot (16 E_1 E_2 E_1' E_2')^{-1} \\
& \cdot \{ \frac{1}{s^2} Tr \, \gamma \cdot p_2' \gamma_\nu \gamma \cdot p_1' \gamma_\mu \cdot Tr \, \gamma \cdot p_2 \gamma_\mu \gamma \cdot p_1 \gamma_\nu \\
& + \frac{1}{st} [Tr \, \gamma \cdot p_1' \gamma_\mu \gamma \cdot p_2' \gamma_\nu \gamma \cdot p_2 \gamma_\mu \gamma \cdot p_1 \gamma_\nu + h.c.] \\
& + \frac{1}{t^2} Tr \, \gamma \cdot p_1' \gamma_\mu \gamma \cdot p_1 \gamma_\nu \cdot Tr \, \gamma \cdot p_2' \gamma_\nu \gamma \cdot p_2 \gamma_\mu \}.
\end{aligned}
\tag{54}
$$

Note that

$$
\begin{aligned}
Tr \, \gamma \cdot & p_2' \gamma_\nu \gamma \cdot p_1' \gamma_\mu \cdot Tr \, \gamma \cdot p_2 \gamma_\mu \gamma \cdot p_1 \gamma_\nu \\
=& 4(p_{2\nu}' p_{1\mu}' + p_{2\mu}' p_{1\nu}' - \delta_{\mu\nu} p_2' \cdot p_1') \\
& \cdot 4(p_{2\mu} p_{1\nu} + p_{2\nu} p_{1\mu} - \delta_{\mu\nu} p_2 \cdot p_1) \\
=& 16(2p_1' \cdot p_2 p_1 \cdot p_2' + 2p_1' \cdot p_1 p_2' \cdot p_2),
\end{aligned}
\tag{55a}
$$

$$Tr\,\gamma\cdot p_1'\gamma_\mu\gamma\cdot p_1\gamma_\nu\cdot Tr\,\gamma\cdot p_2'\gamma_\nu\gamma\cdot p_2\gamma_\mu$$
$$= 16\left(2p_1'\cdot p_2 p_1\cdot p_2' + 2p_1'\cdot p_2' p_1\cdot p_2\right). \tag{55b}$$

Using $(\gamma\cdot a)(\gamma\cdot b) = 2a\cdot b - (\gamma\cdot b)(\gamma\cdot a)$ and $\gamma_\mu\gamma_\mu = 4$, we find

$$Tr\,\gamma\cdot p_1'\gamma_\mu\gamma\cdot p_2'\gamma_\nu\gamma\cdot p_2\gamma_\mu\gamma\cdot p_1\gamma_\nu$$

$$=2p_{2\mu}'Tr\,\gamma\cdot p_1'\gamma_\nu\gamma\cdot p_2\gamma_\mu\gamma\cdot p_1\gamma_\nu - 2\delta_{\mu\nu}Tr\,\gamma\cdot p_1'\gamma\cdot p_2'\gamma\cdot p_2\gamma_\mu\gamma\cdot p_1\gamma_\nu$$

$$+\,2p_{2\mu}Tr\,\gamma\cdot p_1'\gamma\cdot p_2'\gamma_\nu\gamma_\mu\gamma\cdot p_1\gamma_\nu - 4Tr\,\gamma\cdot p_1'\gamma\cdot p_2'\gamma_\nu\gamma\cdot p_2\gamma\cdot p_1\gamma_\nu$$

$$=2p_{2\mu}'\{2p_{1\nu}'Tr\,\gamma\cdot p_2\gamma_\mu\gamma\cdot p_1\gamma_\nu - 4Tr\,\gamma\cdot p_1'\gamma\cdot p_2\gamma_\mu\gamma\cdot p_1\}$$

$$-\,2\{2p_{1\mu}Tr\,\gamma\cdot p_1'\gamma\cdot p_2'\gamma\cdot p_2\gamma_\mu - 4Tr\,\gamma\cdot p_1'\gamma\cdot p_2'\gamma\cdot p_2\gamma\cdot p_1\} \tag{56}$$

$$+\,2p_{2\mu}\{2\delta_{\mu\nu}Tr\,\gamma\cdot p_1'\gamma\cdot p_2'\gamma\cdot p_1\gamma_\nu - 2p_{1\nu}Tr\,\gamma\cdot p_1'\gamma\cdot p_2'\gamma_\mu\gamma_\nu$$

$$+\,4Tr\,\gamma\cdot p_1'\gamma\cdot p_2'\gamma_\mu\gamma\cdot p_1\}$$

$$-4\{2p_{2\nu}Tr\,\gamma\cdot p_1'\gamma\cdot p_2'\gamma\cdot p_1\gamma_\nu - 2p_{1\nu}Tr\,\gamma\cdot p_1'\gamma\cdot p_2'\gamma\cdot p_2\gamma_\nu$$

$$+\,4Tr\,\gamma\cdot p_1'\gamma\cdot p_2'\gamma\cdot p_2\gamma\cdot p_1\}.$$

Using Eq. (53b), we then obtain

$$Tr\,\gamma\cdot p_1'\gamma_\mu\gamma\cdot p_2'\gamma_\nu\gamma\cdot p_2\gamma_\mu\gamma\cdot p_1\gamma_\nu = -32p_1\cdot p_2' p_2\cdot p_1'. \tag{57}$$

Therefore, we find

$$\sum|T_{fi}|^2 = e^4\cdot\frac{1}{4}\cdot(E_1 E_2 E_1' E_2')^{-1}\cdot\{\frac{1}{s^2}(2p_1'\cdot p_2 p_1\cdot p_2' + 2p_1'\cdot p_1 p_2'\cdot p_2)$$

$$+\,\frac{1}{st}(-4)p_1\cdot p_2' p_2\cdot p_1'$$

$$+\,\frac{1}{t^2}(2p_1'\cdot p_2 p_1\cdot p_2' + 2p_1'\cdot p_2' p_1\cdot p_2)\}. \tag{58}$$

Choose the kinematics as in the center-of-mass (CM) frame:

$$\mathbf{p}_1 = -\mathbf{p}_2 = \mathbf{p}, \qquad \mathbf{p}_1' = -\mathbf{p}_2' = \mathbf{p}',$$

$$\mathbf{p}_1\cdot\mathbf{p}_1' = |\,\mathbf{p}_1\,|\,|\,\mathbf{p}_1'\,|\cos\theta;$$

$$E\cong|\,\mathbf{p}\,|, \qquad E'\cong|\,\mathbf{p}'\,|. \tag{59}$$

We obtain, from Eqs. (49) and (58),[1]

---

[1]See, e.g., Bjorken, J.D. and Drell, S.D., *Relativistic Quantum Mechanics* *(McGraw-Hill, New York, 1964).*

$$\frac{d\sigma}{d\Omega} = \frac{\alpha^2}{4s}\{(1+\cos^2\theta) - 4\csc^2\frac{\theta}{2}\cos^4\frac{\theta}{2} + 2\csc^4\frac{\theta}{2}(1+\cos^4\frac{\theta}{2})\}. \qquad (60)$$

Note that the first term in Eq. (60) describes the reation $e^+e^- \to \mu^+\mu^-$ at high energies:[2]

$$\frac{d\sigma}{d\Omega}(e^+e^- \to \mu^+\mu^-) = \frac{\alpha^2}{4s}(1+\cos^2\theta), \qquad (61)$$

so that

$$\sigma(e^+e^- \to \mu^+\mu^-) = \frac{4\pi\alpha^2}{3s}. \qquad (62)$$

Analogously, production cross section for quark-antiquark pairs in e^+e^- collisions in given by

$$\sigma(e^+e^- \to \text{hadrons}) = 3\sum_q Q_q^2 \frac{4\pi\alpha^2}{3s}, \qquad (63)$$

where 3 is the color factor. Accordingly, the ratio R as measured frequently in e^+e^- machines is given by

$$\begin{aligned} R &\equiv \frac{\sigma(e^+e^- \to \text{hadrons})}{\sigma(e^+e^- \to \mu^+\mu^-)} \\ &= 3\sum_q Q_q^2 \\ &= 2 \quad \text{for} \quad \sqrt{s} < m(\psi's); \\ &\quad \frac{10}{3} \quad \text{for} \quad m(\psi's) < \sqrt{s} < m(\Upsilon's); \\ &\quad \frac{11}{3} \quad \text{for} \quad \sqrt{s} > m(\Upsilon's), \end{aligned} \qquad (64)$$

which agrees with the observed values of R as a function of $\sqrt{s}$, supporting the three-color hypothesis.

---

[2] See, e.g., Halzen, F. and Martin, A.D., *Quarks and Leptons: An Introductory Course in Modern Particle Physics (John Wiley & Sons, New York, 1984)*, Ch. 11.

(2) Compton Scattering $\gamma e^- \to \gamma e^-$

The transition amplitude for Compton scattering,

$$\gamma(k, \lambda) + e^-(p, s) \to \gamma(k', \lambda') + e^-(p', s'), \tag{65}$$

is given by Eq. (44); see Figs. 3(a) and 3(b). The differential cross section is determined from Eq. (42f) with the flux factor $f = 1$:

$$d\sigma = \frac{d^3p'}{(2\pi)^3} \frac{d^3k'}{(2\pi)^3} (2\pi)^4 \delta^4(p' + k' - p - k) \overline{\sum} |T_{fi}|^2 . \tag{66}$$

On integrating over d^3k',

$$d\sigma = \frac{1}{(2\pi)^2} d^3p' \delta(p'_0 + k'_0 - p_0 - k_0) \overline{\sum} |T_{fi}|^2 . \tag{67}$$

Choose the kinematics:

$$\mathbf{p} = 0, \qquad \mathbf{p'} + \mathbf{k'} = \mathbf{k}, \qquad \mathbf{p'} \cdot \mathbf{k} = |\mathbf{p'}| |\mathbf{k}| \cos\theta, \tag{68}$$

so that

$$\begin{aligned} d^3p' &= p'^2 dp' d\Omega'_e \\ &= p' p'_0 dp'_0 d\Omega'_e, \end{aligned} \tag{69a}$$

$$\begin{aligned} &\delta(p'_0 + k'_0 - p_0 - k_0) \\ =& (1 + \frac{p'_0 - k_0 \frac{|\mathbf{p'}|}{p'_0} \cos\theta}{p_0 + k_0 - p'_0})^{-1} \delta(p'_0 - p'_{0c}), \end{aligned} \tag{69b}$$

with p'_{0c} the observed final electron energy.

Thus, we obtain

$$\frac{d\sigma}{d\Omega} = \frac{p' p'_0}{(2\pi)^2} \frac{p_0 + k_0 - p'_0}{p_0 + k_0 - k_0 \frac{|\mathbf{p'}|}{p'_0} \cos\theta} \overline{\sum} |T_{fi}|^2 . \tag{70}$$

For the Compton scattering of unpolarized electrons by unpolarized photons, we have

$$\overline{\sum} |T_{fi}|^2 = \frac{1}{4} \sum_s \sum_\lambda \sum_{s'} \sum_{\lambda'} |T_{fi}|^2 . \tag{71}$$

Summation over the photon polarizations λ and λ' may be carried out with the aid of the formula:

$$\sum_\lambda \varepsilon^\lambda(k) \cdot \mathbf{a}\, \varepsilon^{\lambda*}(k) \cdot \mathbf{b} = \mathbf{a} \cdot \mathbf{b} - \mathbf{a} \cdot \hat{k}\, \mathbf{b} \cdot \hat{k}. \tag{72}$$

Here we need to consider only the two physical transverse polizations for the initial (or final) photon.

The rest of the algebra has to do with evaluation of traces of the products of γ-matrices, which is left as an exercise. The final resalt is the well-known Klein-Nishina-Tamm formula:[3]

$$\frac{d\sigma}{d\Omega'_e} = \frac{\alpha^2}{4m_0^2} \left(\frac{k'_0}{k_0}\right)^2 \left\{ 4cos^2\theta + \frac{k_0}{k'_0} + \frac{k'_0}{k_0} - 2 \right\}. \tag{73}$$

[3] See, e.g., Bogoliubov, N.N. and Shirkov, D.V., Introduction to the Theory of Quantized Fields, 3rd Edition (John Wiley & Sons, New York, 1980), Ch. 4.; or, Heitler, W., Quantum Theory of Radiation, 3rd edition (Oxford University, 1956), p. 217.

(3) Electron-Positron Annihilation $e^+e^- \to \gamma\gamma$

We shall also investigate briefly the process of mutual annihilation of an electron and a positron. The simplest diagram which corresponds to this process is shown by Fig. 4, which is the only first-order diagram.

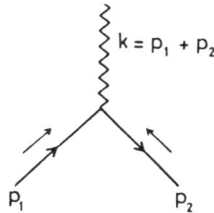

Fig. 4. Annihilation of electron and positron into a single photon.

However, it may readily be seen that the one-photon annihilation described by this diagram is forbidden by the energy-momentum conservation laws. Indeed, the conservation laws yield

$$\mathbf{k} = \mathbf{p}_1 + \mathbf{p}_2,$$
$$|\mathbf{k}| = \sqrt{|\mathbf{p}_1|^2 + m_0^2} + \sqrt{|\mathbf{p}_2|^2 + m_0^2}.$$

By transforming, for example, to the system in which the center of mass of the electron and the positron is at rest $(\mathbf{p}_1 + \mathbf{p}_2 = 0)$, we obtain a clear contradiction.

The two-photon annihilation is described by the two second-order diagrams obtained from Figs. (3a) and (3b) by crossing.

We may carry out the calculation in the system in which the center of mass of the electron and the positron is at rest. Then by setting

$$\mathbf{p}_1 = \mathbf{p}, \qquad \mathbf{p}_2 = -\mathbf{p}, \qquad \mathbf{k}_1 = \mathbf{k}, \qquad \mathbf{k}_2 = -\mathbf{k}, \tag{74}$$

We obtain

$$
\begin{aligned}
electron &: (E = \sqrt{\mathbf{p}^2 + m_0^2}, \mathbf{p}), \\
positron &: (E, -\mathbf{p}), \\
first\,photon &: (k_0 = |\mathbf{k}|, \mathbf{k}), \\
second\,photon &: (k_0, -\mathbf{k}).
\end{aligned}
\tag{75}
$$

The transition amplitude is (see Eq. (44))

$$
\begin{aligned}
T_{fi} = (-i)\{ &\bar{v}(p_2)(e\gamma_\nu) \frac{\varepsilon_\nu'^{\lambda'*}(k_2)}{\sqrt{2k_{20}}} \frac{1}{i} \frac{m - i\gamma \cdot (p_1 - k_1)}{m^2 + (p_1 - k_1)^2 - i\varepsilon} \\
&\cdot (e\gamma_\mu) \frac{\varepsilon_\mu^{\lambda*}(k_1)}{\sqrt{2k_{10}}} u(p_1) \\
+ &\bar{v}(p_2)(e\gamma_\mu) \frac{\varepsilon_\mu^{\lambda*}(k_1)}{\sqrt{2k_{10}}} \frac{1}{i} \frac{m - i\gamma \cdot (p_1 - k_2)}{m^2 + (p_1 - k_2)^2 - i\varepsilon} \\
&\cdot (e\gamma_\nu) \frac{\varepsilon_\nu'^{\lambda'*}(k_2)}{\sqrt{2k_{20}}} u(p_1)\}.
\end{aligned}
\tag{76}
$$

Using

$$
\delta(k_{10} + k_{20} - p_{10} - p_{20}) = \frac{1}{2}\, \delta(k_{10} - k_0), \tag{77}
$$

we find, with the flux factor $f = |\, v_1 - v_2\, | = 2p/E$ in the CM frame,

$$
\frac{d\sigma}{d\Omega} = \frac{E}{2p} \frac{1}{(2\pi)^2} \frac{k_0^2}{2} \overline{\sum} \, |\, T_{fi}\, |^2, \tag{78}
$$

with $p \equiv |\, \mathbf{p}\, |$.

Again, it is straightforward, albeit tedious, to evaluate the traces associated with $\bar\Sigma \, |\, T_{fi}\, |^2$. For an unpolarized experiment, we obtain the well-known formula for the differential cross section:[4]

$$
\frac{d\sigma}{d\Omega} = \frac{\alpha^2}{4k_0 p} \Big\{ \frac{k_0^2 + p^2 + p^2 \sin^2 \theta}{k_0^2 - p^2 \cos^2 \theta} - \frac{2p^4 \sin^4 \theta}{(k_0^2 - p^2 \cos^2 \theta)^2} \Big\}, \tag{79}
$$

with $cos\theta \equiv \mathbf{p} \cdot \mathbf{k}/(|\, \mathbf{p}\, ||\, \mathbf{k}\, |)$.

---

[4] See, e.g., Heitler, W., Quantum Theory of Radiation, 3rd edition (Oxford University, 1956), p. 269.

References

Wu, T.-Y., *Quantum Mechanics* (World Scientific, Singapore, 1986), p. 293; on the evolution operator.

Bjorken, J.D. and Drell, S.D., *Relativistic Quantum Mechanics* (McGraw-Hill, New York, 1964).

Halzen, F. and Martin, A.D., *Quarks and Leptons: An Introductory Course in Modern Particle Physics* (John Wiley & Sons, New York, 1984).

Bogoliubov, N.N. and Shirkov, D.V., *Introduction to the Theory of Quantized Fields*, 3rd Edition (John Wiley & Sons, New York, 1980).

Heitler, W., *Quantum Theory of Radiation*, 3rd edition (Oxford University, 1956).

248

Exercises: *Chapter 9*

1. Prove Eq. (11) from Eq. (8). Discuss possible ambiguities in your proof in light of possible singularites at equal times.

2. Use the anticommutation or commutation relations Eqs. (26) and (31) to obtain Eqs. (27), (32), and (40a)-(40d).

3. On Compton scattering, derive the final formula on the differential cross section, Eq. (73), from Eqs. (44), (70), and (71).

4. On electron-positron annihilation $e^+e^- \rightarrow \gamma\gamma$, derive the differential cross secion, Eq. (79), from Eqs. (76) and (78).

Chapter 10.
Quantum Electrodynamics II: Renormalization

Let us look into the structure of quantum electrodynamics in greater detail. To this end, we consider the three diagrams illustrated in Fig. 1.

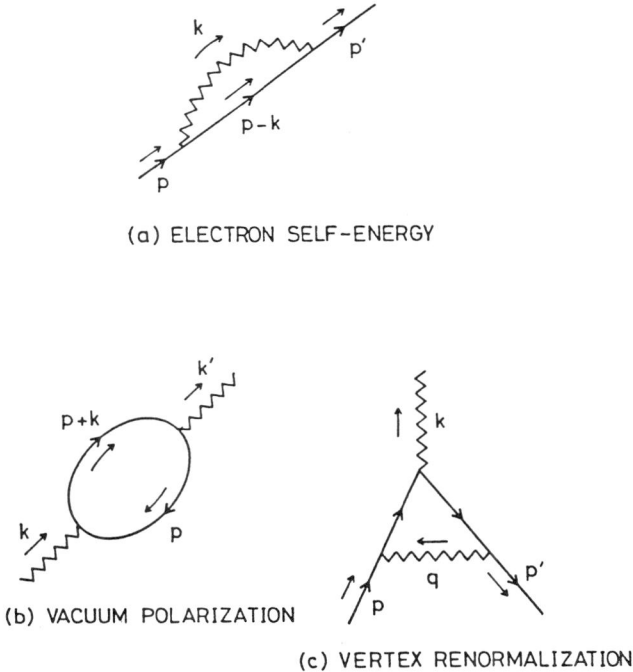

(a) ELECTRON SELF-ENERGY

(b) VACUUM POLARIZATION

(c) VERTEX RENORMALIZATION

Fig. 1. Leading renormalization diagrams in QED: electron self-energy (a), vacuum polarization (b), and vertex renormalization (c).

Specifically, Fig. 1(a) represents the scenario that an electron, as propagating in free space, may emit and reabsorb a virtual photon. Using QED Feynman rules, Eqs. (42a)-(42f) of Ch. 9, we obtain

$$S_a = -i(2\pi)^4 \delta^4(p' - p)\delta_{s',s}\bar{u}(p')\Sigma(p)u(p), \tag{1}$$

with

$$\Sigma(p) = -ie^2 \int \frac{d^4k}{(2\pi)^4} \gamma_\mu \frac{m - i\gamma \cdot (p - k)}{m^2 + (p - k)^2 - i\varepsilon} \gamma_\mu \frac{1}{k^2 - i\varepsilon}. \tag{2}$$

This contribution is often called "electron self-energy." As it stands, however, the $\Sigma(p)$ as given by Eq. (2) is ill-defined since, at large k, the integrand behaves like k^{-3} and $\int d^4k k^{-3}$ diverges. Taking into account the fact that integration over an expression that is odd in k_μ vanishes identically, we still anticipate that $\Sigma(p)$ diverges logarithmically.

Before considering any resolution to this problem, we write down the S-matrix elements in the case of Figs. 1(b) and 1(c).

$$S_b = -i(2\pi)^4 \delta^4(k' - k)\delta_{\lambda\eta}\frac{1}{2k_0}\varepsilon_\nu^{\eta*}(k')\varepsilon_\mu^\lambda(k)\Pi_{\mu\nu}(k), \tag{3}$$

$$S_c = -i(2\pi)^4 \delta^4(p' + k - p)\frac{\varepsilon_\mu^\lambda(k)}{\sqrt{2k_0}}(-ie)\bar{u}(p')\Gamma_\mu u(p), \tag{4}$$

with

$$\Pi_{\mu\nu}(k) = ie^2 \int \frac{d^4p}{(2\pi)^4}T_r\gamma_\mu\frac{m + i\gamma \cdot p}{m^2 + p^2 - i\varepsilon}\gamma_\nu\frac{m + i\gamma \cdot (p + k)}{m^2 + (p + k)^2 - i\varepsilon}, \tag{5}$$

$$\Gamma_\mu = -ie^2 \int \frac{d^4q}{(2\pi)^4}\gamma_\alpha\frac{m - i\gamma \cdot (p' + q)}{m^2 + (p' + q)^2 - i\varepsilon}\gamma_\mu\frac{m - i\gamma \cdot (p + q)}{m^2 + (p + q)^2 - i\varepsilon}\gamma_\alpha\frac{1}{q^2 - i\varepsilon}. \tag{6}$$

Simple power counting indicates that $\Pi_{\mu\nu}(k)$ and Γ_μ are also divergent. It is customary to refer to $\Pi_{\mu\nu}$ and Γ_μ ad "vacuum polarization" and "vertex renormalization", respectively.

What goes wrong? Physics is an empirical science. In principle, we can use only a well-defined mathematical model (without infinities) to model or describe an observed phenomenon. The fact that the diagrams illustrated by Fig. 1, if taken at the face value, are divergent must have critical implications for local field theories as a whole. As stated in the introductory chapter (Ch. 0), serious attempts were made, during the decade from the mid 1950's to the mid 1960's, in searching for an alternative means of describing interactions among elementary particles in terms of the S-matrix approach, in which one tries to determine the S-matrix elements using general principles (such as unitarity and microscopic causality) and a minimal set of dynamical assumptions. However, efforts to find an alternative approach were not very fruitful while gauge field theories have scored an amazing comeback since early 1970's. Successes of the standard model, which is phrased in terms of local gauge field theories (with QED as a prototype), suggest strongly the usefulness of the concept of a local field theory, despite the presence of infinities associated with the diagrams such as those in Fig. 1. The question concerning what goes wrong with these infinities is thus transcended to the questions as to why infinities are there and how we can make sense out of them. These questions are what we wish to investigate in the rest of this chapter, although we shall see that affirmative answers remain to be very elusive and in some sense rather mysterious.

An important clue comes from the observation that the diagrams such as those in Fig. 1. may become well-defined if the assumption of a 4-dimensional Minkowski space-time with usual topology can somehow be altered. As the first illustrative example, consider possible existence of a maximum momentum scale (or a minimum length scale) for which the theory is applicable. The electron self-energy, $\Sigma(p)$ as given by Eq. (2), is finite if the integration over $d^4 k$ is carried out only up to some cut-off momentum k_{max}. This is Feynman's cut-off regularization method. To do it in a Lorentz-covariant manner, we may use the Pauli-Villars regularization method and replace the photon propagator by

$$reg\, D_0^c(k) = \frac{1}{k^2 - i\varepsilon} + \sum_M C_M \frac{1}{M^2 + k^2 - i\varepsilon}, \qquad (7)$$

and the fermion propagator by

$$reg\, S^c(p) = (m - i\gamma \cdot p)\{\frac{1}{m^2 + p^2 - i\varepsilon} + \sum_M C_M \frac{1}{M^2 + p^2 - i\varepsilon}\}. \qquad (8)$$

Here continuity of the regularized function, together with all its derivatives up to order $n - 1$ inclusive, reguires the following conditions:

$$\sum_i C_i = 0, \qquad \sum_i C_i M_i^2 = 0, \, ..., \qquad \sum_i C_i M_i^{2n} = 0, \qquad (9)$$

where the summation over i includes $i = 0$ with $C_0 = 1$ and $M_0 = m$. Eqs. (7)-(9) constitute the basis for the Pauli-Villars regularization scheme, which we wish to elaborate on later in §.10.1. Taking $n = 1$, we find, from Eqs. (2),

(7), (8), and (9),

$$reg\,\Sigma(p) = -ie^2 \int \frac{d^4k}{(2\pi)^4} \gamma_\mu \{m - i\gamma \cdot (p - k)\} \cdot$$

$$\cdot \{\frac{1}{m^2 + (p - k)^2 - i\varepsilon} - \frac{1}{M^2 + (p - k)^2 - i\varepsilon}\}\gamma_\mu \qquad (10)$$

$$\cdot \{\frac{1}{k^2 - i\varepsilon} - \frac{1}{M^2 + k^2 - i\varepsilon}\},$$

which is finite for $0 < m \ll M < \infty$.

An alternative method for regularization of gauge fields was introduced by 't Hooft and Veltman in 1972. They considered an analytic continuation of the S-matrix elements in the complex n-plane, where n is a variable that for positive integer values equals the dimension of the space involved with respect to loop quantities. The physical situation corresponds to $n = 4$. The generalized S-matrix elements so defined are analytic in n and the infinities of perturbation theory manifest as poles at $n = 4$. The "dimensional regularization" of 't Hooft and Veltman will be introduced in some detail in §.10.2.

Of course, there are additional ways to redefine the self-energy $\Sigma(p)$ and other divergences so that the "regularized" expressions, which reproduce original resulsts in a specific limit, are finite. In all cases, there is an additional parameter, say μ^2, which may be chosen to have the dimension of $(mass)^2$. It is clear that the theory is defined as we know μ^2 in addition to the parameters m, e, and others, which appear in the lagrangion. Changing the scale μ^2 from μ_0^2 to μ_1^2, the parameters (m, e) change from (m_0, e_0) to (m_1, e_1). The structure of the theory remains to be the same, as reflected by the renormalization-group analysis. The renormalization-group concept will be introduced in §.10.3. As an application, the concept of a running coupling constant is also described there.

§.10.1. Pauli-Villars Regularization

Consider the electron self-energy in the Pauli-Villars regularization scheme, i.e. Eq. (10). Using $\gamma_\mu \gamma_\mu = 4$ and $\gamma_\mu i\gamma \cdot p\gamma_\mu = -2i\gamma \cdot p$, we obtain

$$
\begin{aligned}
reg\, \Sigma(p) = &-ie^2 \int \frac{d^4k}{(2\pi)^4} \{4m + 2i\gamma \cdot (p-k)\} \\
&\cdot \{\frac{1}{m^2 + (p-k)^2 - i\varepsilon} - \frac{1}{M^2 + (p-k)^2 - i\varepsilon}\} \\
&\cdot \{\frac{1}{k^2 - i\varepsilon} - \frac{1}{M^2 + k^2 - i\varepsilon}\}.
\end{aligned}
\tag{11}
$$

The following mathematical identities are useful:

$$
\frac{1}{m^2 + k^2 - i\varepsilon} = i \int_0^\infty d\alpha e^{-i\alpha(m^2 + k^2 - i\varepsilon)},
\tag{12a}
$$

$$
\int_{-\infty}^\infty dt e^{i(at^2 + 2bt)} = \frac{1+i}{\sqrt{2}} \sqrt{\frac{\pi}{a}} e^{-\frac{ib^2}{a}}, \quad (a > 0);
\tag{12b}
$$

$$
\int_{-\infty}^\infty dt e^{i(at^2 + 2bt)} = \frac{1-i}{\sqrt{2}} \sqrt{\frac{\pi}{|a|}} e^{-\frac{ib^2}{a}}, \quad (a < 0);
\tag{12c}
$$

$$
\int d^4k e^{-i(ak^2 + 2bk)} = \frac{\pi^2}{ia^2} e^{+\frac{ib^2}{a}}, \quad (a > 0);
\tag{12d}
$$

$$
\int d^4k e^{-i(ak^2 + 2bk)} k_\mu = \frac{ib_\mu}{a}(\frac{\pi}{a})^2 e^{+\frac{ib^2}{a}}, \quad (a > 0);
\tag{12e}
$$

$$
\int d^4k e^{-i(ak^2 + 2bk)} k_\mu k_\nu = \frac{a\delta_{\mu\nu} - 2ib_\mu b_\nu}{2a^2}(\frac{\pi}{a})^2 e^{+\frac{ib^2}{a}}, \quad (a > 0);
\tag{12f}
$$

$$
\int d^4k e^{-i(ak^2 + 2bk)} k^2 = \frac{2a - ib^2}{a^2}(\frac{\pi}{a})^2 e^{+\frac{ib^2}{a}}, \quad (a > 0).
\tag{12g}
$$

Eq. (11) becomes

$$
\begin{aligned}
reg\, \Sigma(p) = &-ie^2 \int \frac{d^4k}{(2\pi)^4} \{4m + 2i\gamma \cdot (p-k)\} \\
&\cdot i \int_0^\infty d\beta e^{-i\beta((p-k)^2 - i\varepsilon)} \{e^{-i\beta m^2} - e^{-i\beta M^2}\} \\
&\cdot i \int_0^\infty d\alpha e^{-i\alpha(k^2 - i\varepsilon)} \{1 - e^{-i\alpha M^2}\} \\
= &\frac{e^2}{8\pi^2} \int_0^\infty d\alpha \int_0^\infty d\beta \frac{e^{-\varepsilon(\alpha+\beta)}}{(\alpha+\beta)^2} e^{-i(\frac{\alpha\beta}{\alpha+\beta}p^2)}(2m + i\gamma \cdot p\frac{\alpha}{\alpha+\beta}) \\
&\cdot (1 - e^{-i\alpha M^2})(e^{-i\beta m^2} - e^{-i\beta M^2}).
\end{aligned}
\tag{13}
$$

254

We introduce

$$\alpha = \xi\lambda, \quad \beta = (1-\xi)\lambda, \tag{14}$$

so that

$$\frac{\partial(\alpha, \beta)}{\partial(\xi, \lambda)} = \lambda. \tag{15}$$

We obtain

$$reg\, \Sigma(p) = \frac{e^2}{8\pi^2} \int_0^1 d\xi (2m + i\gamma \cdot p\xi) J_\varepsilon(\xi, M), \tag{16}$$

with

$$J_\varepsilon(\xi, M) = \int_0^\infty \frac{d\lambda}{\lambda} e^{-\lambda\varepsilon - i\lambda\xi(1-\xi)p^2}(1 - e^{-i\xi\lambda M^2}) \\ \cdot (e^{-i(1-\xi)\lambda m^2} - e^{-i(1-\xi)\lambda M^2}). \tag{17}$$

Using

$$\int_0^\infty \frac{d\lambda}{\lambda}(e^{iA\lambda} - e^{iB\lambda})e^{-\varepsilon\lambda} = \ln\frac{B + i\varepsilon}{A + i\varepsilon}, \tag{18}$$

we obtain

$$J_0(\xi, M) = \ln\frac{M^2 + \xi p^2}{m^2 + \xi p^2} + \ln\frac{\xi M^2 + (1-\xi)m^2 + \xi(1-\xi)p^2}{M^2 + \xi(1-\xi)p^2}. \tag{19}$$

Substituting Eq. (19) back into Eq. (16), we write

$$reg\, \Sigma(p) = \Sigma_{div}(p) + \Sigma'(p), \tag{20}$$

where

$$\Sigma'(p) = \frac{e^2}{8\pi^2} \int_0^1 d\xi (2m + i\gamma \cdot p\xi) \ln\frac{m^2}{m^2 + \xi p^2}, \tag{21}$$

$$\Sigma_{div}(p) = \frac{e^2}{8\pi^2} \int_0^1 d\xi (2m + i\gamma \cdot p\xi) \\ \cdot \ln\{\xi\frac{M^2 + \xi p^2}{m^2} \cdot \frac{\xi M^2 + (1-\xi)m^2 + \xi(1-\xi)p^2}{\xi M^2 + \xi^2(1-\xi)p^2}\}. \tag{22}$$

For sufficiently large M, we find

$$\Sigma_{div}(p) = \frac{e^2}{(4\pi)^2}\{\ln\frac{M^2}{m^2}(4m + i\gamma\cdot p) + (-\frac{i}{2}\gamma\cdot p - 4m)\}. \tag{23}$$

By going over to configuration space,

$$\Sigma(x - y) = \frac{1}{(2\pi)^4}\int d^4p\, e^{ip\cdot(x-y)}\Sigma(p), \tag{24}$$

we find, for sufficiently large M,

$$reg\,\Sigma(x) = \frac{e^2}{(4\pi)^2}\{\ln\frac{M^2}{m^2}(4m - \gamma_\mu\partial_\mu)$$
$$+ (\frac{1}{2}\gamma_\mu\partial_\mu - 4m)\}\delta^4(x) + \Sigma'(x), \tag{25}$$

with

$$\Sigma'(x) \equiv \frac{1}{(2\pi)^4}\int d^4p\, e^{ip\cdot x}\Sigma'(p). \tag{26}$$

Note that

$$\lim_{M\to\infty}\,reg\,\Sigma(x) = \Sigma'(x)\quad everywhere\ except\ x = 0. \tag{27}$$

To understand the meaning of these results, we consider the Feynman regularization method. Instead of Eq. (11), we introduce the Feynman cutoff factor,

$$reg\,\Sigma_F(p) = -ie^2\int\frac{d^4k}{(2\pi)^4}\{4m + 2i\gamma\cdot(p - k)\}$$
$$\cdot\frac{1}{m^2 + (p - k)^2 - i\varepsilon}\cdot\frac{1}{k^2 - i\varepsilon}\cdot\frac{M^2}{M^2 + k^2}$$
$$= -ie^2\int\frac{d^4k}{(2\pi)^4}\{4m + 2i\gamma\cdot(p - k)\}$$
$$\cdot\frac{1}{m^2 + (p - k)^2 - i\varepsilon}\{\frac{1}{k^2 - i\varepsilon} - \frac{1}{M^2 + k^2 - i\varepsilon}\}. \tag{28}$$

Following the same procedure, we find, for sufficiently large M,

$$reg\,\Sigma_F(x) = \frac{e^2}{(4\pi)^2}\{\ln\frac{M^2}{m^2}(4m - \gamma_\mu\partial_\mu) + (\frac{1}{2}\gamma_\mu\partial_\mu - 4m)\}\delta^4(x)$$
$$+ \Sigma'_F(x), \tag{29}$$

where Σ'_F differs from Σ' by a quantity,

$$\frac{e^2}{(4\pi)^2}(2m - \frac{3}{4}\gamma_\mu\partial_\mu)\delta^4(x). \tag{30}$$

Accordingly, we may write the general expression for $\Sigma'(p)$ in the form,

$$\Sigma'(p) = \frac{e^2}{8\pi^2}\{\int_0^1 d\xi(2m + i\gamma \cdot p\xi)\ln\frac{m^2}{m^2 + \xi p^2} \\ + C_1(i\gamma \cdot p + m) + C_2 m\}, \tag{31}$$

where C_1 and C_2 are finite constants which depend on the method of regularization. Nevertheless, the expression obtained by subtracting from reg $\Sigma(p)$ the first two terms of Maclaurin's series,

$$reg\ \Sigma(p) - reg\ \Sigma(0) - \frac{\partial reg\ \Sigma(p)}{\partial p_\mu}\big|_{p=0} \cdot p_\mu,$$

does not depend on the method of regularization.

We proceed to consider the regularization of the vacuum polarization $\Pi_{\mu\nu}(k)$ as given by Eq. (5). We obtain

$$reg\ \Pi_{\mu\nu}(k) = ie^2 \int \frac{d^4p}{(2\pi)^4} T_r\gamma_\mu(m + i\gamma \cdot p)\gamma_\nu(m + i\gamma \cdot (p+k)) \cdot \\ \cdot (\frac{1}{m^2 + p^2 - i\varepsilon} - \frac{1}{M^2 + p^2 - i\varepsilon}) \tag{32} \\ \cdot (\frac{1}{m^2 + (p+k)^2 - i\varepsilon} - \frac{1}{M^2 + (p+k)^2 - i\varepsilon}).$$

Evaluating the trace and using Eqs. (12a)-(12g), we find

$$reg\ \Pi_{\mu\nu}(k) = -\frac{ie^2}{4\pi^2}\int_0^\infty d\alpha \int_0^\infty d\beta e^{-\varepsilon(\alpha+\beta)-i\frac{\alpha\beta}{\alpha+\beta}k^2} \cdot \\ \cdot (e^{-i\alpha m^2} - e^{-i\alpha M^2})(e^{-i\beta m^2} - e^{-i\beta M^2}) \\ \cdot \frac{1}{(\alpha+\beta)^2}\{i\frac{\alpha\beta}{(\alpha+\beta)^2}(k^2\delta_{\mu\nu} - 2k_\mu k_\nu) \tag{33} \\ - \delta_{\mu\nu}(im^2 - \frac{1}{\alpha+\beta})\}.$$

Using Eq. (14), we find

$$reg\ \Pi_{\mu\nu}(k) = -\frac{ie^2}{4\pi^2}\int_0^1 d\xi \int_0^\infty d\lambda\frac{e^{-\varepsilon\lambda}}{\lambda} \cdot \\ \cdot \{f(\lambda)[i\xi(1-\xi)(k^2\delta_{\mu\nu} - 2k_\mu k_\nu) - im^2\delta_{\mu\nu}] \tag{34} \\ + \frac{1}{\lambda}f(\lambda)\delta_{\mu\nu}\},$$

with

$$f(\lambda) = e^{-i\lambda\xi(1-\xi)k^2}(e^{-i\xi\lambda m^2} - e^{-i\xi\lambda M^2}).$$
$$\cdot (e^{-i(1-\xi)\lambda m^2} - e^{-i(1-\xi)\lambda M^2}). \tag{35}$$

Noting that, with $f(\lambda) < \infty$ as $\lambda \to \infty$ and $f(\lambda) \to c\lambda^2$ as $\lambda \to 0$,

$$\int_0^\infty \frac{d\lambda}{\lambda^2} f(\lambda) e^{-\varepsilon\lambda} = \int_0^\infty \frac{d\lambda}{\lambda} \frac{\partial}{\partial\lambda}(e^{-\varepsilon\lambda} f(\lambda))$$
$$\to \int_0^\infty \frac{d\lambda}{\lambda} e^{-\varepsilon\lambda} \frac{\partial f(\lambda)}{\partial\lambda} \quad \text{as} \quad \varepsilon \to 0, \tag{36}$$

we obtain

$$reg\, \Pi_{\mu\nu}(k) = -\frac{ie^2}{4\pi^2} \int_0^1 d\xi \int_0^\infty \frac{d\lambda}{\lambda} e^{-\varepsilon\lambda}.$$
$$\cdot \{f(\lambda)[i\xi(1-\xi)(k^2\delta_{\mu\nu} - 2k_\mu k_\nu) - im^2\delta_{\mu\nu}] + \frac{\partial f(\lambda)}{\partial\lambda}\delta_{\mu\nu}\}. \tag{37}$$

A straightforward computation yields, in the limit of sufficiently large M,

$$reg\, \Pi_{\mu\nu}(k) = -\frac{e^2}{8\pi^2}\delta_{\mu\nu}(M^2 - m^2) + (k^2\delta_{\mu\nu} - k_\mu k_\nu)\frac{e^2}{12\pi^2}\ln\frac{M^2}{m^2} + \Pi'_{\mu\nu}(k), \tag{38}$$

with

$$\Pi'_{\mu\nu}(k) = -\frac{e^2}{2\pi^2}(k^2\delta_{\mu\nu} - k_\mu k_\nu)\int_0^1 d\xi\xi(1-\xi)\ln\frac{m^2 + \xi(1-\xi)k^2}{\xi(1-\xi)m^2}. \tag{39}$$

In configuration space, we have

$$reg\, \Pi_{\mu\nu}(x) = -\frac{e^2}{8\pi^2}\delta_{\mu\nu}(M^2 - m^2)\delta^4(x)$$
$$- \frac{e^2}{12\pi^2}\ln\frac{M^2}{m^2}(\delta_{\mu\nu}\partial_\alpha\partial_\alpha - \partial_\mu\partial_\nu)\delta^4(x) + \Pi'_{\mu\nu}(x), \tag{40}$$

so that $reg\, \Pi_{\mu\nu}(x)$ converges to $\Pi'_{\mu\nu}(x)$ everywhere except at $x = 0$.

At this juncture, it is useful to consider the question related to gauge invariance. Eq. (3) yields

$$S_b \propto (2\pi)^4\delta^4(k' - k)A_\mu(k)\Pi_{\mu\nu}(k)A_\nu(k). \tag{41}$$

All physical observables must be invariant under the gauge transformation,

$$A_\mu(x) \;\to\; A'_\mu(k) = A_\mu(x) + \partial_\mu\chi(x), \qquad (42a)$$

or,

$$A_\mu(k) \;\to\; A'_\mu(k) = A_\mu(k) + ik_\mu\chi(k). \qquad (42b)$$

Thus, we have

$$k_\mu\Pi_{\mu\nu}(k) = k_\nu\Pi_{\mu\nu}(k) = 0. \qquad (43)$$

It is clear that Eq. (38) is not in a gauge-invariant form, indicating that the Pauli-Villars regularization method is not gauge invariant. Dropping the first term in Eq. (38) and carrying out the integration associated with $\Pi'_{\mu\nu}(k)$, we obtain

$$reg\,\Pi_{\mu\nu}(k) = (k^2\delta_{\mu\nu} - k_\mu k_\nu)(\frac{e^2}{12\pi^2}\ln\frac{M^2}{m^2} - \frac{e^2}{60\pi^2}\frac{k^2}{m^2}). \qquad (44)$$

To investigate the meaning of Eq. (44), we consider Rutherford scattering as illustrated by Fig. 2.

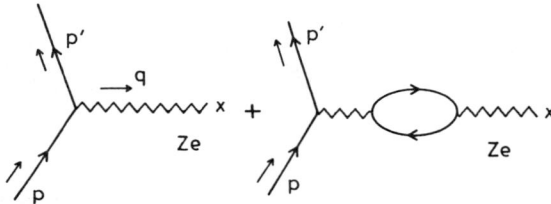

Fig. 2. Rutherford scattering including effects due to vacuum polarization.

The transition amplitude looks like

$$\bar{u}(p')(e\gamma_\mu)u(p) \cdot \frac{1}{i}\frac{1}{q^2} \cdot ZeJ_\mu(q)$$
$$+ \bar{u}(p')(e\gamma_\mu)u(p) \cdot \frac{1}{i}\frac{1}{q^2} \cdot (-i)\Pi_{\mu\nu}(q) \cdot \frac{1}{i}\frac{1}{q^2} \cdot ZeJ_\nu(q). \qquad (45)$$

Using Eq. (44) and $q_\mu J_\mu(q) = 0$, we find

$$T \propto -ei\bar{u}(p')\gamma_\mu u(p) \cdot ZeJ_\mu(q) \cdot \frac{1}{q^2}\{1 - \frac{e^2}{12\pi^2}\ln\frac{M^2}{m^2} + \frac{e^2}{60\pi^2}\frac{q^2}{m^2}\}$$
$$= -Ze_R^2 i\bar{u}(p')\gamma_\mu u(p) \cdot J_\mu(q) \cdot \frac{1}{q^2}\{1 + \frac{e_R^2}{60\pi^2}\frac{q^2}{m^2}\}. \qquad (46)$$

Here we have introduced

$$e_R \equiv e(1 - \frac{e^2}{12\pi^2} \ln \frac{M^2}{m^2})^{\frac{1}{2}}, \tag{47}$$

which is to be identified with the "physical" or "renormalized" charge. The Coulomb potential is obtained by Fourier-transforming the first term of Eq. (46).

$$V_0(\mathbf{r}) = \int \frac{d^3q}{(2\pi)^3} e^{i\mathbf{q}\cdot\mathbf{r}}(-\frac{Ze_R^2}{\mathbf{q}^2}) = -\frac{Ze_R^2}{4\pi r}. \tag{48}$$

Taking into account both terms in Eq. (46), we find

$$V(\mathbf{r}) = -\frac{Ze_R^2}{4\pi r} - \frac{Ze_R^4}{60\pi^2 m^2}\delta^3(\mathbf{r}). \tag{49}$$

The second term contributes $-27\ MHz$ to the total Lamb shift of $+1057\ MHz$ between the $2S_{\frac{1}{2}}$ and $2P_{\frac{1}{2}}$ levels of the hydrogen atom, which has been measured to an accuracy of about 0.01%.

We proceed to consider the vertex renormalization diagram of Fig. 1(c). It is a routine exercise to extract the "finite" contribution out of Eqs. (4) and (6). We may write

$$-ei\bar{u}(p')\{\gamma_\mu[1 - \frac{\alpha}{3\pi}\frac{q^2}{m^2}(\ln\frac{m^2}{\mu^2} - \frac{3}{8})] - \frac{\alpha}{2\pi}\frac{1}{2m}\sigma_{\mu\nu}q_\nu\}u(p), \tag{50}$$

where the leading term refers to the vertex diagram without any loop and the parameter μ is the small cutoff parameter associated with the removal of infrared divergences. Now, the $O(\alpha)$ correction associated with γ_μ can be combined with Eq. (49), yielding a potential:

$$V(\mathbf{r}) = -\frac{Ze_R^2}{4\pi r} + \frac{Ze_R^4}{12\pi^2}\frac{1}{m^2}(\ln\frac{m^2}{\mu^2} - \frac{3}{8} - \frac{1}{5})\delta^3(\mathbf{r}). \tag{51}$$

The second term accounts for the observed value of the Lamb shift.

The Dirac magnetic moment of an electron is given by

$$\mu = -\frac{e}{2m}\sigma = -g\frac{e}{2m}\mathbf{S}, \tag{52}$$

with

$$g = 2. \tag{53}$$

The last term in Eq. (50) yield a correction:

$$\mu = -\frac{e}{2m}\left(1 + \frac{\alpha}{2\pi}\right)\sigma. \tag{54}$$

Or, we have, quoting the most recent result for the reason of accuracy (Kinoshita and Lindquist 1981),

$$\begin{aligned}
\frac{g-2}{2} &= \frac{\alpha}{2\pi} - 0.328478966\left(\frac{\alpha}{\pi}\right)^2 + (1.1765 \pm 0.0013)\left(\frac{\alpha}{\pi}\right)^3 \\
&\quad + (-0.8 \pm 2.5)\left(\frac{\alpha}{\pi}\right)^4 + \ldots \\
&= (1159652.460 \pm 0.127 \pm 0.075) \times 10^{-9},
\end{aligned} \tag{55}$$

where the first error (± 0.127) is due to the uncertainty associated with the fine-structure constant and the second error (± 0.075) is purely theoretical. The experimental value of the electron's anomalous magnetic moment, as from the 1988 publication of Particle Data Group (Phys. Lett. **B204**, p.1), is

$$\left(\frac{g-2}{2}\right)_{exp} = (1159652.193 \pm 0.010) \times 10^{-9}, \tag{56}$$

which is in excellent agreement with Eq. (55).

To sum up, the *finite* contributions associated with Figs. 1(a)-1(b), i.e., those contributions which do not behave like $\delta^4(x)$ or derivatives of $\delta^4(x)$ in configuration space [cf. Eqs. (25) and (40)], are linked to the observable effects such as the Lamb shift and the electron anomalous magnetic moment. These contributions do not depend on the method of regularization. This fact is of specific importance for any future attempt to understand ultraviolet divergences in QED.

§.10.2. Dimensional Regularization

The dimensional regularization method was first adopted in 1972 by 't Hooft and Veltman in their attempt to prove the renormalizability of nonabelian gauge field theories. To describe the central ideas underlying the method, we wish to follow closely the presentation given in their original paper.

The dimensional regularization procedure was based originally on the observation that Ward identities, i.e. identites arising from gauge invariance, hold irrespective of the dimension of the space involved. By introducing a fictitious fifth dimension and attributing a very large momentum inside the loop to the fifth dimension, we may then formulate suitable regulator diagrams which are gauge invariant. It is clear that the procedure breaks down for diagrams containing two or more closed loops since then the "fifth" component of loop momentum may be distributed over the various internal lines. It was suggested that more dimensions would have to be introduced, and thus the idea of continuation in the number of dimensions emerged. In practice, suitable definitions may be introduced such that a slight continuation from dimension 4 into fractal dimensions allows for realization of the dimensional regularization procedure.

Infrared difficulties associated with massless particles are not the subject of our discussion here.

As an example we take a charged scalar field $\phi(x)$ interacting with the electromagnetic field $A_\mu(x)$. The gauge invariant lagrangian is specified by

$$\mathcal{L} = -[D_\mu \phi]^\dagger [D_\mu \phi] - m^2 \phi^\dagger \phi, \tag{57a}$$

with $D_\mu \equiv \partial_\mu - ieA_\mu(x)$ the gauge invariant derivative. Thus the interaction terms are given by

$$\mathcal{L}_{int} = +ie[\partial_\mu \phi]^\dagger A_\mu \phi - ieA_\mu \phi^\dagger [\partial_\mu \phi] - e^2 A_\mu A_\mu \phi^\dagger \phi. \tag{57b}$$

We wish to illustrate the central ideas by considering the one-loop graphs with the lowest order photon self-energy diagrams as the specific example. Extensions to two loops or more will be mentioned at the end. The diagrams are illustrated in Fig. 3.

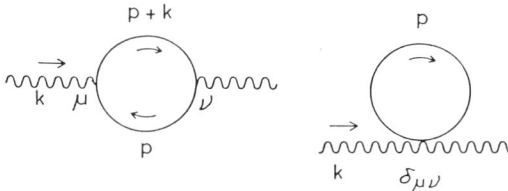

Fig. 3. The lowest order photon self-energy diagrams for a charged scalar particle.

Using the Feynman rules which can easily be obtained from Eq. (57b), we find that the S-matrix element for Fig. 3 is proportional to the integral specified by

$$J_{\mu\nu} = e^2 \int d_n p \left[\frac{(2p+k)_\mu (2p+k)_\nu}{(p^2+m^2)((p+k)^2+m^2)} - \frac{2\delta_{\mu\nu}}{p^2+m^2} \right]. \tag{58}$$

where p and k are the n component loop momentum and the external momentum, respectively. It is straightforward to see that we may recast Eq. (58) in a slightly different form:

$$J_{\mu\nu} = e^2 \int_0^1 dx \int d_n p \frac{4p_\mu p_\nu + 2p_\mu k_\nu + 2k_\mu p_\nu + k_\mu k_\nu - 2((p+k)^2+m^2)\delta_{\mu\nu}}{(p^2+2pkx+k^2x+m^2)^2}.$$

Evaluating the integral with the aid of the formulae listed in the Appendix at the end of the chapter and noting that in the end terms odd in $(1-2x)$ may be dropped, we obtain, with $d_n p = d(ip_0)d_{n-1}p$ used here and in the Appendix,

$$J_{\mu\nu} = e^2 i \pi^{\frac{1}{2}n} \Gamma(2 - \frac{1}{2}n) \int_0^1 dx \frac{(1-2x)^2(k_\mu k_\nu - k^2\delta_{\mu\nu})}{(m^2 + k^2 x(1-x))^{2-\frac{1}{2}n}}. \tag{59}$$

This expression is manifestly gauge invariant, since $k_\mu J_{\mu\nu} = k_\nu J_{\mu\nu} = 0$. In the complex n plane there are simple poles for $n = 4, 6, 8$, etc. Note that gauge invariance holds for any n. This is the property mentioned earlier that Ward identities do not involve the dimensionality of space.

To carry out the regularization procedure, we subtract from (59) the pole and its residue at the dimension $n = 4$:

$$e^2 i \pi^2 \frac{2}{4-n} (k_\mu k_\nu - k^2 \delta_{\mu\nu}) \int_0^1 dx (1-2x)^2, \tag{60}$$

which is a polynomial in the external momentum, and is again gauge invariant. Subtracting (60) from (59) and taking the limit $n = 4$, we obtain the customary result:

$$J_{\mu\nu}^{reg} = - ie^2 \pi^2 (k_\mu k_\nu - k^2 \delta_{\mu\nu}) \int_0^1 dx (1-2x)^2 \ln(m^2 + k^2 x(1-x))$$
$$+ C(k_\mu k_\nu - k^2 \delta_{\mu\nu}), \tag{61}$$

where C is a constant related to the n dependence other than in the exponent of the denominator. The constant C is in fact undetermined, as may be seen as follows: Suppose that in (59) we replace e^2 by $e^2 M^{4-n}$, where M is an arbitrary mass. Thus, the dimension of Eq. (59) is $(mass)^2$ which is independent of the dimension n. However, C in (61) is changed by a term proportional to $\ln M$ (which is arbitrary). It is useful to note that such ambiguity also occurs if we apply the Pauli-Villars regularization scheme to Fig. 3.

The above simple heuristic derivation already displays many of the features of the dimensional regularization method. It is clear that, in practical calculations for one loop diagrams, the method provides a very simple scheme for computing gauge invariant results. It could for instance be used to show cancellation of divergencies in the manifestly unitary set of Feynman diagrams.

However, there are serious objections to the above manipulations. First of all, our starting point Eq. (58) is meaningless for $n \geq 2$. In order to obtain a sensible result, one must (i) extend the Feynman rules such that for non-integer n all diagrams give rise to well-defined expressions, and (ii) define a suitable limiting procedure for $n \to 4$, which restores originally convergent diagrams to their original values while originally divergent diagrams are given a meaning consistent with unitarity, etc.

Thus, we shall first come up with a possible redefinition of the S-matrix. Let us consider again Eq. (58). First we split the n-dimensional space in a 4 dimensional (physical) space and an $n - 4$ dimensional subspace:

$$\int d_n p \to \int d_4 \underline{p} \int d_{n-4} P. \tag{62}$$

Multiplying (58) with two arbitrary physical four vectors $e_{1\mu}$ and $e_{2\nu}$ we see that (58) depends on the direction of $\underline{p}$ but not on the direction of P. Here we have used $(pk) = (\underline{p}k)$, $(e_1 p) = (e_1 \underline{p})$, $(e_2 p) = (e_2 \underline{p})$ and $p^2 = \underline{p}^2 + \omega^2$, with ω the magnitude of P in the $n - 4$ dimensional subspace. Introducing polar coordinates in P space and integrating over angles one finds:

$$J = \int d_4 \underline{p} \int d\omega \omega^{n-5} \frac{2(\pi)^{\frac{1}{2}(n-4)}}{\Gamma(\frac{1}{2}(n-4))} f(\underline{p}, \omega^2). \tag{63}$$

where the dependence on the external vectors e_1, e_2 and k is not shown explicitly. Note that (63) may still quite meaningless, since the second integral in (63) contains an infrared divergence for $n < 4$. Thus, we continue our formal manipulations until we arrive at an expression that can be given a meaning. This divergence is superficial and may be removed by partial integration:

$$\int_0^\infty d\omega \omega^{n-5} f(\underline{p}, \omega^2) = -\frac{2}{n-4} \int_0^\infty d\omega \omega^{n-3} \frac{\partial}{\partial \omega^2} f(\underline{p}, \omega^2),$$

where surface terms have been neglected. Doing this λ times on Eq. (63), we obtain

$$\frac{\pi^{\frac{1}{2}(n-4)} 2}{\Gamma(\frac{1}{2}(n-4)+\lambda)} \int d_4\underline{p} \int_0^\infty d\omega \omega^{n-5+2\lambda} (-\frac{\partial}{\partial \omega^2})^\lambda f(\underline{p}, \omega^2) \qquad (64)$$

For the quadratically divergent diagrams of Fig. 3 (in 4 dimensional space), this is a well defined formula for $4 - 2\lambda < n < 2$. *Eq. (64) with sufficiently large λ defines the contribution of one loop diagrams to the generalized S-matrix elements in a finite region of the complex n-plane. This region is the domain of convergence of the integrals in Eq. (64).*

By taking a sufficiently large λ the domain of convergence extends to arbitrarily small n. Furthermore, the degree of convergence is $2 - n$ as far as ultraviolet behavior is concerned and $n - 4 + 2\lambda$ for the infrared behavior. It is clear that, by choosing a suitable λ and n, one has a representation of the generalized diagrams in some region of the n-plane in terms of arbitrarily convergent integrals.

If a diagram is convergent in 4-dimensional space then the redefinition Eq. (64) exists for $n < n_0$ with $n_0 > 4$. Moreover, for $n = 4$, Eq. (64) equals the result evaluated in the conventional way, as may be seen by taking $\lambda = 1$ and setting $n = 4$. Thus, for finite diagrams our prescription gives the conventional results in the limit $n = 4$. For divergent diagrams, Eq. (64) will be meaningless for $n = 4$. However, as will be shown, Eq. (64) may be continued in the complex n-plane to large n values. The result will in general have a pole at $n = 4$. In order to make sense in the limit $n = 4$, one must introduce counterterms in the perturbation expansion, and those counterterms must be chosen to cancel the poles appearing at $n = 4$. Whether this can be done in a consistent manner is a separate and difficult issue, which is related to the concept of "renormalization" to be discussed in §.10.3.

For values of n outside the domain of convergence of the integrals in Eq. (64), the contribution to the generalized S-matrix elements is to be defined as the analytic continuation of Eq. (64).

It turns out to be possible to construct explicitly this analytic continuation toward larger n values. The method is as follows: By means of partial integration which is valid inside the domain of convergence of Eq. (64) we may derive a new formula, which is identical to Eq. (64) inside the domain of convergence of Eq. (64) but is analytic in n in an enlarged domain. By the principles of analytic continuation, this new formula is then used to define the analytic continuation of Eq. (64) in this enlarged domain.

In view of the importance of this construction, it is of importance to formulate the concept as clearly as possible. We consider the integral specified by

$$I = \int d_k p \frac{p_a^{\lambda_1} p_b^{\lambda_2} \dots p_c^{\lambda_j}}{((p+k_1)^2 + m_1^2)^{\alpha_1} ((p+k_2)^2 + m_2^2)^{\alpha_2} \dots ((p+k_\ell)^2 + m_1^2)^{\alpha_\ell}}. \tag{65}$$

Here p_a, p_b (etc.) are components a, b (etc.) of p. The α_j are positive integers and can be larger than 1. The exponents $\lambda_i \dots \lambda_j$ are not necessarily integers. Eq. (64) is of the form Eq. (65) with $k = 5$, where the integration over p_5 in Eq. (65) is nothing but the ω-integration in Eq. (64). Thus p_5 occurs with an n-dependent exponent in the numerator. Also $\underline{p}_1$, $\underline{p}_2$, etc. may occur in the numerator, they are contained in Eq. (64) in the function f. Note that the differentiations with respect to ω^2 in Eq. (64) have as net effect an increase of the exponents of the factors in the denominator.

The integral in Eq. (65) is convergent if

$$\begin{aligned} &\lambda_1 > -1, \lambda_2 > -1, \dots, \lambda_j > -1; \\ &k + \lambda_1 + \lambda_2 + \dots \lambda_j - 2(\alpha_1 + \alpha_2 + \dots \alpha_\ell) < 0. \end{aligned} \tag{66}$$

Next we insert in Eq. (65) the expression, which is identical to unity:

$$\frac{1}{k} \sum_{i=1}^{k} \frac{\partial p_i}{\partial p_i}. \tag{67}$$

Within the region (66) we may perform partial integrations with respect to p_i. After some trivial algebra we obtain:

$$I = -\frac{\lambda_1 + \lambda_2 + \dots \lambda_j}{k} I + \frac{2(\alpha_1 + \alpha_2 + \dots \alpha_\ell)}{k} I - \frac{1}{k} I',$$

with

$$\begin{aligned} I' = \int d_k p\, p_a^{\lambda_1} \dots p_c^{\lambda_j} \{ &\frac{2\alpha_1 (m_1^2 + k_1^2 + (pk_1))}{((p+k_1)^2 + m_1^2)^{\alpha_1+1} (\,)^{\alpha_2} \dots (\,)^{\alpha_\ell}} \\ + &\frac{2\alpha_2 (m_2^2 + k_2^2 + (pk_2))}{(\,)^{\alpha_1} (\,)^{\alpha_2+1} \dots (\,)^{\alpha_\ell}} + \dots + \frac{2\alpha_\ell (m_\ell^2 + k_\ell^2 + (pk_\ell))}{(\,)^{\alpha_1} (\,)^{\alpha_2} \dots (\,)^{\alpha_\ell+1}} \}, \end{aligned} \tag{68}$$

or

$$I = -\frac{1}{(k + \lambda_1 + \lambda_2 + ...\lambda_j - 2\alpha_1 - 2\alpha_2 - ... - 2\alpha_\ell)} I'. \qquad (69)$$

The integral I' converges if

$$\lambda_1 > -1, \lambda_2 > -1, ..., \lambda_j > -1$$
$$k + \lambda_1 + \lambda_2 + ...\lambda_j - 2(\alpha_1 + \alpha_2 + ...\alpha_\ell) < 1, \qquad (70)$$

which is a domain larger than Eq. (66). The right hand side of (69) is the explicit representation of the analytic continuation of I into this domain.

For one loop diagrams, the variable n appears linearly in some exponent λ, so that one obtains an explicit representation valid in an arbitrarily large domain in the complex n-plane. With the prescription (69) one may now evaluate the integrals in the example Eq. (58). The result is of course precisely Eq. (60).

For diagrams with two closed loops one may proceed in a similar way. There will be two n-fold integrals, and one writes:

$$\int d_n p \int d_n p' \rightarrow \int d_4\underline{p} \int d_4\underline{p}' \int d_{n-4} P \int d_{n-4} P'. \qquad (71)$$

In the P' integral the fifth axis is taken in the direction of the $(n-4)$ vector P:

$$(71) \rightarrow \int d_4\underline{p} \int d_4\underline{p}' \int d_{n-4} P \int dp_5' \int d_{n-5} P'.$$

The integrands will be independent of the P and P' directions. The integration over angles may be performed:

$$(71) \rightarrow \frac{4\pi^{\frac{1}{2}(2n-9)}}{\Gamma(\frac{1}{2}(n-4))\Gamma(\frac{1}{2}(n-5))} \cdot$$
$$\cdot \int d_4\underline{p} \int d_4\underline{p}' \int_0^\infty d\omega\, \omega^{n-5} \int_{-\infty}^\infty dp_5' \int_0^\infty d\omega'\, \omega'^{n-6}. \qquad (72)$$

The argument of such integrals will be a function of the components $\underline{p}$ and $\underline{p}'$, of ω^2, $p_5'^2 + \omega'^2$ and $(p_5' + \omega)^2 + \omega'^2$ (arising from p^2, p'^2 and $(p + p')^2$).

Eq. (72) may be written in an elegant from by introducing a two dimensional space, and the vectors

$$q = \begin{pmatrix} \omega \\ 0 \end{pmatrix}, \qquad q' = \begin{pmatrix} p_5' \\ \omega' \end{pmatrix} \qquad (73)$$

We have, with ϵ_{ij} the completely antisymmetric tensor in two dimensions ($\epsilon_{12} = 1$),

$$\int d_n p \int d_n p'\, f(p^2, p'^2, (p+p')^2)$$
$$= \frac{2\pi^{\frac{1}{2}(2n-9)-1}}{\Gamma(\frac{1}{2}(n-4))\Gamma(\frac{1}{2}(n-5))} \int d_4 \underline{p} \int d_4 \underline{p}' \int d_2 q \int d_2 q'\, (\epsilon_{ij} q_i q_j')^{n-6}. \qquad (74)$$
$$\cdot \theta(\epsilon_{ij} q_i q_j') f(q^2, q'^2, (q+q')^2).$$

The step-function θ is needed because of the lower limit $\omega' > 0$ in the ω' integration in Eq. (72). We have dropped explicit indication of the dependence on the components of $\underline{p}$ and $\underline{p}'$.

The equivalent of Eq. (64) may now be obtained by partial integrations. To this purpose, one observes that

$$(\epsilon_{ij} q_i q_j')^\alpha = \frac{1}{(\alpha+2)(\alpha+1)} \epsilon_{ab} \frac{\partial}{\partial q_a} \frac{\partial}{\partial q_b'} (\epsilon_{ij} q_i q_j')^{\alpha+1}. \qquad (75)$$

Applying Eq. (75) λ times together with subsequent partial integrations, one obtains an expression similar to Eq. (64):

$$\frac{2\pi^{\frac{1}{2}(2n-11)}}{\Gamma(\frac{1}{2}(n-4)+\lambda)\Gamma(\frac{1}{2}(n-5)+\lambda)}$$
$$\times \int d_4 \underline{p} \int d_4 \underline{p}' \int d_2 q \int d_2 q'\, (\epsilon_{ij} q_i q_j')^{n-6+2\lambda}$$
$$\times \theta(\epsilon_{ij} q_i q_j') \left(\frac{\partial^2}{\partial q^2 \partial q'^2} + \frac{\partial^2}{\partial q^2 \partial (q+q')^2} + \frac{\partial^2}{\partial q'^2 \partial (q+q')^2} \right)^\lambda \qquad (76)$$
$$\times f(q^2, q'^2, (q-q')^2).$$

Again, Eq. (76) and its analytic continuation to larger n define the contribution of the two-loop diagrams to the generalized S-matrix elements. Explicit representations for large n may be obtained by operations similar to Eqs. (67)-(70) described earlier. Note that we need four such operations in the two loop case. It is clear that the procedure to redefine the generalized S-matrix elements for the three or more closed loop cases may be worked out in an analogous manner.

The above prescription applies if all loop particles are scalars. To complete our prescription to cover vector fields we note that indices that are part of the propagators contained in the loops now take the values 1 to n for integer

n. This is because polarization vectors corresponding to internal lines become n-component vectors. The only practical consequence of this fact is that in doing the vector algebra of all occurring loop indices one must use the rule, in continuing into larger n (including fractal n),

$$\delta_{\mu\mu} = n. \tag{77}$$

After that, one has expressions that can be used to define the diagrams for non-integer n. In establishing Ward-dentities, one sees that there is an interplay between these factors n and the factors n occurring in association with averaging over all directions in p space of factors like $p_\mu p_\nu$. (See Eqs. (A7) and (A8) in the Appendix.)

Extensions to fermions

The extension to fermions is based on the following observation: Everything may be formulated such that only traces of strings of γ-matrices occur. If there are external fermion lines this may be done through the use of suitable projection operators. These traces must then be evaluated according to the rules generalized to n dimensions:

$$\{\gamma_\mu, \gamma_\nu\} = 2\delta_{\mu\nu}, \tag{78}$$

$$Tr(S) = 0 \quad if \quad S \text{ is an odd string of } \gamma's, \tag{79}$$

$$Tr(I) = 4. \tag{80}$$

Remember $\delta_{\mu\mu} = n$. As far as γ-matrices is concerned, any Ward identity relying only on Eq. (78) (as in quantum electrodynamics) will be satisfied for every n.

Note that there is no place for the pseudo-scalar γ_5 (in conventional notation) in Eq. (78), as there is no place for the pseudo tensor $\epsilon_{\mu\nu\alpha\beta}$. This places certain limitations on the regularization method. (See the orginal paper of 't Hooft and Veltman.)

The rule Eq. (80) can be satisfied by finite matrices only for $n = 4$, but this is of no importance because we are only interested in a consistent algebra for $n \neq 4$. Or, in n-dimensional space one will have

$$Tr(I) = f(n)$$

where f is a function of n only. We need only $f(n) = 4$, and the deviations of $f(n)$ from $f(4)$ are never important for Ward identities because one always compares diagrams with an equal number of traces.

As an expample we consider the lowest order photon self-energy diagram in quantum electrodynamics as illustrated by Fig. 1(b). According to Eq. (5), the contribution is proportional to, in the n dimensional space,

$$I_{\mu\nu} = \int d_n p \frac{Tr[\gamma_\mu(i\gamma(p+k)+m)\gamma_\nu(i\gamma p+m)]}{((p+k)^2+m^2)(p^2+m^2)}$$

The trace may be evaluated using Eqs. (78), (79) and (80). Taking denominators together, one obtains:

$$I_{\mu\nu} = 4 \int_0^1 dx \int d_n p \frac{\delta_{\mu\nu}(m^2+p^2+pk) - 2p_\mu p_\nu - p_\mu k_\nu - k_\mu p_\nu}{(p^2+2pkx+k^2x+m^2)^2}$$

Using the formulae listed in the Appendix, one obtains

$$I_{\mu\nu} = \frac{4i\pi^{\frac{1}{2}n}}{\Gamma(2)}\Gamma\left(2-\frac{1}{2}n\right) \int_0^1 dx \frac{2x(1-x)(k_\mu k_\nu - \delta_{\mu\nu}k^2)}{(m^2+k^2x(1-x))^{2-\frac{1}{2}n}},$$

which is manifestly gauge invariant. It is of some interest to compare this result with what we have obtained earlier in the Pauli-Villars regularization method. (See Eqs. (38) and (39).)

§.10.3. Introduction to Renormalization

Consider a physical process as illustrated by Fig. 4.

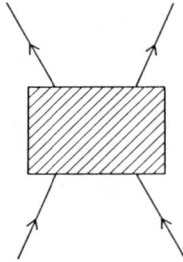

Fig. 4. A schematic view of a typical physical process.

Let $\{p_j\}$, m, and g be, respectively, the external momenta, the mass parameter, and the dimensionless coupling constant. (Our discussion refers to an arbitrary physical process.) The amplitude, i.e. the S-matrix element, can be represented by a vertex function:

$$\Gamma(-\frac{p_j^2}{\mu^2}, \frac{m^2}{\mu^2}, g),$$

where μ^2 is the renormalization parameter required in the treatment of ultraviolet divergent diagrams. For instance, μ^2 can be taken to be the cut-off mass squared M^2 in the Pauli-Villars regularization scheme or μ^2 in the dimensional regularization scheme. Now, suppose that we change the renormalization point from μ^2 to $\bar{\mu}^2$. Note that

$$g \to \bar{g}(\frac{\bar{\mu}^2}{\mu^2}, \frac{m^2}{\mu^2}, g), \tag{81a}$$

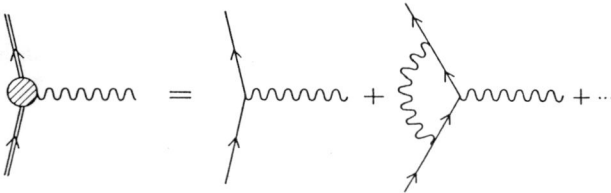

Fig. 5(a). A pictorial representation of Eq. (81a).

$$m \to \bar{m}(\frac{\bar{\mu}^2}{\mu^2}, \frac{m^2}{\mu^2}, g). \tag{81b}$$

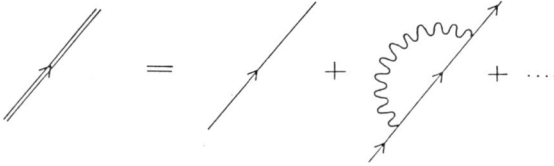

Fig. 5(b). A pictorial representation of Eq. (81b).

Thus, the consistency of the theory demands that the amplitude calculated at the new renormalization point $\bar{\mu}^2$ must be related to that obtained at μ^2 in a simple manner:

$$\Gamma(-\frac{p_j^2}{\bar{\mu}^2}, \frac{\bar{m}^2}{\bar{\mu}^2}, \bar{g}) = Z_\Gamma(\frac{\bar{\mu}^2}{\mu^2}, \frac{m^2}{\mu^2}, g)\Gamma(-\frac{p_j^2}{\mu^2}, \frac{m^2}{\mu^2}, g), \tag{81c}$$

where Z_Γ is a characteristic function for the given vertex Γ.

Differentiating Eq. (81c) with respect to $\bar{\mu}^2$ and then setting $\bar{\mu}^2 = \mu^2$, we obtain

$$\{\mu^2 \frac{\partial}{\partial \mu^2} + \beta(\frac{m^2}{\mu^2}, g)\frac{\partial}{\partial g} + \gamma_m(\frac{m^2}{\mu^2}, g)m^2 \frac{\partial}{\partial m^2} - \gamma_\Gamma(\frac{m^2}{\mu^2}, g)\}\Gamma(-\frac{p_j^2}{\mu^2}, \frac{m^2}{\mu^2}, g) = 0, \tag{82a}$$

with

$$\beta(\frac{m^2}{\mu^2}, g) \equiv \bar{\mu}^2 \frac{\partial}{\partial \bar{\mu}^2}\bar{g}(\frac{\bar{\mu}^2}{\mu^2}, \frac{m^2}{\mu^2}, g) \mid_{\bar{\mu}=\mu}, \tag{82b}$$

$$\gamma_m(\frac{m^2}{\mu^2}, g) \equiv \frac{\bar{\mu}^2}{\bar{m}^2} \frac{\partial}{\partial \bar{\mu}^2}\bar{m}^2(\frac{\bar{\mu}^2}{\mu^2}, \frac{m^2}{\mu^2}, g) \mid_{\bar{\mu}=\mu},$$

$$= \bar{\mu}^2 \frac{\partial}{\partial \bar{\mu}^2}ln\,\bar{m}^2(\frac{\bar{\mu}^2}{\mu^2}, \frac{m^2}{\mu^2}, g) \mid_{\bar{\mu}=\mu}, \tag{82c}$$

$$\gamma_\Gamma(\frac{m^2}{\mu^2}, g) \equiv \bar{\mu}^2 \frac{\partial}{\partial \bar{\mu}^2}ln\,Z_\Gamma(\frac{\bar{\mu}^2}{\mu^2}, \frac{m^2}{\mu^2}, g) \mid_{\bar{\mu}=\mu}. \tag{82d}$$

Eq. (82a) is known by several names: Ovsyannikov's equation, Gell-Mann-Low equation, or Callan-Symanzik equation. Although it is clearly beyond the scope of this book to elucidate how these renormalization-group (RG) equations [Eqs. (82)] should be used in connection with gauge field theories, we wish to stress the point that presence of infinities would render a field theory meaningless if it were not possible to treat these infinities *consistently* in the framework of RG equations. A field theory for which infinities can be handled *consistently* in the framework of RG equations will be referred to as a "renormalizable field theory." Otherwise, it is a non-renormalizable field theory. [*Caution:* This strict definition of "renormalizability", albeit the most appropriate one, is not adopted in field theory books because it may be extremely difficult to prove that a field theory is renormalizable.]

More precisely, presence of infinities, if taken at its face value, already renders the theory meaningless mathematically. However, a renormalizable field theory makes sense because it makes sense for an arbitrary renormalization point $\bar{\mu}^2$ ($< \infty$) and the theories at two different renormalization points, say μ^2 and $\bar{\mu}^2$ (both $< \infty$), are correlated via RG equations. In other words, there is a specific structure, which exists at all finite μ^2 and so at $\mu^2 \to \infty$, that makes a renormalizable field theory a meaningful theory.

It is a rather tedious task to show how, in a given regularization scheme, those ultraviolet infinities which we encounter in QED can be absorbed consistently into redefined couplings including the charge e and the mass m. This task is of course very important since, otherwise, the renormalization step illustrated by Eqs. (45)-(54) to interpret the finite parts of divergent graphs as being physical would be completely unfounded. Since it is not the objective of this book to expose ourselves to the technical complexities associated with the actual renormalization program for QED, we shall close our discussions by turning our attention to another very interesting consequence of the renormalization group equations, namely, that the coupling constant is running, i.e., depends on Q^2 that we are probing.

We may determine the beta function using Eqs. (81a) and (82b). We quote the result for QED (Baker and Johnson 1969):

$$\beta(\alpha) = \frac{\alpha^2}{3\pi} + \frac{\alpha^3}{4\pi^2} + 0(\alpha^4). \tag{83}$$

Note that the first term is just α times the coefficient of the $ln \frac{M^2}{m^2}$ term in Eq. (44). To make the physical meaning of Eq. (83) a little more transparent, we need to consider the RG equation for $\bar{g}(\frac{\bar{\mu}^2}{\mu^2}, g)$ for, say, a massless theory. Let us define

$$t \equiv \frac{\bar{\mu}^2}{\mu^2}. \tag{84}$$

Choosing $t'' = t't$, we have

$$g'' \equiv \bar{g}(t', g') \quad with \quad g' \equiv \bar{g}(t, g),$$

or,

$$\bar{g}(t't, g') = \bar{g}(t', \bar{g}(t, g)).$$

Differentiating with respect to t' and then setting $t' = 1$, we find

$$t \frac{\partial}{\partial t} \bar{g}(t, g) = \beta(\bar{g}). \tag{85}$$

Note that Eq. (85) is more general than Eq. (82b), the latter defining $\beta(g)$.

Eq. (85) can readily be solved for $\bar{g}$. We obtain

$$\int_{t=1}^{t} \frac{d\bar{g}}{\beta(\bar{g})} = ln \; t. \tag{86}$$

Consider the case of QED to lowest order:

$$\beta(\alpha) = \frac{\alpha^2}{3\pi} \tag{87a}$$

We find

$$3\pi \left(\frac{1}{\alpha(\mu^2)} - \frac{1}{\alpha(\bar{\mu}^2)} \right) = ln \frac{\bar{\mu}^2}{\mu^2},$$

or,

$$\alpha(\bar{\mu}^2) = \frac{\alpha(\mu^2)}{1 - \frac{1}{3\pi} \alpha(\mu^2) ln \frac{\bar{\mu}^2}{\mu^2}}. \tag{87b}$$

In an approximate sense, the renormalization parameter μ^2 can be identified also as the momentum squared Q^2 which defines the scale of the physics that we are probing. In other words, we may rewrite Eq. (87b) in a more familiar form:

$$\alpha(Q^2) = \frac{\alpha(Q_0^2)}{1 - \frac{1}{3\pi} \alpha(Q_0^2) ln \frac{Q^2}{Q_0^2}}. \tag{88}$$

At $Q_0^2 \sim (1 GeV)^2$, we have $\alpha(Q_0^2) = \frac{1}{137}$. We note that

$$\alpha(Q^2) \to \infty \quad as \quad 1 - \frac{1}{3\pi} \alpha(Q_0^2) ln \frac{Q^2}{Q_0^2} \to 0, \tag{89}$$

which is the well-known Landau ghost problem.

Appendix
Some Useful Formulae For Dimensional Regularization

The n-dimensional integration is defined by

$$\int d_n x f(x) = \int f(x) r^{n-1} dr \, sin^{n-2} \theta_{n-1} d\theta_{n-1} \, sin^{n-3} \theta_{n-2} d\theta_{n-2} ... d\theta_1, \quad (A1)$$

with $0 \le \theta_i \le \pi$, except $0 \le \theta_1 \le 2\pi$.

If $f(x)$ depends only on $r = \sqrt{x_1^2 + ... x_n^2}$ one may perform the integration over angles using

$$\int_0^\pi sin^m \theta d\theta = \sqrt{\pi} \frac{\Gamma(\frac{1}{2}(m+1))}{\Gamma(\frac{1}{2}(m+2))}, \quad (A2)$$

leading to

$$\int d_n x f(r) = \frac{2\pi^{\frac{1}{2}n}}{\Gamma(\frac{1}{2}n)} \int f(r) r^{n-1} dr. \quad (A3)$$

Another useful formula is given by

$$\int_0^\infty dx \frac{x^\beta}{(x^2 + M^2)^\alpha} = \frac{1}{2} \frac{\Gamma(\frac{1}{2}(\beta+1))\Gamma(\alpha - \frac{1}{2}(\beta+1))}{\Gamma(\alpha)(M^2)^{\alpha - \frac{1}{2}(\beta+1)}} \quad (A4)$$

Keeping the prescriptions and definitions of §.10.2. in mind, the following equations hold for arbitrary n:

$$\int d_n p \frac{1}{(p^2 + 2kp + m^2)^\alpha} = \frac{i\pi^{\frac{1}{2}n}}{(m^2 - k^2)^{\alpha - \frac{1}{2}n}} \frac{\Gamma(\alpha - \frac{1}{2})}{\Gamma(\alpha)}. \quad (A5)$$

$$\int d_n p \frac{p_\mu}{(p^2 + 2kp + m^2)^\alpha} = \frac{i\pi^{\frac{1}{2}n}}{(m^2 - k^2)^{\alpha - \frac{1}{2}n}} \frac{\Gamma(\alpha - \frac{1}{2})}{\Gamma(\alpha)} (-k_\mu). \quad (A6)$$

$$\int d_n p \frac{p_\mu p_\mu}{(p^2 + 2kp + m^2)^\alpha} = \frac{i\pi^{\frac{1}{2}n}}{(m^2 - k^2)^{\alpha - \frac{1}{2}n}} \cdot \frac{1}{\Gamma(\alpha)} \{\Gamma(\alpha - \frac{1}{2}n)k^2 + \Gamma(\alpha - 1 - \frac{1}{2}n)\frac{1}{2}n(m^2 - k^2)\}. \quad (A7)$$

$$\int d_n p \frac{p_\mu p_\nu}{(p^2 + 2kp + m^2)^\alpha} = \frac{i\pi^{\frac{1}{2}n}}{(m^2 - k^2)^{\alpha - \frac{1}{2}n}} \cdot \frac{1}{\Gamma(\alpha)} \{\Gamma(\alpha - \frac{1}{2}n)k_\mu k_\nu + \Gamma(\alpha - 1 - \frac{1}{2}n)\frac{1}{2}\delta_{\mu\nu}(m^2 - k^2)\}. \quad (A8)$$

$$\int d_n p \, \frac{p_\mu p_\nu p_\lambda}{(p^2 + 2kp + m^2)^\alpha} = \frac{i\pi^{\frac{1}{2}n}}{(m^2 - k^2)^{\alpha - \frac{1}{2}n}} \cdot \frac{1}{\Gamma(\alpha)} \{ -\Gamma(\alpha - \frac{1}{2}n) k_\mu k_\nu k_\lambda$$

$$- \Gamma(\alpha - 1 - \frac{1}{2}n) \frac{1}{2} (\delta_{\mu\nu} k_\lambda + \delta_{\mu\lambda} k_\nu + \delta_{\nu\lambda} k_\mu)(m^2 - k^2) \}.$$

$$(A9)$$

$$\int d_n p \, \frac{p^2 p_\mu}{(p^2 + 2kp + m^2)^\alpha} = \frac{i\pi^{\frac{1}{2}n}}{(m^2 - k^2)^{\alpha - \frac{1}{2}n}} \cdot \frac{1}{\Gamma(\alpha)} (-k_\mu) \{ \Gamma(\alpha - \frac{1}{2}n) k^2$$

$$+ \Gamma(\alpha - \frac{1}{2}n - 1) \frac{1}{2} (n+2)(m^2 - k^2) \}.$$

$$(A10)$$

The above equations contain indices μ, ν, λ. These indices are understood to be contracted with arbitrary n-vectors q_1, q_2 etc. In computing the integrals one first integrates over the part of n-space orthogonal to the vectors k, q_1, q_2 etc., using (A1)-(A4). After that, the expressions are meaningful also for non-integer n. Note that formally (A6)-(A10) may be obtained from (A5) by differentiation with respect to k, or by using $p^2 = (p^2 + 2pk + m^2) - 2pk - m^2$.

As is well-known, the Feynman parameter method for non-integer exponents is often very useful:

$$\frac{1}{a^\alpha b^\beta} = \frac{\Gamma(\alpha + \beta)}{\Gamma(\alpha)\Gamma(\beta)} \int_0^1 dx \, \frac{x^{\alpha-1}(1-x)^{\beta-1}}{(ax + b(1-x))^{\alpha+\beta}}, \qquad (A11)$$

valid for $\alpha > 0, \beta > 0$. If one needs this formula for α in the neighbourhood of 0 one may write

$$\frac{1}{a^\alpha b^\beta} = \frac{a}{a^{\alpha+1} b^\beta}$$

and then use (A11). The generalization of eq. (A11) for many factors is also useful:

$$\frac{1}{a_1^{\alpha_1} a_2^{\alpha_2} ... a_m^{\alpha_m}} = \frac{\Gamma(\alpha_1 + \alpha_2 + ...\alpha_m)}{\Gamma(\alpha_1)\Gamma(\alpha_2)...\Gamma(\alpha_m)} \cdot \int_0^1 dx_1 \int_0^{x_1} dx_2 ... \int_0^{x_{m-2}} dx_{m-1}$$

$$\times \frac{x_{m-1}^{\alpha_i - 1}(x_{m-2} - x_{m-1})^{\alpha_2 - 1}...(1 - x_1)^{\alpha_m - 1}}{[a_1 x_{m-1} + a_2 (x_{m-2} - x_{m-1}) + ... + a_m (1 - x_1)]^{\alpha_1 + \alpha_2 + ...\alpha_m}}.$$

$$(A12)$$

References

't Hooft, G. and Veltman, M., Nucl. Phys. **B44**, 189 (1972); on dimensional regularization.

Kinoshita, T. and Lindquist, W.B., Phys. Rev. Lett. **47**, 1573, (1981); on higher-order calculation of the $g - 2$.

Baker, M. and Johnson, K., Phys. Rev. **183**, 1292 (1969); on the β function for QED.

Bjorken, J.D. and Drell, S.D., *Relativistic Quantum Fields* (McGraw-Hill, New York, 1965).

Halzen, F. and Martin, A.D., *Quarks and Leptons: An Introductory Course in Modern Particle Physics* (John Wiley & Sons, New York, 1984).

Bogoliubov, N.N. and Shirkov, D.V., *Introduction to the Theory of Quantized Fields*, 3rd Edition (John Wiley & Sons, New York, 1980).

Itzykson, C., and Zuber, J., *Introduction to Quantum Field Theory* (McGraw-Hill, New York, 1980).

Cheng, T.-P., and Li, L.-F., *Gauge Theory of Elementary Particle Physics* (Clarendon Press, Oxford, 1984).

Exercises: *Chapter 10*

10.1. On the Pauli-Villars regularization scheme, work out the following problems.

(a) Prove Eqs. (12a)-(12g).

(b) Examine the step leading to Eq. (37) starting from Eq. (32).

(c) Derive Eq. (38) from Eq. (37).

10.2. On the dimensional regularization scheme, work out the following problems.

(a) Prove Eqs. (A3)-(A6), (A8), and (A11).

(b) Fill the missing steps leading to Eq. (60).

(c) Obtain the last equation in §.10.2.

10.3. Consider possible renormalization effects due to the diagrams illustrated in Fig. 1. Follow the step leading to Eq. (82a) and try to construct the set of renormalization group equations for the charge e, the mass m, etc. Could you justify Eqs. (45)-(54)?

278

Part C. The Standard Model: Quantum Chromodynamics and Glashow-Salam-Weinberg Electroweak Theory

Chapter 11.

Symmetries, Transformations, and Invariants

In the last part of this book, we wish to describe in a pedagogical manner the so-called "Standard Model", which consists of quantum chromodynamics (QCD), an $SU(3)$ gauge field theory of strong interactions, and the Glashow-Salam-Weinberg (GSW) $SU(2) \times U(1)$ gauge field theory of electroweak interactions. Hermann Weyl in 1919 introduced the concept of "scale invariance", or "gauge invariance" as Weyl himself called it, in his attempt to unify gravity and eletromagnetism. In the quantum mechanical description of a charged particle, one makes the substitution:

$$p_\mu - \frac{e}{c} A_\mu \rightarrow -i\hbar \{ \frac{\partial}{\partial x_\mu} - i\frac{e}{\hbar c} A_\mu \}.$$

In 1927, Fock observed that one could obtain quantum electrodynamics using this operator. London (1927) soon pointed out that Fock's observation amounts to adding an additional phase i to Weyl's notion of "gauge invariance". Thus, the concept of "gauge invariance", as it is called nowadays, is in fact "phase invariance" of some sort. Another breakthrough along this line had to wait until 1954 as Yang and Mills[1] attempted to describe nucleon-nucleon interactions in terms of *local* $SU(2)$ isospin phase symmetry. The idea was revolutionary since, in the proposed non-abelian gauge theory as often referred to as "Yang-Mills theory" nowadays, the way of identifying, or distinguishing between, protons and neutrons, varies from one space-time point to the other. Analogously, the color of a quark, as described by an $SU(3)$ Yang-Mills theory (i.e., QCD), varies with the space-time point under local $SU(3)$ gauge transformations.

In the preceding chapters, we have seen, to a certain extent, the important role played by the concepts of the invariance of physical laws under certain transformations, as may be summarized by Noether's theorem formulated in §.6.3. According to Noether's theorem, these concepts are related to the concepts of symmetry and laws, known as "conservation laws".

It is often useful to differentiate between *exact symmetries* and *approximate symmetries*. Invariance under Lorentz transformations (which form a Lie group called "Lorentz group") is believed to be an "exact symmetry". Lorentz group

---

[1] *Yang, C.N., and Mills, R. L., Phys. Rev.* **96**, *191 (1954).*

contains the rotational $SU(2)$ symmetry group as a subgroup. On the other hand, the isospin symmetry, an $SU(2)$ symmetry which arises from the almost equal mass for the proton and the neutron, is regarded as an "approximate symmetry". In the framework of present-day gauge field theories, gauge principle is taken as an exact symmetry whereas conservation of "lepton number" or "baryon number" in fundamental reactions is assumed to be only "approximate", although such distinction is yet to be confirmed further by experiments.

The revolutionary suggestion to extend the notion of "gauge invariance" from the $U(1)$ group to a non-abelian group such as $SU(2)$ or $SU(3)$ is to be elucidated in §.12.1. To this end, we wish in this chapter to introduce Lie groups $SU(2)$ and $SU(3)$ as groups of symmetries of some kind. In particular, we review in §.11.1., respectively, rotation symmetry as an *exact* $SU(2)$ symmetry and isospin symmetry as an *approximate* $SU(2)$ symmetry. In §.11.2., we consider flavor $SU(3)$ symmetry, a generalization of isospin symmetry to include strangeness, as an example of *approximate* $SU(3)$ symmetry. As a by-product, we introduce the näive quark model for low-lying mesons and baryons. In §.14.3., we mention briefly additional symmetries often encountered in particle and nuclear physics. In the Appendix, we discuss briefly potentials and phase in quantum mechanics.

§.11.1. $SU(2)$ Symmetries in Particle Physics

(1) Invariance under Spatial Rotations as an Exact $SU(2)$ Symmetry

Consider rotations in configuration space of a physical system that is characterized by the Hamiltonian H. The states of the system are labeled by $\mid \psi >$, $\mid \phi >$, etc.

The set of rotations of a system form a group, each rotation being an element of the group. Two successive rotations R_1 followed by R_2 (written as the "product" $R_2 R_1$) are equivalent to a single rotation (that is, to another group element). The set of rotations is closed under "group multiplication." There is an identity element (no rotation), and every rotation has an inverse (rotate back again). The "product" is not necessarily commutative, $R_1 R_2 \neq R_2 R_1$, but the associative law $R_3(R_2 R_1) = (R_3 R_2)R_1$ always holds. The rotation group is a continuous group in that each rotation can be labeled by a set of continuously varying parameters $(\theta_1, \theta_2, \theta_3)$. These can be regarded as the components of a vector θ directed along the axis of rotation with magnitude given by the angles of rotation. See §.4.5. (2) for the specification of a rotation.

Thus, the theory of groups is the mathematics that arise naturally in treatment of symmetries in physics.

The rotation group is a Lie group. Every rotation can be expressed as the product of a succession of infinitiesimal rotations (rotations arbitarily close to the identity). The group is thus completely defined by the "neighborhood of the identity."

To ensure that an experimental result does not depend on the specific laboratory orientation of the system which we are measuring, rotations must form a symmetry group of a system. The rotations must leave the transition probabilities of the system invariant. Suppose that under a rotation R the states of a system transform as

$$| \psi > \to | \psi' > = U \mid \psi > . \tag{1}$$

The probability that a system described by $\mid \psi >$ will be found in state $\mid \phi >$ must be unchanged by R,

$$|< \phi \mid \psi >|^2 = |< \phi' \mid \psi' >|^2 = |< \phi \mid U^\dagger U \mid \psi >|^2, \tag{2}$$

so that U must be a unitary operator. The operators $U(R_1)$, $U(R_2)$ ...., form a group with exactly the same structure as the original group R_1, R_2, ... That is, they form a unitary representation of the rotation group.

It is assumed that the Hamiltonian H is unchanged by a symmetry operator R of the system and the matrix elements are preserved:

$$< \phi' \mid H \mid \psi' > = < \phi \mid U^\dagger H U \mid \psi > = < \phi \mid H \mid \psi >,$$

so that

$$H = U^\dagger H U, \quad \text{or,} \quad [U, H] \equiv U H - H U = 0. \tag{3}$$

The equation of motion,

$$i \frac{d}{dt} \mid \psi(t) > = H \mid \psi(t) >, \tag{4}$$

is not affected by the designated rotation. Accordingly, the expectation value of U is a constant of motion:

$$i \frac{d}{dt} < \psi(t) \mid U \mid \psi(t) > = < \psi(t) \mid UH - HU \mid \psi(t) > = 0, \qquad (5)$$

where it is assumed that U does not have an explicit time dependence.

As a basic property of a Lie group, all the rotation group properties are in fact defined through infinitesimal rotations in the neighborhood of the identity. To see this, we consider a rotation through an infinitesimal angle ε about the 3- (or z) axis. We may write, to first order in ε,

$$U = 1 - i\varepsilon J_3. \qquad (6)$$

The operator J_3 is called the generator of rotations about the 3-axis. We have

$$\begin{aligned} 1 = U^\dagger U &= (1 + i\varepsilon J_3^\dagger)(1 - i\varepsilon J_3) \\ &= 1 + i\varepsilon(J_3^\dagger - J_3) + O(\varepsilon^2). \end{aligned}$$

so that J_3 is hermitian (an important property for a physical observable). Note that the factor i has been introduced in Eq. (6) to ensure the hermiticity of the operator J_3.

To understand the physical significance of the operator J_3, we consider the effect of a rotation on the wave function $\psi(\mathbf{r})$ describing the system. As already familiar in quantum mechanics, one should distinguish between two points of view: Either we can keep the axes fixed and rotate the system (the active viewpoint) or we may rotate the axes and keep the physical system fixed (the passive viewpoint). The viewpoints are equivalent; a rotation of the axes through an angle θ is the same as a rotation of the physical system by $-\theta$. We adopt the active viewpoint and rotate the physical system. The wave function ψ' describing the rotated state at $\mathbf{r}$ is then equal to the original function ψ at the point $R^{-1}\mathbf{r}$, which is transformed into $\mathbf{r}$ under the rotation R:

$$\psi'(\mathbf{r}) = \psi(R^{-1}\mathbf{r}). \qquad (7)$$

This determines the correspondence between ψ' and ψ:

$$\psi' = U\psi. \qquad (8)$$

Thus, we obtain, for an infinitesimal rotation ε about the z axis,

$$
\begin{aligned}
U\psi(x,y,z) = \psi(R^{-1}\mathbf{r}) &= \psi(x + \varepsilon y, y - \varepsilon x, z) \\
&\simeq \psi(x,y,z) + \varepsilon\left(y\frac{\partial\psi}{\partial x} - x\frac{\partial\psi}{\partial y}\right) \\
&= (1 - i\varepsilon(xp_y - yp_x))\psi.
\end{aligned}
\tag{9}
$$

Comparing Eq. (9) with Eq. (6), we identify the *generator*, J_3, of rotations about the 3- (or z) axis with the third-component of the angular momentum operator.

Eq. (5) indicates that the eigenvalues of the observable J_3 are constants of motion. A symmetry of the system has led to a conservation law. The fact that experiments performed with an apparatus of different orientations yield identical physics results (rotational symmetry) has led to the conservation of angular momentum. This is in fact a special case of Noether's theorem introduced earlier in §.6.3. Of course, Noether's theorem may be realized for an arbitrary symmetry transformation that leaves the system invariant.

A rotation through a finite angle θ may be built up from a succession of n infinitesimal rotations

$$
U(\theta) = (U(\varepsilon))^n = \left(1 - i\frac{\theta}{n}J_3\right)^n \to e^{-i\theta J_3}, \quad \text{as } n \to \infty.
\tag{10}
$$

Upon introducing similar hermitian generators of rotations, J_1 and J_2, respectively about the 1- and 2-axes, we obtain the commutator algebra of the generators as follows:

$$
[J_i, J_j] = i\varepsilon_{ijk}J_k,
\tag{11}
$$

where $\varepsilon_{ijk} = +1(-1)$ if ijk is a cyclic (anticyclic) permutation of 1 2 3 and $\varepsilon_{ijk} = 0$ otherwise. The set of relations (13), which is said to form a Lie algebra, completely define the group properties; the ε_{ijk} coefficients are called the structure constants of the group.

For the rotation group, the combination

$$
J^2 = J_1^2 + J_2^2 + J_3^2,
\tag{12}
$$

commutes with all three generators of the group,

$$[J^2, J_i] = 0 \qquad \text{with } i = 1, 2, 3. \tag{13}$$

Such nonlinear function of the generators which commute with all the generators is called an *invariant* or *Casimir operator*. It follows that we can construct simultaneous eigenstates $| jm >$ of J^2 and one of the generators, say J_3. Using only Eq. (11), we may show that

$$\begin{aligned} J^2 \mid jm > &= j(j+1) \mid jm >, \\ J_3 \mid jm > &= m \mid jm >, \end{aligned} \tag{14}$$

with $m = -j, -j+1, ..., j$, and where j can take one of the values $0, \frac{1}{2}, 1, \frac{3}{2}, ...$ Introducing the "raising" and "lowering" operators:

$$J_\pm = J_1 \pm iJ_2, \tag{15}$$

we find

$$J_\pm \mid jm >= (j(j+1) - m(m \pm 1))^{1/2} \mid j, m \pm 1 > . \tag{16}$$

A state $| jm >$ is transformed under a rotation through an angle θ about the 2-axis into a linear combination of the $2j + 1$ states $| jm' >$, with $m' = -j, -j+1, ..., j$:

$$e^{-i\theta J_2} \mid jm >= \sum_{m'} d^j_{m'm}(\theta) \mid jm' >, \tag{17}$$

where the coefficients $d^j_{m'm}$ are frequently called *rotation matrices*. Eq. (17) indicates that, although all the $2j + 1$ states are mixed by rotations, the states with the same j and all possible m transform among themselves under rotations. That is, they form the basis of a $(2j + 1)$-dimensional *irreducible representation* of the rotation group. The set of such states is called a *multiplet*.

The generators in the lowest-dimensional nontrivial representation of the rotation group $(J = \frac{1}{2})$ may be written as follows:

$$J_i = \frac{1}{2}\sigma_i, \qquad \text{with } i = 1, 2, 3, \tag{18}$$

where σ_i are the Pauli matrices

$$\sigma_1 = \begin{pmatrix} 0 & 1 \\ 1 & 0 \end{pmatrix}, \qquad \sigma_2 = \begin{pmatrix} 0 & -i \\ i & 0 \end{pmatrix}, \qquad \sigma_3 = \begin{pmatrix} 1 & 0 \\ 0 & -1 \end{pmatrix}. \tag{19}$$

The basis for this representation is conventionally chosen to be the eigenvectors of σ_3,

$$\begin{pmatrix} 1 \\ 0 \end{pmatrix} \quad \text{and} \quad \begin{pmatrix} 0 \\ 1 \end{pmatrix},$$

which describe a spin-$\frac{1}{2}$ particle of spin projection up ($m = +\frac{1}{2}$ or $\uparrow$) and spin projection down ($m = -\frac{1}{2}$ or $\downarrow$) along the 3-axis, respectively.

The Pauli matrices σ_i are hermitian, and the transformation matrices

$$U(\theta_i) = e^{-i\theta_i \sigma_i/2}, \tag{20}$$

are unitary. The set of all unitary 2×2 matrices is known as the group $U(2)$. However, $U(2)$ is larger than the group of matrices $U(\theta_i)$, since the generators σ_i all are traceless. For any hermitian traceless matrix σ, we can show that

$$det\left(e^{i\sigma}\right) = e^{iTr\,\sigma} = 1. \tag{21}$$

Since the unit determinant is preserved in matrix multiplication, the set of traceless unitary 2×2 matrices form a special subgroup, $SU(2)$, of $U(2)$. $SU(2)$ is used to denote the special unitary group in two dimensions. The set of transformation matrices $U(\theta_i)$ therefore form an $SU(2)$ group. The $SU(2)$ algebra is just the algebra of the generators J_i, as given by Eq. (11). There are $1, 2, 3, 4, ...$ dimensional representations of $SU(2)$ corresponding to $j = 0, \frac{1}{2}, 1, \frac{3}{2}, ...$, respectively. The two-dimensional representation specified by the σ-matrices is called the *fundamental representation* of $SU(2)$, since from which all other representations can be constructed.

(2) Isospin Symmetry

Owing to the fact that the proton and neutron masses are nearly equal ($m_p = 938.27\,MeV$ versus $m_n = 939.57\,MeV$), we may choose to view the two particles as two substates of one and the same particle called the "nucleon," in analogy with the two spin substates of an electron. We note that electrons of spin *up* and *down* are thought of as one, not two, particles. Accordingly, the mathematical structure used to discuss the similarity of the neutron and the proton is just a xerox copy of spin, and is called "*isospin*".

Consider the description of the two-nucleon system. Each nucleon has spin $\frac{1}{2}$ (with spin states $\uparrow$ and $\downarrow$) and, following the rules for the addition of angular momenta, the composite system may have total spin $S = 1$ or $S = 0$. The composition of these spin triplet and spin singlet states is given by

$$\begin{cases} | S = 1, M_s = 1 > = \uparrow\uparrow \\ | S = 1, M_s = 0 > = \sqrt{\frac{1}{2}}(\uparrow\downarrow + \downarrow\uparrow) \\ | S = 1, M_s = -1 > = \downarrow\downarrow \end{cases} \qquad (22a)$$

$$| S = 0, M_s = 0 > = \sqrt{\frac{1}{2}}(\uparrow\downarrow - \downarrow\uparrow). \qquad (22b)$$

Analogously, each nucleon is postulated to have isospin $I = \frac{1}{2}$, with $I_3 = \pm\frac{1}{2}$ for protons and neutrons, respectively. $I = 1$ and $I = 0$ states of the nucleon-nucleon system are then given by

$$\begin{cases} | I = 1, \quad I_3 = 1 > = pp \\ | I = 1, \quad I_3 = 0 > = \sqrt{\frac{1}{2}}(pn + np) \\ | I = 1, \quad I_3 = -1 > = nn \end{cases} \qquad (23a)$$

$$| I = 0, I_3 = 0 > = \sqrt{\frac{1}{2}}(pn - np). \qquad (23b)$$

The three-dimensional isospin space is so defined that its relation to isospin is identical to that of configuration space (x, y, z) to spin. A rotation in isospin space induces transformations on isospin states, called "isospin transformations". There is much evidence to show that the nuclear force is invariant under isospin transformations. In other words, the interaction is independent of the value of I_3 in the $I = 1$ multiplet of Eq. (23a).

Nowadays, we may attribute the success of $SU(2)$ isospin symmetry to the fact that the u and d quarks have almost equal masses while strong interactions (or QCD) are flavor independent.

§.11.2. Flavor $SU(3)$ symmetry: isospin and strangeness

(1) $SU(3)$ Group

The group $SU(3)$ consists of all unitary 3×3 matrices with $det\, U = 1$. The generators for the group may be taken to be any $3^2 - 1 = 8$ linearly independent traceless hermitian 3×3 matrices. Following Gell-Mann,[2] we may take

$$\lambda_i = \begin{pmatrix} \tau_i & 0 \\ 0 & 0 \end{pmatrix}, \quad i = 1, 2, 3; \qquad (24a)$$

$$\lambda_4 = \begin{pmatrix} 0 & 0 & 1 \\ 0 & 0 & 0 \\ 1 & 0 & 0 \end{pmatrix}, \quad \lambda_5 = \begin{pmatrix} 0 & 0 & -i \\ 0 & 0 & 0 \\ i & 0 & 0 \end{pmatrix}; \qquad (24b)$$

$$\lambda_6 = \begin{pmatrix} 0 & 0 & 0 \\ 0 & 0 & 1 \\ 0 & 1 & 0 \end{pmatrix}, \quad \lambda_7 = \begin{pmatrix} 0 & 0 & 0 \\ 0 & 0 & -i \\ 0 & i & 0 \end{pmatrix},$$

$$\lambda_8 = \frac{1}{\sqrt{3}} \begin{pmatrix} 1 & 0 & 0 \\ 0 & 1 & 0 \\ 0 & 0 & -2 \end{pmatrix}. \qquad (24c)$$

with τ_i $(i = 1, 2, 3)$ three 2×2 Pauli matrices. It is possible to have only two of these traceless matrices diagonal which is the maximum number of mutually commuting generators. This number is called the *rank* of the group so that $SU(3)$ is of rank 2. For a Lie group, the number of Casimir operators is equal to the rank of the group. In the case of $SU(3)$, the Casimir operators are taken to be F^2 and G^3. (It is J^2 in the case of $SU(2)$.)

The fundamental representation of $SU(3)$ is a triplet. The three color substates of a quark, red (R or x), yellow (Y or y), and blue (B or z) [as introduced in §.0.1.], constitute the basis for the triplet fundamental representation of $SU(3)$.

$$x = \begin{pmatrix} 1 \\ 0 \\ 0 \end{pmatrix}, \quad y = \begin{pmatrix} 0 \\ 1 \\ 0 \end{pmatrix}, \quad z = \begin{pmatrix} 0 \\ 0 \\ 1 \end{pmatrix}. \qquad (25)$$

An arbitrary state φ ,

---

[2] *Gell-Mann, M., Phys. Rev.* **125**, *1067 (1962)*.

$$\varphi = ax + by + cz = \begin{pmatrix} a \\ b \\ c \end{pmatrix}, \qquad (26a)$$

transforms under $SU(3)$ as follows:

$$\varphi \to \varphi' = U(\theta)\varphi, \qquad (26b)$$

with

$$U(\theta) = exp(-i\sum_{j=1}^{8} \theta_j \lambda_j/2). \qquad (26c)$$

Here $\{\theta_j\}$ are eight real numbers which characterize some "rotation" in the 8-dimensional $SU(3)$ adjoint space. [Using $SU(2)$ as an analogue, the adjoint space may be viewed as the ordinary space and $\{\theta_1, \theta_2, \theta_3\}$ characterize an ordinary rotation so that the transformation $\varphi \to \varphi'$ describes the effect of an ordinary rotation on the electron spinor (which is a linear combination of Pauli spinors).]

The generators $\{\lambda_i\}$ of the $SU(3)$ group satisfy the Lie algebra,

$$[\lambda_i, \lambda_j] = 2if_{ijk}\lambda_k, \qquad (27)$$

with f_{ijk} the completely antisymmetric structure coefficients. Using the Gell-Mann matrices listed in Eqs. (24), we have, for nonvanishing f_{ijk}'s,

$$\begin{aligned} f_{ijk} =&1 \quad \text{for } (ijk) = (123); \\ &\frac{1}{2} \quad \text{for } (ijk) = (147), (246), (257), \text{ or } (345); \\ &-\frac{1}{2} \quad \text{for } (ijk) = (156) \text{ or } (367); \\ &\frac{\sqrt{3}}{2} \quad \text{for } (ijk) = (458) \text{ or } (678). \end{aligned} \qquad (28)$$

Finally, we have

$$\begin{aligned} \{\lambda_i, \lambda_j\} &\equiv \lambda_i\lambda_j + \lambda_j\lambda_i \\ &= 2d_{ijk}\lambda_k + \frac{4}{3}\delta_{ij}, \end{aligned} \qquad (29)$$

with

$$d_{ijk} = \frac{1}{2}, \text{ for } (ijk) = (146), (157), (256), (344), \text{ or } (355);$$

$$-\frac{1}{2}, \text{ for } (ijk) = (247), (366), \text{ or } (377);$$

$$\frac{1}{\sqrt{3}}, \text{ for } (ijk) = (118), (228), \text{ or } (338); \tag{30}$$

$$-\frac{1}{\sqrt{3}}, \text{ for } (ijk) = (888);$$

$$-\frac{1}{(2\sqrt{3})}, \text{ for } (ijk) = (448), (558), (668), \text{ or } (778).$$

Note that the structure constants d_{ijk} are completely symmetric. Note also that those d_{ijk} which are not listed in Eq. (30) vanish identically.

(2) Flavor SU(3) Symmetry

Since the pion was discovered in 1947, the nucleon has lost its unique role in particle physics. Subsequently, many more strongly interacting particles (hadrons) have been identified. Some of the new particles were surprisingly long-lived on the time scale of strong interactions, despite being massive enough to decay into lighter objects without violating the conservation of charge or baryon number. For instance, a Σ^- is readily produced by the strong interaction $\pi^- p \to K^+ \Sigma^-$ and yet decays only weakly via $\Sigma^- \to n\pi^-$. Gell-Mann and, independently, Nishijima, took this as the hint for the existence of a new additive quantum number called "strangeness" S. They assigned to each hadron an integer value of strangeness,

$$S = 0 : \qquad \pi, N, \Delta, ...,$$
$$S = 1 : \qquad K^+, ..., \tag{31}$$
$$S = -1 : \qquad \Lambda, \Sigma, ...,$$

with $-S$ for their antiparticles. Assuming that strong and electromangnetic interactions are forbidden unless S is conserved by the reaction, Gell-Mann and Nishijima were able to account for the strong production and the weak decay of the Σ. Indeed, the total strangeness is conserved in the reaction $\pi^- p \to K^+ \Sigma^-$, accounting for th production of the Σ-particle by the strong interaction. Subsequently, the Σ^- can only decay by the strangeness-violating weak interaction $\Sigma^- \to n\pi^-$, yielding its long lifetime. The Gell-Mann and Nishijima scheme was confirmed by observations of the properties of the large number of strange particles that were subsequently discovered.

Now, owing to the existence of a second additive quantum number S in addition to I_3, it was natural to attempt to enlarge isospin symmetry to a larger group, say, a group of rank 2. This new symmetry group must allow for grouping of the hadrons with similar properties naturally into its multiplet representations. Since no strange particles exist that are close in mass to the nucleon, the appropriate grouping was relatively difficult to identify and the choice of group was not obvious at all. As $SU(3)$ was originally proposed in 1961, it groups the n, p, Σ^+, Σ^0, Σ^-, Λ, Ξ^0, and Ξ^-, with a mass spread of nearly $400\,MeV$, into an $SU(3)$ octet representation. In addition, it groups the lightest mesons into an octet, with the K-meson belonging to the same representation as the much lighter π. The extra symmetry linking strange and non-strange particles is thus much more approximate than is isospin.

Noting that the $SU(3)$ multiplet structure of elementary particles was quite similar to the grouping of chemical elements in Mendeleev's periodic table, successes of the $SU(3)$ classification strongly hinted at the existence of a substructure. The $SU(3)$ incorporating the three lightest flavors of quark can only be an approximate symmetry, which is called "flavor $SU(3)$". We may use such symmetry to enumerate low-lying hadronic states. The concept of flavor $SU(3)$ symmetry was not firmly established until 1964. Note that "flavor $SU(3)$" has nothing to do with the concept of "color $SU(3)$", which is believed to be an exact symmetry of fundamental origin. (See Chapter 12.)

Nowadays, the role of the $SU(3)$ group of isospin and strangeness is mostly historical: It sets the stage for the entry of quarks into particle physics.

(3) Quark Model

According to näive quark models, hadrons are made up of a small variety of more basic entities, called quarks and gluons, bound together in different ways. The fundamental representation of flavor $SU(3)$, the multiplet (u, d, s) from which all other multiplets can be built, is a triplet. Each quark is assigned spin $\frac{1}{2}$ and baryon number $B = \frac{1}{3}$. In näive quark models, baryons are treated as systems of three quarks (qqq) with quarks confined to the region characterized by the baryon size while mesons are quark-antiquark pairs $(q\bar{q})$ with the quark (antiquark) content restricted to a small region. The additive quantum number "hypercharge" Y ($\equiv B + S$) rather than the strangeness S is introduced so that the charge Qe is given by

$$Q = I_3 + \frac{Y}{2} \tag{32}$$

Mesons as Quark-Antiquark Pairs

Mesons are quark-antiquark pairs confined to within the region defined by the meson size. Considering the subspace defined by only two flavors, $q = u$ or d, we obtain the $q\bar{q}$ bound-state wave functions by making the substitutions $p \to u$ and $n \to d$ in Eqs. (2). This results in an isotriplet and an isosinglet of mesons:[3]

$$\begin{cases} \mid I = 1, \quad I_3 = 1 >= -u\bar{d} \\ \mid I = 1, \quad I_3 = 0 >= \sqrt{\tfrac{1}{2}}(u\bar{u} - d\bar{d}) \ , \\ \mid I = 1, \quad I_3 = -1 >= d\bar{u} \end{cases}$$

$$\mid I = 0, I_3 = 0 >= \sqrt{\frac{1}{2}}(u\bar{u} + d\bar{d}).$$

Adding the s quark to the picture, we have five more states:

$$u\bar{s}, d\bar{s};$$
$$s\bar{d}, -s\bar{u};$$
$$s\bar{s}.$$

Thus, there are nine possible $q\bar{q}$ combinations. We may divide the nine states into an $SU(3)$ octet and an $SU(3)$ singlet. Under operations of the $SU(3)$ group, the eight states transform among themselves and do not mix with the singlet state.

Among the nine $q\bar{q}$ states, we note that there are three states which have $I_3 = Y = 0$. These are linear combinations of $u\bar{u}$, $d\bar{d}$, and $s\bar{s}$ states. The singlet combination, η_1, must contain each quark flavor on an equal footing:[4]

$$\eta_1 = \sqrt{\frac{1}{3}}(u\bar{u} + d\bar{d} + s\bar{s}).$$

---

[3] *Note that, if (u, d) is an isospin doublet, then $(-\bar{d}, \bar{u})$ is also an isospin doublet. The extra sign in $-\bar{d}$ ia essential here.*

[4] *Here we use particle names to denote the various $q\bar{q}$ combinations.*

The second state may taken to be a member of the isospin triplet,

$$\pi^0 = \sqrt{\frac{1}{2}}(u\bar{u} - d\bar{d}).$$

By requiring orthogonality to both η_1 and π^0, the isospin singlet state is found to be

$$\eta_8 = \sqrt{\frac{1}{6}}(u\bar{u} + d\bar{d} - 2s\bar{s}).$$

Like any quantum-mechanical bound system, the $q\bar{q}$ pair should have a discrete energy level spectrum corresponding to the different modes of $q\bar{q}$ excitations. Since the quark has spin $\frac{1}{2}$, the total intrinsic spin of the $q\bar{q}$ pair can be either $S = 0$ or 1. The spin J of the composite meson is then the vector sum of this spin S and the relative orbital angular momentum L between q and $\bar{q}$. The parity of the meson, which specifies the meson property under space inversion, is thus given by

$$P = -(-1)^L. \tag{33}$$

Here the minus sign arises beacuse the q and $\bar{q}$ have opposite intrinsic parity and $(-1)^L$ is due to the space inversion property of the $q\bar{q}$ wavefunction $Y_{LM}(\theta, \phi)$. A neutral $q\bar{q}$ system is an eigenstate of the particle-antiparticle conjugation operator C.

$$C = -(-1)^{S+1}(-1)^L = (-1)^{L+S}, \tag{34}$$

where S is the total intrinsic spin of the $q\bar{q}$ pair. Note that, in Eq. (34), the minus sign arises from interchanging fermions, the factor $(-1)^{S+1}$ is determined from the symmetry of the $q\bar{q}$ spin states [cf. Eqs. (22)], and the factor $(-1)^L$ from the symmetry of $Y_{LM}(\theta, \phi)$.

For the $J^P = 0^-$ $q\bar{q}$, we have two isospin doublets

$$K^0(d\bar{s}), \qquad K^+(u\bar{s}) \quad \text{with } Y = 1,$$

$$K^-(s\bar{u}), \qquad \bar{K}^0(s\bar{d}) \quad \text{with } Y = -1.$$

These pseudoscalar mesons form an octet along with Y=0 isotriplet (the π^+, π^0, π^- states) and the $(Y = 0, I = 0)$ state (the η meson). The $SU(3)$ singlet state is identified with η' meson.[5]

---

[5] *In practice, the $\eta_1 - \eta_8$ mixing is of importance since flavor $SU(3)$ symmetry breaking is not negligible.*

To sum up, we have, in group notations,

$$3 \otimes \bar{3} = 8 \oplus 1, \tag{35a}$$

where η_0 is the only member forming the singlet representation ("nonet") while the eight members of the octet representation are listed below:

$$\eta_8, \pi^+, \pi^0, \pi^-, K^+, K^0, \bar{K}^0, K^-. \tag{35b}$$

Looking over the particle data table, there are also nine lowlying vector mesons $(J^P = 1^-)$ which may be grouped into an octet and nonet representations of flavor $SU(3)$ symmetry:

$$\omega, \rho^+, \rho^0, \rho^-, K^{*+}, K^{*0}, \bar{K}^{*0}, K^{*-}, \phi. \tag{36}$$

Here ω and ϕ are states obtained by mixing "ideally" the ω_1 and ω_8 (analogues of η_1 and η_8 described above) such that ϕ is an almost pure $s\bar{s}$ state.

Finally, we remark on the color wave function for mesons. In view of color confinement, it was conjectured earlier that all observed hadrons are colorless, or in the singlet representation of the color $SU(3)$ group. Using the result which we have just learned for flavor $SU(3)$, we find that the color wave function for a meson is unique:

$$\Phi_M^c = \sqrt{\frac{1}{3}}(x\bar{x} + y\bar{y} + z\bar{z}), \tag{37}$$

where x, y, and z are three color substates introduced earlier in Eq. (3), §.0.1.

Baryons as Three-Quark Systems

The flavor $SU(3)$ decomposition of the 27 possible qqq combinations is a little more involved than that for mesons. Nevertheless, the quark content of baryons can be readily obtained using the same techniques. We first combine the first two quarks and arrange the nine qq combinations into two $SU(3)$ multiplets,

$$3 \otimes 3 = 6 \oplus \bar{3}, \tag{38}$$

where the 6 is symmetric and the $\bar{3}$ is antisymmetric under interchange of the two quarks. The members of the 6 representation ("sextet") are listed below:

$$uu, \qquad dd, \qquad ss,$$

$$(ds)_S \equiv \frac{1}{\sqrt{2}}(ds + sd), \qquad (su)_S, \qquad (ud)_S. \tag{39a}$$

The members of the antitriplet representation are given by

$$(ds)_A \equiv \frac{1}{\sqrt{2}}(ds - sd), \qquad (su)_A, \qquad (ud)_A. \tag{39b}$$

Adding the third quark to the triplet, we obtain the decomposition:

$$3 \otimes 3 \otimes 3 = (6 \otimes 3) \oplus (\bar{3} \otimes 3)$$
$$= 10 \oplus 8 \oplus 8 \oplus 1. \tag{40}$$

The decuplet states are totally symmetric under interchange of quarks, as evidenced by the uuu, ddd, and sss members. The symmetric combination of "uud" is given by

$$\Delta_S \equiv \frac{1}{3}[uud + (ud + du)u]. \tag{41}$$

The two other uud states orthogonal to Δ are

$$p_s = \sqrt{\frac{1}{6}}[(ud + du)u - 2uud], \tag{42}$$

$$p_a = \sqrt{\frac{1}{2}}(ud - du)u. \tag{43}$$

The states p_s and p_a have mixed symmetry, where the subscripts are used to indicate the symmetry property under interchange of the first two quarks. The quark structrue of the other states can be obtained in a similar manner. The completely antisymmetric three-quark state is

$$(qqq)_{singlet} = \frac{1}{6}(uds - usd + sud - sdu + dsu - dus). \tag{44}$$

Replacing $u \rightarrow \uparrow$ and $d \rightarrow \downarrow$ to Eqs. (41)-(43), we immediately obtain three possible spin states for a quark triplet:

$$\chi(S) = \sqrt{\frac{1}{3}}(\uparrow\uparrow\downarrow + \uparrow\downarrow\uparrow + \downarrow\uparrow\uparrow)$$

$$\chi(M_s) = \frac{1}{6}(\uparrow\downarrow\uparrow + \downarrow\uparrow\uparrow - 2\uparrow\uparrow\downarrow) \tag{45}$$

$$\chi(M_A) = \sqrt{\frac{1}{2}}(\uparrow\downarrow\uparrow - \downarrow\uparrow\uparrow).$$

To enumerate the baryons expected in the quark model, we must combine the $SU(3)$ flavor wave function such as Eqs. (41)-(43) with the spin wave function, Eq. (45).

There is a problem with this symmetry of the ground state. For example, a Δ^{++} of $J_3 = \frac{3}{2}$ is described by the symmetric wave function,

$$u \uparrow u \uparrow u \uparrow, \tag{46}$$

in violation with Pauli exclusion principle for a system of identical fermions. As already noted earlier, the explanation is that the quarks possess an additional internal degree of freedom, called "color", which can take three possible values, x, y, or z. The quarks form a fundamental triplet of an $SU(3)$ color symmetry which is believed to be exact. All observed hadrons are postulated to be colorless. That is, they belong to singlet representations of the $SU(3)$ color group. The color wavefunction for a baryon is given by[6]

$$(qqq)_{color\ singlet} = \sqrt{\frac{1}{6}}(xyz - xzy + yzx - yxz + zxy - zyx). \tag{47}$$

The required antisymmetric property of the total wavefunction under particle exchange is then ensured: It is overall symmetric in space, spin, and flavor structure and completely antisymmetric in color space. Note that this precludes the choice of Eq. (44) as the flavor wave function.

As an example of an explicit quark model wavefunction, we have, for a spin-up proton,

$$
\begin{aligned}
\mid p \uparrow > =& \sqrt{\frac{1}{18}}[uud(\uparrow\downarrow\uparrow + \downarrow\uparrow\uparrow -2\uparrow\uparrow\downarrow) + udu(\uparrow\uparrow\downarrow + \downarrow\uparrow\uparrow -2\uparrow\downarrow\uparrow) \\
&+ duu(\uparrow\downarrow\uparrow + \uparrow\uparrow\downarrow -2\downarrow\uparrow\uparrow)] \\
=& \sqrt{\frac{1}{18}}[u \uparrow u \downarrow d \uparrow +u \uparrow u \downarrow u \uparrow d \uparrow -2u \uparrow u \uparrow d \downarrow + permutations].
\end{aligned} \tag{48}
$$

The three quarks have zero orbital angular momentum in ground-state baryons, so that the parity of the state, $(-1)^{\ell+\ell'}$, is positive for lowlying baryons.

We may now look over the particle data table on lowlying baryons. There are eight $J^P = \frac{1}{2}^+$ baryons which may be grouped together to form an octet representation:

---

[6] *See Eq. (44).*

$$p, \quad n,$$

$$\Sigma^+, \quad \Sigma^0, \quad \Sigma^-, \quad \Lambda, \tag{49}$$

$$\Xi^+, \quad \Xi^-.$$

In addition, there are ten lowlying $J^P = \frac{3}{2}^+$ baryons which can be grouped into a decuplet representation:

$$\Delta^{++}, \quad \Delta^+, \quad \Delta^0, \quad \Delta^-,$$

$$\Sigma^{*+}, \quad \Sigma^{*0}, \quad \Sigma^{*-},$$

$$\Xi^{*+}, \quad \Xi^{*-}, \tag{50}$$

$$\Omega^-.$$

As a matter fact, the existence of the last member in the decuplet, namely Ω^-, was a prediction made by Gell-Mann as he tried to identify members for a possible decuplet representation under flavor $SU(3)$. The prediction was soon confirmed by experiments, strengthening considerably the idea of flavor $SU(3)$ symmetry.

There are quark models which provide detailed descriptions of the hadron wave functions, which include wave functions in configuration space, spin space, flavor space, and color space and sometime even with specific background fields related to confinement. A general survey of the various quark models may be found in the literature.

§.11.3. Additional Symmetries in Particle Physics

In addition to $SU(2)$ and $SU(3)$ symmetries which were elucidated in §.11.1. and §.11.2., similar discussions may easily be extended to many other symmetries.

(1) For a system possessing translational symmetry in time, i.e., invariant under translation in time, there is the law of conservation of energy.[7]

---

[7] *We wish to insert a remark here to emphasize the real significance of the use of tensors and vectors. The Lorentz transformation (or the general coordinate transformation in the general relativity theory) defines the tensors (and vectors)*

(2) For a dynamical system possessing translational symmetry in a certain direction in space, dynamical laws are invariant under linear translations of coordinate systems in that direction, leading to the law of conservation of momentum in that direction.

(3) For systems in uniform relative motion, the physical laws are invariant under the Lorentz transformation. This gives (or, is the expression of) the relativity principles.

Note that the ten dynamical operators, including (1) the generator for time translation ("*Hamiltonian*"), i.e., for item (1) above, (2) the three generators for spatial translations ("*momenta*"), i.e., for item (2) above, (3) the three *Lorentz boost* operators, as in item (3) above, and (4) the *angular momenta* (for generating rotaions in configuration space, as discussed earlier in §.11.1.), form a Lie group, called "proper Lorentz group". The Lie algebra may be called "Poincaré algebra". (Dirac 1949)

(4) Parity symmetry, charge conjugation symmetry, and time-reversal symmetry. See §.4.4. for an introduction to these important concepts.

It has been intuitively taken for granted that physical laws have a symmetry with respect to a mirroring operation (i.e., a reflection of the coordinate system on the $X - Y$ plane, say.). This intuitive feeling has been justified by the discovery of the empirical law (Laporte, 1924) for dipole radiation from the analysis of the spectrum lines of iron atom. That this left-right symmetry is not universal was proposed by T. D. Lee and C. N. Yang in 1956 and soon experimentlally verified by C. S. Wu and others for the interactions involved in nuclear and neutron β decays, $\pi \rightarrow \mu$ decay, and $\mu \rightarrow e$ decay.[8]

It was found that the combined symmetry of charge conjugation and space inversion (CP) was broken in the case of the 2π decay of the K_L^0 meson, by

---

by their transformation properties, so that a tensor equation is automatically invariant in form under the transformation, and any physical law expressible in the form of a tensor equation is automatically guaranteed to obey the relativity theory.

---

[8] See, e.g., Commins, E.D. and Bucksbaum, P.H., Weak Interactions of Leptons and Quarks (Cambridge University Press, 1983).

Christenson et al. in 1964.[9] If the CTP theorem is assumed (as is the case for a local field theory), then the symmetry of time (with respect to reversal) was also broken.

(5) The invariance under gauge transformation as introduced in §.1.1. for a Klein-Gordon particle and in §.4.2. for a Dirac particle.

It can be shown that from the requirement of the invariance of the electromagnetic laws under the gauge transformation follows the conservation of electric charge, which is of course not a new result but a familiar empirical fact.

As already mentioned earlier in this chapter, the concept of "gauge invariance" was first introduced by Weyl (1919), revised slightly by Fock (1927) and London (1927),[10] and finally generalized to the case of a non-abelian gauge group by Yang and Mills (1954). Generalization of "$U(1)$ gauge symmetry to a non-abelian group will be described, in the case of QCD or an $SU(3)$ Yang-Mills theory, at the beginning of the next chapter (§.12.1.). Here we note that the notion of "gauge symmetry" under a non-abelian gauge group has become the backbone of the modern theory of particle physics, namely, the "Standard Model" consisting of an $SU(3)$ Yang-Mills theory of strong interactions and the Glashow-Salam-Weinberg $SU(2) \times U(1)$ Yang-Mills theory of eletroweak interactions.

(6) There is a permutation symmetry for a system of many particles (fermions or bosons). As elucidated in detail in Ch. 7, permutation symmetry determines the statistics of the many-body system.

To sum up, it is fair to say that, in the present form of elementary particle physics, the concepts of symmetry has assumed a very fundamental role.

---

[9] *Christenson, J.H., Cronin, J.W., Fitch, V.L., and Turlay, R., Phys. Rev. Lett. **13**, 138 (1964).*

[10] *In ordinary quantum mechanics, the equation of an electron in an electromagnetic field is invariant under the gauge transformation if it is accompanied by a suitable transformation in the phase of the wave function. For the meaning of the phase in quantum mechanics and its possible relation to gauge symmetry, consult the Appendix for further discussions.*

Appendix

Potentials and Phase in Quantum Mechanics

To shed light on the meaning of the gauge symmetry, we append here a few notes[11] which we believe may be relevant for the problem. The notes are organized as follows: In §.A.1. and §.A.2., we contrast the role of field and potential, respectively, in classical physics and in quantum mechanics. In §.A.3., we introduce Dirac's theory of the magnetic monopole. In §.A.4., aspects related to the Aharonov-Bohm experiment are briefly discussed. For geometrization of gauge field theories, the lecture note given by C. N. Yang (1975) at the Sixth Hawaii Topical Conference in Particle Physics provides an in-depth treatment of the problem while, for an introduction to the interesting question of Berry's phase (Berry 1984), the articles by Berry himself (1988) and by R. Jackiw (1988) may be helpful.

§.A.1. Field and potential in classical physics[12]

In classical dynamics, the force $\mathbf{F}\ (F_x, F_y, F_z)$ appears explicitly in the equations of motion of a particle,

$$m\frac{d^2\mathbf{r}}{dt^2} = \mathbf{F}(\mathbf{r}).$$

It is true that a potential can be introduced

$$\mathbf{F} = -\nabla V,$$

but all potentials differing from V by a constant give the same $\mathbf{F}$ so that V is not uniquely determined by $\mathbf{F}$.

---

[11] *This appendix is prepared using the colloquium lectures given by one of us (T.-Y. Wu) some ten years ago both at the State University of New York at Buffalo and at National Tsing-hua University on Dirac's theory of the magnetic monopole and in that connection also on the Aharonov-Bohm theory.*

[12] *In view of the nature of the materials to be discussed, the natural unit system ($\hbar = c = 1$) is not adopted in this specific appendix.*

In classical electromagnetic theory the basic laws (Maxwell's equations, together with Lorentz's force expression) are all explicitly expressed in terms of the fields **E**, **D**, **B**, **H**. One may introduce the potentials $(\mathbf{A}, \phi)$ such that

$$\mathbf{E} = -\nabla\phi - \frac{1}{c}\frac{\partial \mathbf{A}}{\partial t}, \qquad \mathbf{H} = \nabla \times A, \qquad (A1)$$

but under the gauge transformation to new potentials $(\mathbf{A}', \phi')$

$$\mathbf{A}' = \mathbf{A} + \nabla\chi, \qquad \phi' = \phi - \frac{1}{c}\frac{\partial \chi}{\partial t}, \qquad \partial_\mu \partial_\mu \chi = 0, \qquad (A2)$$

the fields **E**, **H** and the Maxwell equations (together with the Lorentz gauge condition) remain invariant.

Thus, in classical dynamics and electromagnetic theory, the potentials do not have definite meaning; they are "mathematical tools"; only the fields **F**, **E**, **H**, etc. are measurable and have physical meaning.

§.A.2. Field and potential in quantum mechanics

(i) In quantum mechanics, the situation regarding the roles of the fields and the potentials are different!

In the non-relativistic Schrödinger equation,

$$(\frac{1}{2m}\mathbf{p}^2 + V - E)\psi = 0,$$

$$-\frac{\hbar}{i}\frac{\partial}{\partial t}\psi = (\frac{1}{2m}\mathbf{p}^2 + V)\psi,$$

or the relativistic Dirac equation,

$$\gamma_\mu(\frac{\hbar}{i}\frac{\partial}{\partial x^\mu} - \frac{e}{c}A_\mu)\psi = imc\psi, \qquad (A3)$$

$$x^\mu = (x, y, z, ict), \qquad A_\mu = (A_x\, A_y\, A_z, i\phi),$$

the potentials V, A, ϕ appear and not the fields.

The next question is "how do the potentials affect the wave function of a system?" Or, do the potentials have observable effects?

(ii) Introduction of potential A_μ; non-integrable phase

Let us start with a free electron whose equation is

$$\left(\gamma_\mu \frac{\hbar}{i} \frac{\partial}{\partial x_\mu} - imc\right)\psi = 0. \tag{$A4$}$$

Let

$$\psi = \psi' \, exp\left(-\frac{ie}{\hbar c} \int A_\mu dx_\mu\right). \tag{$A5'$}$$

Then ψ' satisfies the equation

$$exp\left(-\frac{ie}{\hbar c} \int A_\mu dx_\mu\right)\left[\gamma_\mu\left(\frac{\hbar}{i} \frac{\partial}{\partial x_\mu} - \frac{e}{c} A_\mu\right) - imc\right]\psi' = 0,$$

i.e., ψ' satisfies the equation for an electron in the potential $(A, i\phi)$, and ψ' is related to ψ by

$$\psi' = \psi \, exp\left(\frac{ie}{\hbar c} \int A_\mu dx_\mu\right). \tag{$A5$}$$

Now the phase

$$\beta = \frac{e}{\hbar c} \int A_\mu dx_\mu, \tag{$A6$}$$

is not an integrable function of the point x_μ, but depends on the path of integration C from some initial point to x_μ.[13]

Note that the β as specified by Eq. (A6) is a property of the potential alone, and is independent of ψ.

The difference in phase between two points $x_\mu^{(1)}$ and $x_\mu^{(2)}$ depends on the path from $x_\mu^{(1)}$ to $x_\mu^{(2)}$!

The phase $\beta = \frac{e}{\hbar c} \int A_\mu dx_\mu$ is then said to be non-integrable.

Thus we obtain

---

[13] *By Stokes' theorem,*

$$\oint A_\mu dx_\mu = \int\int_\sigma curl\,\mathbf{A} \cdot d\mathbf{S}. \tag{$A7$}$$

Unless Curl $\mathbf{A} = 0$, *i.e., unlesss*

$$A_\mu = \frac{\partial\chi}{\partial x_\mu}, \qquad \chi = a \; scalar \; function, \tag{$A8$}$$

the integral $\oint A_\mu dx_\mu$ *does not vanish.*

Theorem I: The effect of a potential is to introduce a non-integrable phase in the wave function ψ of the potential-free particle.

The relationship between gauge transformation of the electromagnetic potentials and the phase of wave functions is first shown by H. Weyl in 1929. The possibility of using non-integrable phases to characterize geometrically a gauge theory was explored by C. N. Yang in 1974.

(iii) Gauge invariance and phase

The Dirac equation for an electron in the potential $(\mathbf{A}, i\phi)$ is (A3),

$$[\gamma_\mu(\frac{\hbar}{i}\frac{\partial}{\partial x_\mu} - \frac{e}{c}A_\mu) - imc]\psi = 0. \qquad (A3)$$

On making the gauge transformation (A2)

$$\mathbf{A}' = \mathbf{A} + \nabla\chi, \qquad \phi' = \phi - \frac{1}{c}\frac{\partial\chi}{\partial t}, \qquad \partial_\mu\partial_\mu\chi = 0, \qquad (A2)$$

and at the same time the transformation

$$\psi' = \psi\, exp(-\frac{ie}{\hbar c}\chi), \qquad (A9)$$

Eq. (A3) becomes

$$[\gamma_\mu(\frac{\hbar}{i}\frac{\partial}{\partial x_\mu} - \frac{e}{c}A'_\mu) - imc]\psi' = 0. \qquad (A10)$$

Thus the effect of a gauge transformation (A2) is to introduce a phase factor $exp(-\frac{ie}{\hbar c}\chi)$ to ψ. The phase

$$\beta = \frac{e}{\hbar c}\chi(x_\mu), \qquad (A11)$$

is a function of the 4-point x_μ, and the change in phase in going from $x_\mu^{(1)}$ to $x_\mu^{(2)}$ is

$$\beta^{(1)} - \beta^{(2)} = \frac{e}{\hbar c}(\chi(x_\mu^{(2)}) - \chi(x_\mu^{(1)})), \qquad (A12)$$

which is independent of the path joining $x_\mu^{(1)}$ and $x_\mu^{(2)}$. The change in phase in going around a closed curve is therefore zero.

Such a phase as (A11) is said to be integrable.

Theorem II: In quantum mechanics, a gauge transformation of the potential $A_\mu(\mathbf{A}, i\phi)$ causes a phase change which is integrable.

§.A.3. Dirac's theory of the magnetic monopole

(i) Phase of wave function in a magnetic field

As seen from Eqs. (A5') and (A6), it is possible to generate the desired equation for a Dirac particle moving in an electromagnetic field from the free Dirac equation by substituting the wave function ψ by ψ':

$$\psi' = \psi\, e^{i\beta}, \qquad \beta = \frac{e}{\hbar c} \int A_\mu dx_\mu. \qquad (A13)$$

The phase β is determined by the potential A_μ and is the same for all wave functions. Thus, even though β is not integrable, it still has the following property, namely, the change $\Delta\beta$ when a wave function is carried around any given closed curve (in space-time) is the same for all ψ. This property follows from the following considerations. Let φ_m, ψ_n be any two arbitrary wave functions. The integral

$$\int \varphi_m^* \psi_n d\tau = \text{a complex number}, \qquad (A14)$$

is a measure of the "identity" (sameness) of the two states φ_m and ψ_n. If both φ_m and ψ_n are carried around an arbitrary closed curve Γ, the change $\Delta(-\beta_m + \beta_n)$ in $\varphi_m^* \psi_n$ must be zero (or $2\pi n$) in order that the integral in Eq. (A14) have a definite value. From this it follows that

$$\Delta\beta_m = \Delta\beta_n, \qquad \text{or} \quad (\pm 2\pi n). \qquad (A15)$$

The indeterminacy $2\pi n$ can be eliminated on the following consideration: Let the closed curve Γ shrink continuously. On account of the continuous character of the $\psi's$, the phases $\Delta\beta_m$ and $\Delta\beta_n$ must decrease continuously to zero, respectively. Hence $\Delta\beta_m = \Delta\beta_n$; there cannot be any difference $2\pi n$.

But the argument based on continuity would fail in the following case: The wave function ψ has a *nodal line* on which $\psi = 0$. When $\psi = 0$, the phase itself has no meaning (undefined). In this case, taking ψ around a nodal line (in 3-dimensional space), we have no continuity argument for requiring $\Delta\beta$ to be "small"; we may only say that $\Delta\beta$ approximates to $2\pi n$, with n some integer.[14]

---

[14]*n must be an integer because of the single-valued nature of the quantum mechanical wave function.*

The numerical value of n is a property of the nodal line; the sign of n depends on the sense of revolution around the nodal line.

At this point, Dirac introduced an important new idea, namely:

Let ψ_0 be the wave function that has a nodal line. Along a small closed curve Γ around its nodal line, the change in β_0 of ψ_0 is $\Delta\beta_0$. The difference between $\Delta\beta_0$ and the nearest value $2\pi n$ (n being a positive or negative integer) is the same as the change $\Delta\beta$ in phase β, around the same closed curve Γ, of any wave function ψ not having that nodal line,

$$\Delta\beta_0 - 2\pi n = \Delta\beta. \qquad (A16)$$

This is a new idea, an assumption, and cannot be proved, nor derived. It is a conjecture of the overall consistency among all the solutions to the same equation. As we shall see immediately, it leads to the quantization rule for the strength of the Dirac magnetic monopole.

From (A13) or (A6), we then obtain, for Γ in 3-space,

$$\Delta\beta_0 - 2\pi n = \frac{e}{\hbar c} \int_\sigma \mathbf{H} \cdot d\sigma \qquad (A17)$$

where σ is any surface whose boundary is Γ.

Let Γ shrink so that σ becomes a "bottle"; in the limit of Γ shrinking to a point, σ becomes a closed surface, a bubble, and $\Delta\beta_0$ approaches zero, so that

$$-2\pi n = \frac{e}{\hbar c} \oint_\sigma \mathbf{H} \cdot d\sigma \qquad (A18)$$

The righthand side $\frac{e}{\hbar c} \oint \mathbf{H} \cdot d\sigma$ has nothing to do with the nodal line, by the original assumption (A16). Hence the value of n is independent of the wave function, and can only be a property of the field.

For example, take $n = \pm$ integer and $n \neq 0$. The righthand side ordinarily vanishes,

$$\oint \mathbf{H} \cdot d\sigma = \iiint \nabla \cdot \mathbf{H} \, d\tau,$$
$$= 0, \quad \text{since} \quad \nabla \cdot \mathbf{H} = 0.$$

There is then a contradiction!

This contradiation can be avoided by assuming the presence of a magnetic pole of magnetic charge g (in the same unit as electric charge e), so that

$$\oint \mathbf{H} \cdot d\sigma = 4\pi g, \tag{A19}$$

and Eq. (A18) becomes

$$2\pi n = \frac{e}{\hbar c} 4\pi g,$$

or

$$g = \frac{\hbar c}{2e} n, \qquad n = 0, \pm 1, \pm 2, \ldots \tag{A20}$$

If $n = 1$, then the smallest magnetic monopole g is

$$\frac{g}{e} = \frac{\hbar c}{2e^2} = \frac{1}{2\alpha} = \frac{137}{2}. \tag{A21}$$

The importance of the relation (A20) is this: If the quantization of electric charge (the universal unit e) is accepted, then (A20) is the law of quantization of the magnetic pole strength.

Then, for $n \neq 0$, the nodal line have an endpoint inside the closed surface σ; for, otherwise, a nodal line cutting the surface σ an even number of times will have their total contributions to $2\pi n$ equal to zero, n being positive or negative depending on the direction of the nodal line and the sense of the curve Γ around it. The endpoint of the nodal line is the seat of the magnetic monopole as we shall see in subsection (ii) below.

Following Dirac's conjecture, we see that, unless there exists a magnetic monopole (or something else that is similar), it is not possible to have a wave function that has a nodal line.

If ψ_0 is taken along a closed curve Γ around a number of nodal lines, we shall make small closed curves around each of them. The assumption (A16) is now

$$\triangle \beta_0 - 2\pi \sum n = \triangle \beta, \tag{A16a}$$

where $\triangle \beta$ is $\frac{e}{\hbar c} \iint \mathbf{H} \cdot d\sigma$ calculated for any surface whose boundary is Γ, and $\sum n$ in the sum of n over the contributions of the various nodal lines. (A18) is now replaced by

$$-2\pi \sum n = \frac{e}{\hbar c} \oint_{\sigma} \mathbf{H} \cdot d\sigma. \tag{A18a}$$

Each nodal line having an endpoint inside the closed surface σ contributes a $2\pi n$.

These endpoints being the same for all wave functions are, according to the conclusion following (A18), not the properties of the wave function; they are the singular properties of the potential **A**.

(ii) Singularities of potential and monopole

We have seen, following (A21), that the endpoints in σ are properties of the potential and not of the wave functions.

It remains to show that the vector potential **A** due to a monopole precisely has a line of singularities. [And along the line of singularities in **A**, there is then a nodal line in the wave function of an electron!]. Assume that a monopole of strength g in placed at the origin. The magnetic field (in Gauss system) is

$$\mathbf{H} = \frac{g}{r^3}\mathbf{r} \qquad (A22)$$

and the vector potential A such that $\mathbf{H} = curl\ \mathbf{A}$ is

$$A_x = -\frac{gy}{r(r+z)}, \qquad A_y = \frac{gx}{r(r+z)}, \qquad A_z = 0, \qquad (A23)$$

or, in polar coordinates

$$A_r = 0, \qquad A_\vartheta = 0, \qquad A_\varphi = \frac{g}{r}\tan\frac{\vartheta}{2} \qquad (A24)$$

Let C be a latitude circle with latitude θ, and S be the surface of the cap above **C**. Then

$$\begin{aligned}
\oint_C \mathbf{A}\cdot ds = \oint_S \mathbf{H}\cdot d\sigma &= g\oint_S \frac{1}{r^3}\mathbf{r}\cdot d\sigma, \\
&= 2\pi g(1 - cos\theta), \qquad (A25) \\
&= \begin{cases} 0, & \text{for } \theta = 0, \\ 4\pi g, & \text{for } \theta = \pi. \end{cases}
\end{aligned}$$

When C shrinks to an infinitesimally small circle as $\theta \to \pi$, the last equality of (A25) becomes $4\pi g$ and the l.f.s. would approach zero if **A** is finite. In that case, we have a contradiction $0 = 4\pi g$.

But from (A24), it is seen that at $\vartheta = \pi$, A_φ is infinite!

Since the radius r in Eqs. (A23)-(A25) is entirely arbitrary, it is seen that that the half-line from $(x, y, z) = (0, 0, 0)$ to $(0, 0, -\infty)$ is a line of singular $\mathbf{A}$. The magnetic monopole is at the *end* $(0, 0, 0)$ of the line of singularities.

§.A.4. The Aharonov-Bohm experiment

(i) The principle of the Aharonov-Bohm experiment

The principle of the experiment is as follows:

Since $A_\mu \, dx_\mu$ is Lorentz invariant, its value can be calculated in a Lorentz frame in which $dx_4 = 0$, and

$$A_\mu \, dx_\mu \to \mathbf{A} \cdot d\mathbf{x},$$

$$\int A_\mu \, dx_\mu \to \sum \mathbf{A} \cdot d\mathbf{x} \to \int \mathbf{A} \cdot d\mathbf{x},$$

$$\oint A_\mu \, dx_\mu \to \oint \mathbf{A} \cdot d\mathbf{x} = \oint \mathbf{H} \cdot d\sigma.$$

Imagine a long thin solenoid (perpendicular to the plane of this paper). A coherent beam of electrons is split at A into two beams, along path C_1 and C_2 (both on the plane of this paper, encircling the long thin solenoid), which are reunited at B. It is assumed that there is a nonzero magnetic field lines $\mathbf{H}$ cutting through the surface (which is the surface of this paper) expanded by the two paths C_1 and C_2 but $\mathbf{H} = 0$ on the paths and elsewhere outside the solenoid.

The difference in phase of the two beams at B is[15]

$$\int_{C_2} \mathbf{A} \cdot ds - \int_{C_1} \mathbf{A} \cdot ds = \oint \mathbf{A} \cdot ds = \oint \mathbf{H} \cdot d\sigma \neq 0.$$

The difference in phase can be detected by the interference effect at B. This experiment was carried out by Chambers, and a positive result was found!

---

[15] *The changes in phase due to the bending of the electron beams can be calculated and allowed for in the change in the interference fringe due to the enclosed $\mathbf{H}$,*

$$\frac{e}{\hbar c} \oint \mathbf{H} \cdot d\sigma,$$

which is a "quantum effect".

(ii) The Essence of the Aharonov-Bohm experiment

The important point, as emphasized by Aharonov and Bohm, is the following: The electrons, along the paths C_1 and C_2, move in a region of space where the field $\mathbf{H}$ is zero, so that the electrons nowhere have experienced any "force" (Lorentz force). On the other hand, the potential A cannot be zero everywhere; otherwise $\oint \mathbf{A} \cdot ds$ would have been zero. Hence the experiment shows that, in quantum mechanics, the potential plays a role (i.e., in determining some observable effect such as interference) not expected on classical theories! As stressed by C.N. Yang (1974), the knowledge on the field strengths $F_{\mu\nu}$ under-describes electromagnetism while the knowledge on the potentials A_μ over-describes the situation, leading to the attempt (C. N. Yang 1974) of using the non-integrable phases to characterize a gauge theory.

(iii) Remarks

(1) It is to be noted that, while the potential A affects the phase $\frac{e}{\hbar c} \int A_\mu \, dx_\mu$, this effect is gauge invariant so that all potentials connected by the gauge transformation (A2) are indistinguishable in the experiment of Chambers. This must be remembered when one is thinking of regarding the potentials as quantum mechanical observables.

(2) In the experimental arrangement of Chambers, the phase difference $\oint A_\mu dx_\mu = \oint_C \mathbf{A} \cdot dx$ depends on the curve C. If C does not enclose the solenoid, then $\oint_C \mathbf{A} \cdot ds = \oint \mathbf{H} \cdot d\sigma = 0$. Thus the space around a solenoid (or a magnet) is not simply-connected; it can be made simply-connected by a cut. That this property of space in connection with a magnetic field towards the phase of a wave function is not the first instance of its kind.

In 1931, Dirac, on consideration of the phase of the wave function, arrived at the new idea of a magnetic monopole. We have given a report on the theory in Subsection **A.3.** above.

In 1984, M. V. Berry made an interesting new observation by noting that there may be nontrivial phases for a quantum mechanical system that is influenced adiabatically by the background system with (periodic) slow motion. The surprising discovery of Berry's phases in the well-known framework of the adiabatic approximation adds a new dimension to the question of phases and angles in quantum mechanics. The articles by Berry himself (1988) and by R. Jackiw (1988) provide an excellent introduction to the subject.

References

Halzen, F., and Martin, A.D., *Quarks and Leptons: An Introductory Course in Modern Particle Physics* (John Wiley & Sons, New York, 1984), Ch. 2.; on a very nice introduction to symmetries and quarks.

Wu, T.Y., *Physics: Its Development and Philosophy* (The Physical Society of the Republic of China, Taipei, 1989), Ch. V.; on symmetry, transformation, and invariance.

Dirac, P.A.M., *Rev. Mod. Phys.* **21**, 392 (1949); on the Lie algebra for the proper Lorentz group.

Weyl, H., *Ann. Physik* **59**, 101 (1919).

Weyl, H., *Zeits. f. Physik* **56**, 330 (1929). This is the classic paper on gauge invariance (Eichinvariance).

Fock, V., *Zeits. f. Physik* **39**, 226 (1927).

London, F., *Zeits. f. Physik* **42**, 375 (1927).

Dirac, P.A.M., *Proc. Royal Soc. London* **A133**, 60 (1931). This is the first paper on the magnetic monopole. It is followed by a second paper in **Phys. Rev. 74**, 817 (1948).

Ehrenberg, W., and Siday, R.E., *Proc. Phys. Soc. London* **B62**, 8 (1949). This is the first paper pointing out the essential idea of the experiment of Chamber (of Aharonov & Bohm's). This paper has received no attention.

Aharonov, Y., & Bohm, D., *Phys. Rev.* **115**, 485 (1959).

Chambers, R.G., *Phys. Rev. Letters* **5**, 3 (1960).

Yang, C.N., *Gauge Fields*, in the Proceedings of the Sixth Hawaii Topical Conference in Particle Physics (1975), Eds. P.N. Dobson, Jr., S. Pakvasa, V.Z. Peterson, and S.F. Tuan, p. 488; *Phys. Rev. Lett.* **33**, 445 (1974).

Wu, T.-Y., *Theoretical Physics* (in Chinese), Vol. 3, p. 164; Vol. 4, p. 75 and p. 68; on singularites of the potential **A** of a magnetic monopole.

M. V. Berry, Proc. R. Soc. **A392**, 45 (1984); Sci. Am. **259**, No. 6, 46 (1988); R. Jackiw, Comments At. Mol. Phys. **21**, 71 (1988); Intl. Jnl. Mod. Phys. **A3**, 285 (1988).

Exercises: *Chapter 11*

1. In analogy with our discussion on the $SU(2)$ algebra, define

$$F^a = \frac{\lambda^a}{2}, \qquad a = 1, ..., 8.$$

(a) Write down the Lie algebra for F^a and prove that $F^2 \equiv F^a F^a$ is a Casimir operator.

(b) Generalize the definition for the raising and lowering operators (such as Eqs. (17)-(19) for $SU(2)$) to the case of $SU(3)$.

(c) Discuss how to label the members of a given irreducible representation of $SU(3)$.

2. *Use the results obtained from Exercise 1.*

(a) Write down the wave function for $\Delta^{++}(J_3 = +3/2)$ and generate the wave functions of the highest weight $(J_3 = +3/2)$ for all other decuplet baryons by applying successively suitable raising and lowering operators. Fix the relative phases of the wave functions as well as the normalization factors of your raising and lowering operators.

(b) Repeat the exercise for the case of low-lying octet baryons, now starting from the proton wave function given by Eq. (48) in the text.

3. Assume that the color wave function for a three-quark system is given uniquely by Eq. (47) in the text. Also assume that quarks are spin-$\frac{1}{2}$ particles satisfying Pauli exclusion principle.

(a) Working in the space of the first three flavors of quarks, i.e., u, d, and s with all quarks sitting at the lowest S-orbitals, show that there is only one set of octet wave functions *and* that there is no nonet baryon.

(b) By substituting the s quark by the c quark in the flavor $SU(3)$ octet and decuplet baryons, describe the spectrum for charmed baryons.

4. Classify the states of mesons which contain at least one charm quark or antiquark. Your scheme should contain (1) those well-known $c\bar{c}$ bound states which have been observed experimentally and (2) those low-lying charmed mesons which contain one charm quark or antiquark. Try to give some description on the wave functions of those mesons which you have considered.

Chapter 12.
Quantum Chromodynamics

Quantum chromodynamics, or QCD, is an $SU(3)$ gauge field theory of strong interactions among quarks and gluons. The word "chromodynamics" means "color-dynamics", meaning the dynamics among colors carried by quarks and gluons. As explained earlier in Ch. 11, the concept of "gauge invariance" was first introduced by Weyl (1919), revised slightly by Fock (1927) and London (1927), and finally generalized to the case of a non-abelian gauge group by Yang and Mills (1954). Thus, a non-abelian gauge theory is also referred to as "Yang-Mills theory". QCD is just a Yang-Mills theory with $SU(3)$ as the gauge group. In contrast, quantum electrodynamics (QED), which was treated in detail previously in Ch. 9 and Ch. 10, is an abelian gauge theory with $U(1)$ as the gauge group.

§.12.1. QCD is an $SU(3)$ Gauge Theory.

Consider the free lagrangian for a quark of a given flavor,

$$\mathcal{L}_0(x) = -\bar{\psi}(x)\gamma_\mu \partial_\mu \psi(x) - m\bar{\psi}(x)\psi(x), \tag{1}$$

which is almost identical to the lagrangian for a free eletron, except that, in the present case, the quark field $\psi(x)$ has three components in color space. Following the important suggestion made by Yang and Mills (1954), we attempt to introduce a local phase transformation in color space:

$$\psi(x) \to \psi'(x) = U(\theta)\psi(x), \tag{2a}$$

with

$$U(\theta) = exp(-i\sum_{a=1}^{8} \theta_a(x)\lambda_a/2), \tag{2b}$$

where λ_a are eight hermitian, traceless Gell-Mann 3×3 matrices introduced earlier in §.11.2. Note that $\{\theta_a(x)\}$ are eight real functions to ensure that $U(\theta)$ is unitary.

Accordingly, the "gauge principle" may be generalized as follows:

$$\partial_\mu \to D_\mu \equiv \partial_\mu - i\frac{g}{2}\lambda_a G_\mu^a(x), \tag{3}$$

so that the lagrangian (1) becomes

$$\mathcal{L}(x) = -\bar{\psi}(x)\gamma_\mu D_\mu \psi(x) - m\bar{\psi}(x)\psi(x). \tag{4}$$

Here $G_\mu^a(x)$ are eight gauge fields which are called "gluons", meaning the bosons that glue quarks together. To ensure that the lagrangian (4) is invariant under the gauge transformation of Eqs. (2a) and (2b), we require

$$\frac{\lambda_a}{2}G_\mu^a \to \frac{\lambda_a}{2}G_\mu^{\prime a} = U(\theta)\{\frac{\lambda_a}{2}G_\mu^a - \frac{i}{g}U^{-1}(\theta)\partial_\mu U(\theta)\}U^{-1}(\theta). \tag{5}$$

The task to write down the lagrangian for the gauge field, which must be invariant under the gauge transformation Eq. (5), was first accomplished by Yang and Mills in 1954. First, we note that the QED analogue, $\partial_\mu G_\nu^a - \partial_\nu G_\mu^a$, is *not* gauge invariant. This is where the non-abelian nature of $SU(3)$ starts to manifest itself in a certain way. Nevertheless, a suitable gauge-invariant expression may be inferred from the gauge principle as follows:

$$\begin{aligned}
&\partial_\mu(\frac{1}{2}\lambda_a G_\nu^a) - \partial_\nu(\frac{1}{2}\lambda_a G_\mu^a) \\
&\to D_\mu(\frac{1}{2}\lambda_a G_\nu^a) - D_\nu(\frac{1}{2}\lambda_a G_\mu^a) = \frac{\lambda_a}{2}G_{\mu\nu}^a,
\end{aligned} \tag{6a}$$

with

$$G_{\mu\nu}^a \equiv \partial_\mu G_\nu^a - \partial_\nu G_\mu^a + gf_{abc}G_\mu^b G_\nu^c. \tag{6b}$$

It is then straightforward to show that the expression specified by

$$T_r\frac{\lambda_a}{2}G_{\mu\nu}^a\frac{\lambda_b}{2}G_{\mu\nu}^b = \frac{1}{2}G_{\mu\nu}^a G_{\mu\nu}^a, \tag{7}$$

is invariant under the gauge transformation Eq. (5). Fixing up the normalization factor as in QED, we obtain the QCD lagrangian,

$$\begin{aligned}
\mathcal{L}_{QCD}(x) = &-\frac{1}{4}G_{\mu\nu}^a(x)G_{\mu\nu}^a(x) \\
&- \bar{\psi}(x)\gamma_\mu D_\mu\psi(x) - m\bar{\psi}(x)\psi(x).
\end{aligned} \tag{8}$$

Strictly speaking, this lagrangian defines the so-called "classical chromodynamics". The quantized version of it is called "guantum chromodynamics" or QCD.

Note that the first term in Eq. (8) contains a piece that describes interactions among three gluons and another piece that requires four gluons to interact. This brings in the nonlinearity which makes QCD very difficult to comprehend (or to solve numerically) as a theory of strong interactions.

Quantization of a non-abelian gauge field theory such as Eq. (8) is by no means trivial since application of Dirac's correspondence principle, which maps the Poisson algebra of a classical theory into the commutator algebra of the corresponding quantized theory, is obscured by the difficulty in choosing an appropriate set of independent generalized coordinates. In the case of QED (with $U(1)$ as the gauge group), it is possible to identify the transversal components of the gauge field, i.e., $\mathbf{A}_\perp(x)$, as the generalized coordinates and then to work out a quantized theory. This was explained previously in Chapter 8. It is in fact the procedure followed by many authors.[1] The price which we have to pay in this program is that Lorentz covariance [and gauge invariance] is no longer manifest. In addition, it is not clear whether such program can be of any practical use in the non-abelian case.

In Chapter 8, we outlined another quantization procedure in which one sets out to treat the four components of $A_\mu(x)$ as though they are linearly independent. To this end, one must introduce a term proportional to $(\partial_\mu A_\mu(x))^2$ to ensure that the generalized momentum field conjugate to $A_0(x)$ does not vanish identically. To quantize the electromagnetic field, one imposes an appropriate set of elementary commutation relations. The Lorentz condition is then imposed as an operator condition that ensures exact cancellation between "unphysical" timelike and longitudinal photons. Gauge invariance ensures that such cancellation occurs for all cases, i.e., for all gauges and for an arbitrary number of unphysical photons. Manifest Lorentz covariance is maintained by this quantization procedure, but application of the procedure to the non-abelian case such as QCD is nontrivial.

Thus, it is important to consider the covariant quantization through the path-integral formulation, which can easily be generalized from the $U(1)$ case (such as QED) to the non-abelian case (such as QCD). In the path-integral formulation, one sets out to deal with the partition function as a path integral

---

[1] See, e.g., Lee, T.D., "Particle physics and introduction to field theory" (Harwood, New York, 1981).

and attempts to deduce all physical quantities from it. In view of the extensive mathematical nature of the formalism, we wish to relegate introduction and discussion of this important subject to Appendix A. Here we wish to say a few words about the underlying ideas.

Conceptually, in the path-integral formulation we treat $\partial_\mu G_\mu^a(x) = C^a(x)$ as additional independent degrees of freedom but attempt to maintain the constraint that physical quantities are invariant under an arbitrary gauge transformation, Eqs. (2) and (5). Gauge invariance requires that two sets of $\{G_\mu^a(x)\}$ which are linked by a gauge transformation are completely equivalent, so that only one set in each equivalent class of $\{G_\mu^a(x)\}$ is needed for the determination of the partition function. Qualitatively speaking, fixing the gauge is equivalent to that the gauge is not fixed but the extra degrees of freedom are removed by suitable cancelling fields (which is "Fadde'ev-Popov ghost fields"). This procedure has been referred to as "gauge fixing". It is clear that gauge fixing and treating $C^a(x)$ as independent degrees of freedom are two separate issues in quantization of a gauge field theory. For technical details, consult Appendix A at the end of this chapter.

As we shall see in Ch. 14, experimental tests of QCD rely primarily on the asymptotical free nature of QCD, which says that, for sufficiently large Q^2, the strong coupling $\alpha_s(Q^2)$ ($\equiv g^2/(4\pi)$) becomes sufficiently small to warrent perturbative treatments. The idea of a "running" coupling constant was already introduced in §.10.3. for QED and we shall consider it again in §.12.2. to deduce the asymptotically free property of QCD. For the purpose of carrying out perturbative QCD calculations, it suffices to summarize final results on Feynman rules ('t Hooft, 1971). Qualitatively speaking, these Feynman rules may be obtained from an effective lagrangian,

$$\mathcal{L}_{eff}(x) = \mathcal{L}(x) - \partial_\mu \varphi^a(x) D_\mu \varphi^a(x), \tag{9}$$

where $\mathcal{L}(x)$ is specified by Eq. (8) and $D_\mu \varphi^a(x)$ is given by

$$D_\mu \varphi^a(x) \equiv \partial_\mu \varphi^a(x) + g f_{abc} G_\mu^b(x) \varphi^c(x). \tag{10}$$

$\varphi^a(x)$ is the scalar field of color a that serves as a "cancelling field" and so satisfies Fermi statistics. The fields $\varphi^a(x)$ are the "Fadde'ev-Popov ghost fields" mentioned above. The gluon propagator takes the form,

$$\frac{1}{i}\frac{\delta_{ab}}{k^2 - i\varepsilon}(\delta_{\mu\nu} - \frac{k_\mu k_\nu}{k^2}), \tag{11}$$

which vanishes identically upon contraction by k_μ or k_ν. This corresponds to $\partial_\mu G_\mu^a(x) = 0$, which defines the Landau gauge. It is always possible to modify the effective action by adding a suitable functional $S'(\partial_\mu G_\mu^a)$ such that the new effective lagrangian no longer contains terms involving $\partial_\mu G_\mu^a(x)$ explicitly. The gluon propagator is then given by

$$\frac{1}{i}\frac{\delta_{ab}}{k^2 - i\varepsilon}\delta_{\mu\nu}, \tag{12}$$

which defines the 't Hooft-Feynman gauge.

Feynman rules for carrying out perturbative QCD calculations are summarized immediately below. The reader who is interested in the derivation of these rules may consult Appendix A.

(a) *Fermion propagator*

$$\Leftrightarrow \quad \frac{1}{i}\frac{m - i\gamma \cdot p}{m^2 + p^2 - i\varepsilon} \tag{13a}$$

(b) *Gluon propagator*

$$\Leftrightarrow \quad \frac{1}{i}\frac{\delta_{ab}}{p^2 - i\varepsilon}[\delta_{\mu\nu} - (1-\xi)\frac{p_\mu p_\nu}{p^2}] \tag{13b}$$

$$\xi = 0, \qquad \text{Landau gauge;}$$
$$1, \qquad \text{'t Hooft-Feynman gauge.}$$

(c) *Ghost propagator*

$$\Leftrightarrow \quad \frac{1}{i}\frac{\delta_{ab}}{p^2 - i\varepsilon} \tag{13c}$$

(d) *Fermion vertex*

$$\Leftrightarrow \quad -g\gamma_\mu \frac{\lambda_{\alpha\beta}^a}{2} \tag{13d}$$

(e) *Triple gluon vertex*

$\leftrightarrow$ $-gf_{abc}\{\delta_{\mu\nu}(k-q)_\sigma + \delta_{\nu\sigma}(q-r)_\mu + \delta_{\sigma\mu}(r-k)_\nu\}$ $\qquad$ (13e)

(f) *Quartic gluon vertex*

$\leftrightarrow$ $\begin{aligned} & -ig^2\{f_{abd}f_{cde}(\delta_{\mu\sigma}\delta_{\nu\rho} - \delta_{\mu\rho}\delta_{\nu\sigma}) \\ & + f_{ace}f_{bde}(\delta_{\mu\nu}\delta_{\sigma\rho} - \delta_{\mu\rho}\delta_{\nu\sigma}) \\ & + f_{ade}f_{cde}(\delta_{\mu\sigma}\delta_{\nu\rho} - \delta_{\mu\nu}\delta_{\sigma\rho})\} \end{aligned}$ $\qquad$ (13f)

(g) *Ghost vertex*

$\leftrightarrow$ $+gf_{abc}\gamma_\mu$ $\qquad$ (13g)

(h) *External lines*

$$u(\mathbf{p},s) \quad \text{for incoming spinor,}$$
$$\bar{u}(\mathbf{p},s) \quad \text{for outgoing spinor;}$$
$$\bar{v}(\mathbf{p},s) \quad \text{for incoming antispinor,}$$
$$v(\mathbf{p},s) \quad \text{for outgoing antispinor;}$$
$$\frac{\varepsilon_\mu^{a,\lambda}(k)}{\sqrt{2k_0}} \quad \text{for a gluon in the initial state,}$$
$$\frac{\varepsilon^{a,\lambda *}(k)}{\sqrt{2k_0}} \quad \text{for a gluon in the final state.}$$

$\qquad$ (13h)

(i) Every closed fermion loop or closed ghost loop induces a minus sign. Counting factors are required for a set of identical particles to ensure that the state is suitably normalized. There is a factor of $-i$ in going from the S-matrix element S_{fi} to the T-matrix element T_{fi}. Cross sections are computed from T_{fi} as follows:

$$d\sigma = \frac{1}{|v_1 - v_2|} \{\prod_f \frac{d^3 k_f}{(2\pi)^3}\}(2\pi)^4 \delta^4(\sum_f k_f - \sum_i k_i) \cdot \overline{\sum} |T_{fi}|^2, \qquad (13i)$$

where $\bar{\Sigma}$ indicates suitable averaging (over the initial discrete indices) and summation (over the final discrete degress of freedom).

§.12.2. QCD is Asymptotically Free.

Consider the coupling of a gluon to the quark. Such coupling may be described pictorially as in Figure 1.

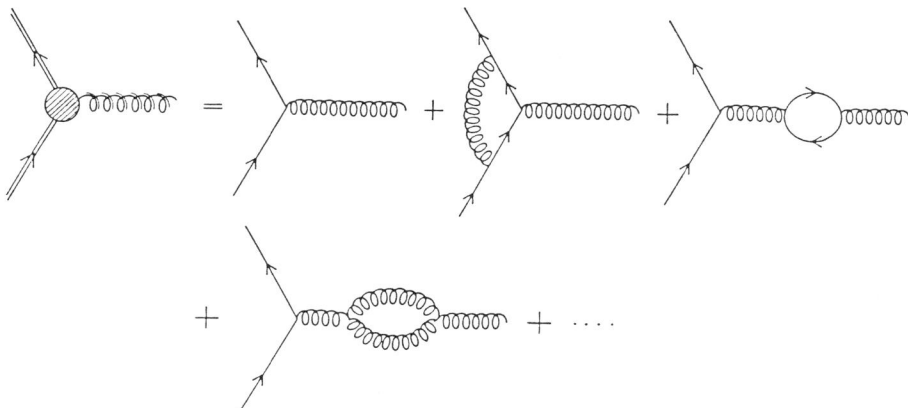

Fig. 1.

We may interpret this figure in two different manners:

(A) We may consider the coupling on the left-hand side (LHS) of Fig. 1 as the coupling when the probing scale is characterized by $\bar{\mu}^2$ with $\bar{\mu}$ of the dimension of a mass. In other words, the coupling is $\bar{g}$ when the distance which we are probing is $1/\bar{\mu}$. Now, suppose that we manage to probe the same coupling at a distance $1/\mu$ smaller than $1/\bar{\mu}$. The coupling at μ^2 is g. It is then clear that the coupling $\bar{g}$ at $\bar{\mu}^2$, as looked upon at a smaller distance, is a sum of diagrams as illustrated by the RHS of Fig. 1. We have

$$\bar{g} \equiv \bar{g}(\frac{\bar{\mu}^2}{\mu^2}, \frac{m^2}{\mu^2}, g), \tag{14}$$

provided that the basic parameters of the theory at the scale μ^2 are m^2 and g. In this view, a hierarchy of theories at different values of μ^2 is envisioned and the notion of the "bare coupling" (or "bare mass") is rejected (as μ^2 is arbitrary and the smallest probing distance does not exist).

(B) Suppose that we write down a theory with g and m^2 as the basic parameters. We set out to compute higher order diagrams as described by the RHS of Fig. 1. However, most of them are divergent unless we introduce a regularization of some sort. In other words, we need an additional parameter μ^2 related to the method of regularization. Therefore, the theory makes sense only when the set of the basic parameters include μ^2 in addition to $\{g, m^2\}$. Eq. (14) comes out when we relate theories at two different values of μ^2.

The variation of the coupling constant with respect to the renormalization point μ^2 is charaterized by the beta function:

$$\beta(\frac{m^2}{\mu^2}, g) \equiv \bar{\mu}^2 \frac{\partial}{\partial \bar{\mu}^2} \bar{g}(\frac{\bar{\mu}^2}{\mu^2}, \frac{m^2}{\mu^2}, g) \mid_{\bar{\mu}^2 = \mu^2}, \tag{15}$$

which was considered earlier in §.10.3 for the case of QED.

The β-function in the case of QED is given by (Baker and Johnson, 1969)

$$\beta(\alpha) = \frac{\alpha^2}{3\pi} + \frac{\alpha^3}{4\pi^2} + O(\alpha^4), \tag{16}$$

which was used previously in §.10.3.

The β-function for an $SU(n)_{color}$ gauge theory with N_f flavors of fermions may be obtained from Fig. 1 and is given by (Belavin and Migdal 1974; Caswell 1974; Jones 1974)

$$\begin{aligned}
\beta(\alpha_s) &= A\frac{\alpha_s^2}{4\pi} + B\frac{\alpha_s^3}{16\pi^2} + O(\alpha_s^4), \\
A &= -\frac{11}{3}n + \frac{2}{3}N_f, \\
B &= -\frac{34}{3}n^2 + N_f(\frac{13n}{3} - \frac{1}{n}).
\end{aligned} \tag{17}$$

To make sense out of Eqs. (15) and (17), we consider again the renormalization group (RG) equation for $\bar{g}(\bar{\mu}^2/\mu^2, g)$, neglecting the mass parameter m^2 for the sake of simplicity. (See §.10.3. for a similar derivation.) We define

$$t \equiv \bar{\mu}^2/\mu^2. \tag{18}$$

Choosing $t'' = t't$, we have

$$g'' \equiv \bar{g}(t', g') \quad with \quad g' \equiv \bar{g}(t, g),$$

or,

$$\bar{g}(t't, g) = \bar{g}(t', \bar{g}(t, g)). \tag{19}$$

Differentiating Eq. (19) with respect to t' and then setting $t' = 1$, we find

$$t\frac{\partial}{\partial t}\bar{g}(t, g) = \beta(\bar{g}), \tag{20}$$

which was just Eq. (85), Ch. 10. Note that Eq. (20) can readily be solved for $\bar{g}$. We obtain

$$\int_{t=1}^{t} \frac{d\bar{g}}{\beta(\bar{g})} = \ln t. \tag{21}$$

Considering the QED case to lowest order, we find

$$3\pi\Big(\frac{1}{\alpha(\mu^2)} - \frac{1}{\alpha(\bar{\mu}^2)}\Big) = \ln\frac{\bar{\mu}^2}{\mu^2},$$

or,

$$\alpha(\bar{\mu}^2) = \frac{\alpha(\mu^2)}{1 - \frac{1}{3\pi}\alpha(\mu^2)\ln\frac{\bar{\mu}^2}{\mu^2}}. \tag{22}$$

Analogously, we obtain, for QCD with N_f flavors,

$$\alpha_s(\bar{\mu}^2) = \frac{\alpha_s(\mu^2)}{1 + \frac{1}{4\pi}(11 - \frac{2}{3}N_f)\alpha_s(\mu^2)\ln\frac{\bar{\mu}^2}{\mu^2}}. \tag{23}$$

It is customary to identify the renormalization point μ^2 with the momentum squared Q^2 which defines the scale of the physics that we are probing. In other words, we may rewrite Eqs. (22) and (23) in a more familiar form:

$$\alpha(Q^2) = \frac{\alpha(Q_0^2)}{1 - \frac{1}{3\pi}\alpha(Q_0^2)\ln\frac{Q^2}{Q_0^2}}, \quad for \quad QED; \tag{24a}$$

$$\alpha_s(Q^2) = \frac{\alpha_s(Q_0^2)}{1 + \frac{1}{4\pi}(11 - \frac{2}{3}N_f)\alpha_s(Q_0^2)\ln\frac{Q^2}{Q_0^2}}, \quad for \quad QCD. \qquad (24b)$$

It is known that, at $Q_0^2 \sim (1GeV)^2$, $\alpha(Q_0^2) = \frac{1}{137}$ and $\alpha_s(Q_0^2) \sim O(1)$. Accordingly, we find

$$\alpha(Q^2) \to \infty \quad as \quad \frac{1}{3\pi}\alpha(Q_0^2)\ln\frac{Q^2}{Q_0^2} \to 1, \qquad (25)$$

which occurs at an enormous value of Q^2/Q_0^2. The divergent behavior of $\alpha(Q^2)$ as indicated by Eq. (25) is related to the so-called "Landau ghost problem" which becomes irrelevant as QED gets unified with other interactions well before the Landau ghost appears. On the other hand, we also have

$$\alpha_s(Q^2) \to 0 \quad as \quad Q^2 \to \infty, \qquad (26)$$

which means that the interaction becomes weak as Q^2 is large ($\gg Q_0^2$). This phenomenon has been referred to as "asymptotic freedom". It can be shown that inclusion of the next-to-leading order terms, the B term in Eq. (17), does not modify significantly the asymptotically free nature of QCD.

It is important to know that QCD is asymptotically free. If we use very high energy probes to probe a hadron [which is a system consisting of quarks, antiquarks, and gluons], the large Q^2 implies the smallness of $\alpha_s(Q^2)$ so that interactions among constituents can be treated perturbatively, thereby explaining why parton models are so successful at high energies. Perturbative treatment (Mueller 1981; Reya 1981) of QCD effects at high energies allows for specific tests of QCD in experiments such as production of jets, decay of quarkonium states, scaling violation in deep inelastic scattering, and Drell-Yan processes.

§.12.3. Color Confinement

As an important empirical fact, hadrons observed so far are *colorless*, i.e., in a color singlet configuration. This suggests that color fields exist only inside the region specified by the extent of a given hadron. A quark or gluon carries color so that confinement of color explains why quarks and gluons are not observed in isolation. Color confinement also explains why observed hadrons, such as baryons and mesons, appear in the way they are. For example, four-quark clusters (Q^4) must have a net color and so cannot exist in isolation. But, it does not preclude the possibility of having another color-singlet configuration for a multiquark cluster such as $(Q^2\bar{Q}^2)$ or Q^6, or having systems consisting only of gluons ("glueballs"), or having systems consisting of both quarks and gluons such as $(Q^3 g)$ ("hermaphrodites"), etc., so long as these hadrons are in an overall color-singlet configurations. Search for an unambiguous manifestation of "color" is of critical importance, but it is often obscured by uncertainties related to strong interactions.

A natural way to realize color confinement is to conjecture that the color fields, $\{\mathbf{E}^a, \mathbf{B}^a\}$, exist only in the quark-gluon phase [as characterized by the QCD vacuum or the trivial QCD ground state] that is different from the hadron phase [the physical vacuum or the true QCD ground state]. This has been referred to as the "two-phase picture". In the strict two-phase picture, formation of flux tubes is required if a color constituent is pulled away from a hadron.

By discretizing $SU(2)$ or $SU(3)$ color gauge theory on a lattice [lattice gauge theory], it is possible to investigate whether the two-phase picture emerges. Indeed, extensive early lattice gauge calculations[2], in the so-called "quenched approximation" (where the role of fermions is suppressed) suggested the presence of a deconfinement phase transition at a certain transition temperature T_c, indicating the occurrence of a phase transition from the hadron phase (at $T < T_c$) to the quark-gluon phase (at $T > T_c$). Meanwhile, the restoration phase transition for chiral symmetry is found to take place simultaneously. Although such a simple and nice picture may be only a gross simplification to what is really going on in the real world, it is truly amazing that a non-abelian gauge

---

[2] *For a recent review, see, e.g., Kogut, J.B., in "Nuclear and Particle Physics on the Light Cone", Eds. Johnson, M.B., and Kisslinger, L.S. (World Scientific, Singapore, 1989), p. 239 and references therein.*

field theory has so many nontrivial properties.

Of course, the definitive results obtained from lattice gauge calculations do not carry over directly to the case of continuum QCD but this is by far the most convincing evidence that QCD is compatible with the two-phase picture, thereby suggesting confinement of color. The compatibility between QCD and color confinement must be considered as another important reason why QCD is being regarded as the candidate theory of strong interactions among quarks.

The question of how to discretize a gauge field theory on a lattice is a technical issue which is not critical for understanding the subject of our discussion. For gaining some basic insights on the problem, however, an interested reader may consult Appendix B at the end of the present chapter.

To sum up, QCD describes strong interactions among quarks and gluons in a satisfactory manner: At high energies $(Q^2 \gg m_p^2)$, it explains why parton models are so successful while, at low energies $(Q^2 \leq m_p^2)$, it is compatible with the empirical fact that color is confined. Just like the Glashow-Salam-Weinberg $SU(2) \times U(1)$ electroweak theory, QCD is phrased on the basis of both gauge principle and renormalizability. If these theories sustain future experimental tests, then one must seek for the answer to the basic question as to why the present specific conceptual framework, in which fundamental laws of the nature can be phrased in terms of a gauge field theory, is so powerful and successful.

Appendix A
Method of Path Integrals

Feynman's method of path integrals is a method of formulating a quantum theory without explicit reference to operators and states in Hilbert space. It is a method that can be used for both quantum mechanics and quantum field theory. The method becomes a powerful tool for treating quantized gauge field theories.

§.A.1. Quantum Mechanics and Path Integrals

As an illustrative example, we consider the scattering of a spinless particle from a potential $V(x)$.[3]

$$i\hbar\frac{\partial\psi}{\partial t} = -\frac{1}{2m}\hbar^2\nabla^2\psi + V(x)\psi \equiv H\psi. \qquad (A1)$$

The solution to Eq. (A1) is specified by the wavefunction $\psi(x,t)$ which is a function of x and t. Alternatively, we may relate the solutions at two space-time points, $\psi(x',t')$ and $\psi(x,t)$ $(t' > t)$, through an integral equation:

$$\psi(x',t') = \int K(x',t';x,t)\psi(x,t)dx. \qquad (A2)$$

Here the function $K(x',t';x,t)$ is called the "Green's function" or "kernel". It is possible to express K as a path integral. To do so, we first divide the time interval $t' - t$ into $n + 1$ small segments of magnitude ϵ. To each value of t, we associate a value of the coordinate x:

$$
\begin{aligned}
t = t_0 &: x_0 \\
t_1 &: x_1 \\
t_2 &: x_2 \\
&\vdots \\
t_n &: x_n \\
t' = t_{n+1} &: x_{n+1} = x'
\end{aligned}
\qquad (A3)
$$

Note that the values x_0 and x' fix the endpoints of a "path" while the intermediate values $x_1, ..., x_n$ determine the rest of the path.

---

[3] In §. A.1., we do not set $\hbar = 1$.

For infinitesimal ϵ, the time evolution is determined from Eq. (A1). For example, we have

$$
\begin{aligned}
\psi(x_1, t_1) &\equiv\; < x_1 \mid t_1 > \\
&=\; < x_1 \mid U(t_1, t_0) \mid t_0 > \\
&=\; < x_1 \mid exp\{-\frac{i}{\hbar} H(x, p)\epsilon\} \mid t_0 > \\
&= \int \frac{dx_0 dp}{2\pi\hbar} exp\{-\frac{i}{\hbar} V(\frac{x_1 + x_0}{2})\epsilon\}\cdot \\
&\quad \cdot < x_1 \mid exp\{-\frac{i}{\hbar}\frac{p^2}{2m}\epsilon\} \mid p >< p \mid x_0 >< x_0 \mid t_0 > \\
&= \int \frac{dx_0 dp}{2\pi\hbar} exp\{-\frac{i}{\hbar} V(\frac{x_1 + x_0}{2})\epsilon\}\cdot \\
&\quad \cdot exp\{\frac{i}{\hbar} p x_1\} exp\{-\frac{i}{\hbar}\frac{p^2}{2m}\epsilon\} exp\{-\frac{i}{\hbar} p x_0\}\psi(x_0, t_0).
\end{aligned}
\tag{A4}
$$

Using the integration formula,

$$
\int_{-\infty}^{\infty} dz \exp(iaz^2) = (i\pi/a)^{1/2},
\tag{A5}
$$

we obtain, by comparing the result with Eq. (A2),

$$
\begin{aligned}
K(x_1, t_1; x, t_0) \\
= (\frac{m}{2\pi\hbar i\epsilon})^{\frac{1}{2}} exp\{\frac{i}{\hbar} L(\frac{x_1 + x_0}{2}, \frac{x_1 - x_0}{\epsilon})\epsilon\}.
\end{aligned}
\tag{A6}
$$

where $L = \frac{m}{2}\dot{x}^2 - V(x)$ is the classical Lagrangian. The operation leading to Eq. (A6) can easily be carried out step by step for the rest of the path. The net result is the desired Green's function:

$$
K(x', t'; x, t_0) = \lim_{n \to \infty, \epsilon \to 0} (\frac{m}{2\pi\hbar i\epsilon})^{(n+1)/2} \int \prod_{i=1}^{n} dx_i \exp(\frac{i}{\hbar} \int L dt).
\tag{A7}
$$

Eq. (A7) is often abbreviated as follows:

$$
K = \int [dx] \exp(iS/\hbar),
\tag{A8}
$$

with $S = \int_{t_0}^{t'} L dt$ the action for the path. Note that Eq. (A7) is not adequate if the potential is velocity-dependent. In fact, in the case that there are a number of quantized systems corresponding to the same classical mechanical system, it is not clear which quantized system the method of path integrals actually selects out.

As a simple example for obtaining the kernel, we consider the case of a free particle, where $L = L_0 = \frac{m}{2}\dot{x}^2$. The kernel is

$$
\begin{aligned}
K_0^{(n)}(x',t';x,t) =& (\frac{m}{2\pi\hbar i\epsilon})^{(n+1)/2} \int_{-\infty}^{\infty} dx_1 ... \int_{-\infty}^{\infty} dx_n \\
& \times \exp[\frac{i}{\hbar}\frac{m}{2\epsilon}(x_1 - x_0)^2] \exp[\frac{i}{\hbar}\frac{m}{2\epsilon}(x_2 - x_1)^2]... \qquad (A9) \\
& \times \exp[\frac{i}{\hbar}\frac{m}{2\epsilon}(x_{n+1} - x_n)^2].
\end{aligned}
$$

Introducing the variables,

$$
\begin{aligned}
& x_1 - x_0 = x_1', \quad x_2 - x_1 = x_2', \quad ..., \quad x_n - x_{n-1} = x_n', \\
& x_{n+1} - x_n = (x_{n+1} - x_0) - z = (x' - x) - z, \qquad (A10) \\
& z = x_1' + x"_2 + ... + x_n',
\end{aligned}
$$

we rewrite Eq. (9) as follows:

$$
\begin{aligned}
K_0^{(n)}(x',t';x,t) =& (\frac{m}{2\pi\hbar i\epsilon})^{(n+1)/2} \int dz \int \frac{dp}{2\pi\hbar} \int_{-\infty}^{\infty} dx_1' ... \int_{-\infty}^{\infty} dx_n' \\
& \times \exp[\frac{i}{\hbar}p \cdot (z - x_1' - x_2' - ... - x_n')] \qquad (A9') \\
& \times \exp[\frac{i}{\hbar}\frac{m}{2\epsilon}\{x_1'^2 + x_2'^2 + ... + x_n'^2 + (z - (x' - x))^2\}].
\end{aligned}
$$

Use Eq. (A5) to carry out the integrations in Eq. (A9') and take into account the fact that in the Schrödinger theory waves propagate forward in time (giving rise to a step function $\theta(t' - t)$ as defined by Eq. (A11), Ch. 8):

$$
K_0(x',t';x,t) = [\frac{m}{2\pi\hbar i(t'-t)}]^{1/2} \exp[\frac{im}{2\hbar(t'-t)}(x'-x)^2] \theta(t'-t), \qquad (A11)
$$

which is a well-known formula in elementary quantum mechanics.

§.A.2. The Path Integral in Field Theory

It is clear that Eq. (A7) may be generalized readily to field theory. Consider a field ϕ which may have many components. We now divide space-time into four-dimensional cells of volume ϵ^4 and define a "path" by specifying the value of ϕ in each cell. The path integral now consists of summing over all possible values of the field in each cell. Setting $\hbar = 1$, we find a kernel:

$$K = \int [d\phi] \exp(i \int \mathcal{L}(x) d^4 x) \qquad (A12)$$

where $\mathcal{L}$ is the Lagrangian density. In practice, it is convenient to define a generating functional $Z[J]$ by the path integral:

$$Z[J] = \int [d\phi] \exp\{i \int [\mathcal{L}(x) + J(x)\phi(x)] d^4 x\}, \qquad (A13)$$

which is the same as Eq. (A12) except that it contains an additional "source" term in the action. $Z[J]$ plays a central role in the development of quantum field theory according to the method of path integrals since, with the aid of this generating functional, all the standard results of field theory can be obtained. In particular, the n-point Green's function is given by

$$\frac{\delta^n Z[J]}{\delta J(x_1) \delta J(x_2) ... \delta J(x_n)}|_{J=0} = i^n < 0 \mid T[\phi(x_1)\phi(x_2)...\phi(x_n)] \mid 0 >$$
$$\equiv i^n G(x_1, ..., x_n) \qquad (A14)$$

where $< 0 \mid T[\phi(x_1)...\phi(x_n)] \mid 0 >$ is the vacuum expectation value of the time-ordered product of n fields $\phi(x_i)$.

We may illustrate the procedure by considering the case of a real scalar field ϕ, with Lagrangian $\mathcal{L} = \mathcal{L}_0 + \mathcal{L}_I$, where $\mathcal{L}_0 = -\frac{1}{2}(\partial_\mu \phi \partial_\mu \phi + \mu^2 \phi^2)$ and $\mathcal{L}_I = \mathcal{L}_I(\phi)$ is an interaction term which we ignore here but may take into account if necessary. Using integration by parts, we find

$$Z_0[J] = \int [d\phi] \exp\{i \int d^4 x [-\frac{1}{2}\phi(-\partial_\mu \partial_\mu + \mu^2 - i\epsilon)\phi + J\phi]\}, \qquad (A15)$$

where the factor $-i\epsilon$ has been introduced on the ground that the resultant Green function is equivalent to what is expected from the Euclidean-space fields through the imaginary time prescription. Note that the integral in the exponent is the limit of a sum over four-dimensional cells. By labeling the field in cell α as ϕ_α, that in neighboring cell β as ϕ_β, and so on, $\frac{1}{2}(-\partial_\mu \partial_\mu + \mu^2 - i\epsilon)$ may be considered as the limit of a symmetric matrix $A_{\alpha\beta}$ that connects neigboring cells. Note that we have, for symmetric A,

$$\int_{-\infty}^{\infty} \prod_{i=1}^{N} dx_i \exp(-x_i A_{ij} x_j + 2S_k x_k) = (\pi/\det A)^{1/2} \exp(S_i A_{ij}^{-1} S_j) \qquad (A16)$$

Employing this result with $S_i = J(x)/2$ and dropping an (infinite but non-essential) multiplicative constant, we obtain

$$Z_0[J] = \exp\{-\frac{1}{2}i \int d^4x d^4y J(x)[-\partial_\mu \partial_\mu + \mu^2 - i\epsilon]^{-1} J(y)\} \qquad (A17)$$

Using Eq. (14), we then obtain the two-point Green's function:

$$\triangle_F(x) = \int \frac{d^4k}{(2\pi)^4} e^{ik\cdot x} \left(\frac{1}{k^2 + \mu^2 - i\epsilon}\right), \qquad (A18)$$

which yields $(-i)(k^2 + \mu^2 - i\epsilon)^{-1}$ as the Feynman propagator for a scalar meson of mass μ.

§.A.3. **Path Integrals and Quantum Electrodynamics**

In the case of quantum electrodynamics (QED), the Green's function for the photon must satisfy

$$(\partial_\nu \partial_\mu - \delta_{\mu\nu} \partial_\eta \partial_\eta)\triangle_{\mu\sigma}(x - y) = \delta_{\nu\sigma} \delta^4(x - y). \qquad (A19)$$

If we differentiate both side by applying ∂_ν, the left-hand side becomes zero but the right-hand side becomes $\partial_\sigma \delta^4(x - y)$. The problem arises because the inverse of the operator $K_{\mu\nu} = \partial_\nu \partial_\mu - \delta^{\mu\nu} \partial_\eta \partial_\eta$ does not exist. This fact is closely linked to gauge invariance, which states that $F_{\mu\nu}$ is invariant under an arbitrary gauge transformation: $A_\nu \rightarrow A_\nu + \partial_\nu \Lambda$. Thus, $K_{\mu\nu} \partial_\nu \Lambda = 0$; or, $K_{\mu\nu}$ has zero eigenvalues and so its inverse is singular.

To obtain the path integral $\int [dA] e^{iS(A)}$ for QED, we integrate over all possible vector potential values in each cell ϵ^4, which include, for each distinct A, all values of A equivalent up to a gauge transformation. That is, we must integrate over all values of Λ. However, for these distinct values of Λ, the lagrangian $\mathcal{L}$ and thus the action S are invariant. Therefore, we are carrying out an integral with a constant integrand over an infinite number of paths. To avoid the problem, we need to select just one path for each gauge-inequivalent A. That this can be done consistently for non-abelian gauge theories such as QCD was first demonstrated by Faddeev and Popov in 1967. Here we wish to illustrate how the solution to the problem results in the introduction of "ghost" fields.

Specifically, we divide the field A_μ into subsets such that, in each subset, none of the members is related to the other members by a gauge transformation. For each member $\tilde{A}_\mu$, we consider all possible gauge transformation Λ. In this way, we have

$$Z = \int [dA]\exp(i\int d^4x\mathcal{L}) = \int [d\tilde{A}]\exp(i\int d^x\mathcal{L})\int [d\Lambda], \qquad (A20)$$

where we may factor out the integrand for $d\Lambda$ because $\mathcal{L}$ is invariant under gauge transformations.

To remove the unphysical infinite constant $\int [d\Lambda]$, we introduce a factor $exp(-i\int d^4xC^2/2)$, with C some function of A_μ (which helps to fix the gauge), and make the resulting integral independent of the choice of C by multiplying it with the Jacobian $det(\partial C/\partial\Lambda)$. That is, we have

$$\begin{aligned}\int [d\Lambda] \quad &\rightarrow \quad \int [d\Lambda]\,det(\partial C/\partial\Lambda)\,exp(-i\int d^4x\frac{1}{2}C^2) \\ &= \int dC\,exp(-i\int d^4x\frac{1}{2}C^2),\end{aligned} \qquad (A21)$$

which amounts to multiplying Z by an overall constant. This is permissible since it affects only normalization factors. Thus, the new partition function is given by

$$Z = \int [d\tilde{A}]\int [d\Lambda]\,det(\frac{\partial C}{\partial\Lambda})\,exp[i\int d^4x(\mathcal{L} - \frac{1}{2}C^2)]. \qquad (A22)$$

Under an infinitesimal gauge transformation

$$A_\mu \quad \rightarrow A_\mu \quad + \partial_\mu\Lambda, \qquad (A23)$$

the function C is altered as follows:

$$C \quad \rightarrow \quad C + M\Lambda \qquad (A24)$$

where M is some operator that may include derivatives. Thus,

$$det(\partial C/\partial \Lambda) = det\, M. \tag{25}$$

We note that, for any Hermitian matrix A, we have

$$(det\, A)^{-1} = \pi^{-n} i^{-n} \int dz_1 dz_2 ... \int dz_n\, exp(i < z \mid A \mid z >), \tag{A26}$$

where $< z \mid A \mid z >= z_1^* A z_1 + z_2^* A z_2 + ... + z_n^* A z_n$.
Letting $z = x + iy$ and $dz \equiv dx dy$, we obtain

$$\int dx \exp(-az^*z) = \pi/a,$$
$$\prod_{i=1}^{n} \int dz_i \exp(-a_i z_i^* z_i) = \pi^n/(a_1...a_n). \tag{A27}$$

where the a_i are real.

For a diagonal $n \times n$ matrix A,

$$A = \begin{pmatrix} a_1 & & & \\ & \cdot & & \\ & & \cdot & \\ & & & \cdot \\ & & & & a_n \end{pmatrix}.$$

we have $\sum a_n z_n^* z_n =< z \mid A \mid z >$. Substituting each a_j by $-ia_j$, we find

$$\int dz_1 \int dz_2 ... \int dz_n \exp(i < z \mid A \mid z >) = i^n \pi^n/a_i...a_n$$
$$= i^n \pi^n/det(A),$$

which leads to Eq. (26) quoted above. On the other hand, for a nondiagonal hermitian matrix A, we can find a unitary transformation such that

$$z' = Uz, \qquad A' = UAU^{-1}$$

so that $dz_1', ..., dz_n' = dz_1, ..., dz_n$ and Eq. (A26) follows.

Apart from an irrelevant constant factor, Eq. (A26) is the cell equivalent of the equation

$$(det\,M)^{-1} = \int [d\psi] \exp(i \int d^4 x \psi^* M \psi). \qquad (A28)$$

However, what we really want to know is $det\,M$, whereas Eq. (A28) is a formula for $(det\,M)^{-1}$. The problem may be fixed by assuming that it is possible to represent $det\,M$ by an equation similar to Eq. (A28):

$$det\,M = \int [d'\phi] \exp(i \int d^4 x \phi^* M \phi), \qquad (A29)$$

where ϕ is also a complex scalar field and the symbol $[d'\phi]$ is yet to be clarified. We then have

$$det\,M (det\,M)^{-1} = 1 = \int [d\psi] \exp(i \int d^4 x \psi^* M \psi)$$
$$\times \int [d'\phi] \exp(i \int d^4 x \phi^* M \phi). \qquad (A30)$$

Since ψ is a complex scalar field not connected to any sources and $\psi^* M \psi$ appears in every term in the Green's function expansion, the ψ lines must appear in closed loops in all Feynman diagrams. The ϕ lines appear in closed loops for exactly the same reason. However, since the left-hand side of (A30) is unity, the contributions of the ψ and ϕ loops must cancel order by order in the perturbation expansion. This can be achieved if we associate a minus sign with each ϕ loop to cancel the plus sign associated with each ψ loop. This is the significance of the notation $[d'\phi]$. Accordingly, Eq. (A29) may be interpreted as a path integral over the complex scalar ϕ field with a factor of (-1) assigned to every closed loop. This yields Fermi-Dirac statistics, or, the "wrong" statistics for a complex scalar field (leading to the name "ghost").

The net result of the above standard discussion is that the unphysical infinite constant factor associated with gauge invariance has been removed from $Z[J]$. A specific choice of gauge fixes the factor C, which implies the ghost factor $exp(i \int d^4 x \phi^* M \phi)$, where M depends on the choice of C. For example, we may choose $C = \partial_\mu A_\mu$. Thus, a gauge transformation $A_\mu \to A_\mu + \partial_\mu \Lambda$ yields $M = \partial_\mu \partial_\mu$ by definition. Therefore, the ghost field in QED does not couple to matter fields and so its role may be ignored.

Consider the $SU(2)$ Yang-Mills field $\mathbf{A}_\mu$. As we recall, the Lagrangian for the gauge field is $\mathcal{L}_{YM} = -\frac{1}{4} \mathbf{F}_{\mu\nu} \cdot \mathbf{F}_{\mu\nu}$, where $\mathbf{F}_{\mu\nu} = \partial_\mu \mathbf{A}_\nu - \partial_\nu \mathbf{A}_\mu - g(\mathbf{A}_\mu \times \mathbf{A}_\nu)$. Note that $\mathcal{L}_{YM}$ is invariant under local gauge transformations of the form

$$\delta\psi = i\epsilon \cdot \frac{\tau}{2}\psi, \qquad \delta\bar{\psi} = -\bar{\psi}i\epsilon \cdot \frac{\tau}{2};$$

$$\delta\mathbf{A}_\mu = -(1/g)\partial_\mu\epsilon - (\epsilon \times \mathbf{A}_\mu).$$

Or, the gauge-invariant derivative is

$$D_\mu = \partial_\mu + igA_\mu \cdot \frac{\tau}{2}.$$

Now, we consider a function corresponding to the Landau gauge condition $\partial_\mu A_\mu = 0$:

$$\mathbf{C} = \partial_\mu\mathbf{A}_\mu. \qquad (A31)$$

Thus, we have

$$\partial_\mu\mathbf{A}'_\mu = \partial_\mu\mathbf{A}_\mu - (1/g)\partial_\mu(\partial_\mu\epsilon) - \partial_\mu(\epsilon \times \mathbf{A}_\mu)$$
$$= \partial_\mu\mathbf{A}_\mu - (1/g)\partial_\mu D_\mu\epsilon,$$

which yields

$$C' = C - (1/g)\partial_\mu(D_\mu\epsilon), \qquad (A32)$$

or,

$$M = -(1/g)\partial_\mu(D_\mu).$$

Therefore, we have a complex scalar ghost field ϕ with an additional factor:

$$\int [d'\phi]\exp(i\int d^4x\phi^*\partial_\mu D_\mu\phi), \qquad (A33)$$

where $[d'\phi]$ signifies that a factor of -1 is associated with each closed loop. The new feature in the massless non-abelian Yang-Mills case is that there is a coupling of the ghost field to the field A_μ, the existence of which would not have been suspected from the original Lagrangian. This coupling cannot be ignored, since otherwise the theory is inconsistent and not renormalizable.

Appendix B
Method of Lattice Gauge Fields

In this appendix, we use Hamiltonian lattice formulation to illustrate how a gauge field theory can be treated on a lattice.

§.B.1. Formulation

We discretize space into a simple cubic lattice with lattice spacing a. The lattice sites are vectors of the form

$$\mathbf{x} = a(i\hat{\mathbf{e}}_x + j\hat{\mathbf{e}}_y + k\hat{\mathbf{e}}_z), \qquad (B1)$$

where i, j, k are integers and $\hat{\mathbf{e}}_i$ are unit vectors along the lattice directions. The directed link from $\mathbf{x}$ to $\mathbf{x} + a\hat{\mathbf{e}}$ will be denoted by $(\mathbf{x}, \hat{\mathbf{e}})$, which may be distinguished from the oppositely directed link $(\mathbf{x} + a\hat{\mathbf{e}}, -\hat{\mathbf{e}})$.

By introducing a lattice, we are forced to give up various space symmetries. Nevertheless, it is possible to maintain gauge invariance. To do so, we wish to require that the theory be invariant under the discrete generalization of the quark field gauge transformation.

It is natural to associate the vector gauge fields with the links. Indeed, we note that, on a lattice, geometry scalars are associated with lattice sites, vectors with directed links, antisymmetric tensors with oriented areas, etc. Our basic unit will be the string operator:

$$U_{\mathbf{x};\hat{\mathbf{e}}} = U(\mathbf{x} + a\hat{\mathbf{e}}, \mathbf{x}; c) = P \exp\ i[\int_{\mathbf{x}}^{\mathbf{x}+a\hat{\mathbf{e}}} \tilde{A} \cdot dz], \qquad (B2)$$

with

$$\tilde{A}_\mu \equiv \frac{\lambda^\alpha}{2} g A_\mu^\alpha. \qquad (B2a)$$

The integral runs along the link from $\mathbf{x}$ to $\mathbf{x} + a\hat{\mathbf{e}}$. An obvious property, as due to unitarity of U, is

$$U_{\mathbf{x};\hat{\mathbf{e}}}^{-1} = U_{\mathbf{x}+a\hat{\mathbf{e}};-\hat{\mathbf{e}}}.$$

The continuum limit may be achieved by letting a approach zero so that we have, at the classical level,

$$U_{x;\hat{e}} \approx \exp\{ia\tilde{A}(\mathbf{x}) \cdot \hat{e}\} \approx 1 + ia\tilde{A}(x) \cdot \hat{e}, \qquad (B3)$$

which may be used to justify the lattice versions of various operators.

U is a unitary 3×3 matrix and thus can be parametrized as an element of $SU(3)$, namely

$$U_{x,\hat{e}} = exp\{i\frac{\lambda^{\alpha}}{2} \cdot b^{\alpha}_{x,\hat{e}}\}. \qquad (B4)$$

Comparing Eq. (B4) with Eq. (B3), we see that the group parameters b are related to the vector potentials A. It is important to note that the b's vary over a compact manifold in distinction to the continuum A's. Integration over the group denoted by $[dU]$ will, in reality, be an integration over the b's. We are now in a position to define a gauge invariant operator whose continuum limit will be related to the color magnetic field B^a. Consider the trace of a product of four link operators, $U_{\mathbf{x},\hat{e}}$, along a fundamental lattice square, or plaquette, illustrated in Fig. 2.

$$U_{\mathbf{x};\hat{e}_1,\hat{e}_2} = \{U_{\mathbf{x}+a\hat{e}_2;-\hat{e}_2} U_{\mathbf{x}+a(\hat{e}_1+\hat{e}_2);-\hat{e}_1} U_{\mathbf{x}+a\hat{e}_1;\hat{e}_2} U_{\mathbf{x};\hat{e}_1}\}. \qquad (B5)$$

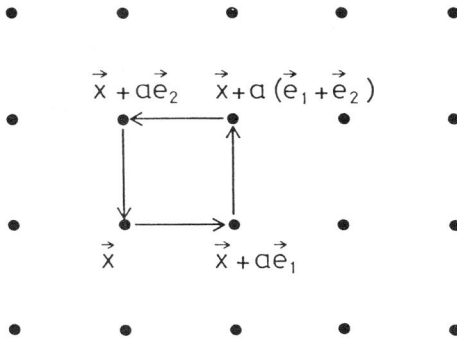

Fig. 2: A fundamental plaquette.

Using Eq. (B3), we obtain, to order a^2,

$$Tr\, U_{x;\hat{e}_i;\hat{e}_j} = \epsilon_{ijk} Tr\, \exp[ia^2 \tilde{B} \cdot \hat{e}_k] \approx \epsilon_{ijk} Tr\,[1 - \frac{1}{2}a^4(\tilde{B} \cdot \hat{e}_k)^2], \qquad (B6)$$

with

$$F_{\mu\nu} \equiv \frac{1}{g}(\partial_\mu \tilde{A}_\nu - \partial_\nu \tilde{A}_\mu - i[\tilde{A}_\mu, \tilde{A}_\nu]), \qquad (B7a)$$

$$\tilde{B}_k \equiv \frac{1}{2}g\epsilon_{ijk}F_{ij}. \qquad (B7b)$$

Unless ambiguity may arise, we replace the combination $\mathbf{x}, \hat{\mathbf{e}}_i, \hat{\mathbf{e}}_j$ by p and let

$$U_p = U_{\mathbf{x};\hat{\mathbf{e}}_i,\hat{\mathbf{e}}_j}. \qquad (B8)$$

In view of Eq. (B6) a lattice version of

$$H_M = Tr \int d^3x \frac{1}{g^2} B^2(x) \qquad (B9)$$

is

$$H_M = \sum_p \frac{1}{g^2 a} Tr\, [2 - U_p - U_p^\dagger]. \qquad (B10)$$

We now need a lattice generalization of the electric field. To this end, we note that

$$[\mathbf{E}^\alpha(\mathbf{y}) \cdot \hat{\mathbf{e}}_i, U_{\mathbf{x};\hat{\mathbf{e}}_j}] = \delta_{ij}\delta^2(y_\perp)U(\mathbf{x} + a\hat{\mathbf{e}}, \mathbf{y}; c)\frac{\lambda^\alpha}{a}U(\mathbf{y}, \mathbf{x}; c), \qquad (B11)$$

if $\mathbf{y}$ coincides with any of the points in the interval $(\mathbf{x}, \mathbf{x} + a\hat{\mathbf{e}})$ and zero otherwise. In the lattice formulation, $U_{\mathbf{x},\hat{\mathbf{e}}}$ must be considered in total and cannot be broken up along the link. We can define an electric field at the start or end of a link. The two fields associated with the link $(\mathbf{x}, \hat{\mathbf{e}})$ are $E^\alpha_{\mathbf{x},\hat{\mathbf{e}}}$ and $E^\alpha_{\mathbf{x}+a\hat{\mathbf{e}};-\hat{\mathbf{e}}}$, and are determined from their commutation relations with link variables $U_{\mathbf{x},\hat{\mathbf{e}}}$. For $\hat{\mathbf{e}}$ positive, we may postulate

$$\begin{aligned}[E^\alpha_{\mathbf{x};\hat{\mathbf{e}}_i}, U_{\mathbf{y};\hat{\mathbf{e}}_j}] &= \delta_{ij}\delta_{\mathbf{x},\mathbf{y}}U_{\mathbf{x};\hat{\mathbf{e}}_i}\frac{\lambda^\alpha}{2}, \\ [E^\alpha_{\mathbf{x}+a\hat{\mathbf{e}}_i;-\hat{\mathbf{e}}_j}, U_{\mathbf{y};\hat{\mathbf{e}}_j}] &= -\delta_{ij}\delta_{\mathbf{x},\mathbf{y}}\frac{\lambda^\alpha}{2}U_{\mathbf{x},\hat{\mathbf{e}}_i}.\end{aligned} \qquad (B12)$$

The fact that $U^\dagger_{\mathbf{x};\hat{\mathbf{e}}} = U_{\mathbf{x}+a\hat{\mathbf{e}};-\hat{\mathbf{e}}}$ yields the commutation relations for negative $\hat{\mathbf{e}}$'s. Use of the Jacobi identities yields the electric field commutation relations:

$$\begin{aligned}[E^\alpha_{\mathbf{x};\hat{\mathbf{e}}_i}, E^\beta_{\mathbf{y};\hat{\mathbf{e}}_j}] &= i\delta_{ij}\delta_{\mathbf{x},\mathbf{y}}f^{\alpha\beta\gamma}E^\gamma_{\mathbf{x};\hat{\mathbf{e}}_i}, \\ [E^\alpha_{\mathbf{x};-\hat{\mathbf{e}}_i}, E^\beta_{\mathbf{y};-\hat{\mathbf{e}}_j}] &= -i\delta_{ij}\delta_{\mathbf{x},\mathbf{y}}f^{\alpha\beta\gamma}E^\gamma_{\mathbf{x};-\hat{\mathbf{e}}_i}, \\ [E^\alpha_{\mathbf{x};\hat{\mathbf{e}}_i}, E^\beta_{\mathbf{y},-\hat{\mathbf{e}}_j}] &= 0.\end{aligned} \qquad (B13)$$

The two electric fields are not independent of each other; the defining commutation relations provide the connection:

$$\lambda^\alpha E^\alpha_{\mathbf{x}+a\hat{e};-\hat{e}} = -U_{\mathbf{x};\hat{e}}\lambda^\beta E^\beta_{\mathbf{x};\hat{e}}U^\dagger_{\mathbf{x};\hat{e}},$$

which implies that

$$E^2_{\mathbf{x};\hat{e}} = E^2_{\mathbf{x}+a\hat{e};-\hat{e}}. \tag{B14}$$

The electric energy

$$H_E = g^2 Tr \int d^3x \tilde{E}^2 \tag{B15}$$

goes over to

$$H_E = \frac{g^2}{2a}\sum_{links} E^\alpha_{\mathbf{x};\hat{e}}E^\alpha_{\mathbf{x};\hat{e}}. \tag{B16}$$

Thus, the total lattice QCD Hamiltonian is given by

$$H = \frac{g^2}{2a}\sum_{links} E^2_{\mathbf{x};\hat{e}} + \frac{1}{ag^2}Tr\sum_p (2 - U_p - U^\dagger_p) + \sum_{\mathbf{x}} m_0\bar{q}_{\mathbf{x}}q_{\mathbf{x}}, \tag{B17}$$

where the last term describes the quark mass effect.

The generator of infinitesimal time-independent gauge transformation is specified by

$$\sum_{\mathbf{x}}\{\sum_{\hat{e}}(E^\alpha_{\mathbf{x};\hat{e}} - E^\alpha_{\mathbf{x};-\hat{e}}) + q^\dagger_{\mathbf{x}}\frac{\lambda^\alpha}{2}q_{\mathbf{x}}\}, \tag{B18}$$

which annihilates all physical states.

We may rewrite the Hamiltonian of Eq. (B17) as follows

$$H = \frac{g^2}{2a}\{\sum_{links} E^2_{\mathbf{x};\hat{e}} + xTr\sum_p (2 - U_p - U^\dagger_p)\} + \sum_{\mathbf{x}} m_0\bar{q}(\mathbf{x})q(\mathbf{x}),$$
$$x = 2/g^4. \tag{B19}$$

§.B.2. An Illustrative Example

In what follows, we wish to consider the quark-antiquark potential, to zeroth, first and second order in x. As we will be interested in the difference in the energies of this configuration and in the vacuum energy, we first calculate the spectrum *without* quarks.

Due to Gauss' law, electric flux lines must close. To zeroth order in x the two lowest lying states are the vacuum, with no electric flux anywhere, and the states where the electric flux is in a 3 or $\bar{3}$ representation along the links of some fundamental plaquette. The first state we identify by $\mid 0 >$ with energy $E_0^{(0)} = 0$. The second class of states consists of

$$\begin{aligned} \mid P > &= U_p \mid 0 >, \\ \mid \tilde{P} > &= U_p^\dagger \mid 0 >. \end{aligned} \qquad (B20)$$

Using the definition of U, Eq. (B5), and the commutation relations, Eq. (B12), we obtain

$$\begin{aligned} H_E \mid P > &= E_p^{(0)} \mid P >, \\ H_E \mid \tilde{P} > &= E_p^{(0)} \mid \tilde{P} >, \end{aligned}$$

with

$$E_p^{(0)} = \frac{g^2}{2a} \{4C^{(3)}\} = \frac{16}{3} \frac{g^2}{2a}. \qquad (B21)$$

Here $C^{(3)}$ is the quadratic Casimir operator for the 3 or $\bar{3}$ representation. To first order in x all energies shift by the amount

$$E_{0,p}^{(1)} = 3x \frac{g^2}{2a} N(P), \qquad (B22)$$

where $N(P)$ is the number of plaquettes in our system. (Here we assume for the moment a finite world.) The order x^2 correction may be obtained from second-order perturbation theory,

$$E_0^{(2)} = -\frac{g^2}{2a} x^2 \sum_p \frac{|< 0 \mid U_p^\dagger \mid P >|^2 + |< 0 \mid U_p \mid \tilde{P} >|^2}{(16/3)} = -\frac{g^2}{2a} \left(\frac{3}{8} N(P)\right) x^2. \qquad (B23)$$

We may now turn to states with a heavy quark at the origin and a heavy antiquark at $R = na\hat{e}_1$. Again due to Gauss' law, we must have an electric flux joining the two particles. We expect that the lowest-energy configuration consists of the shortest flux path possible, of which the energy is

$$E^{0a}(x) = \frac{g^2}{2a}\frac{4}{3}n + 2m_0 = \frac{g^2}{2a^2}\frac{4}{3}R + 2m_0. \tag{B24}$$

Accordingly, *strong coupling lattice gauge theories confine quarks.*

To first order in x the perturbing Hamiltonian connects the state with the shortest length to those with an additional plaquette.

Calculating the result and subtracting from it the second order correction to the vacuum, Eq. (B24), we obtain

$$E(R) = \frac{g^2}{2a^2}R\left[\frac{4}{3} - x^2\frac{11}{153} + ...\right] + 2m_0. \tag{B25}$$

§.B.3. Euclidean Lattice Gauge Theories

Lagrangian theories in a path integral formalism may also be transcribed to a lattice, which is four-dimensional:

$$< \prod O_i(\mathbf{x}_i; \hat{\mathbf{e}}_i) > = \frac{1}{Z}\int \prod_{\mathbf{x};\hat{\mathbf{e}}} dU_{\mathbf{x};\hat{\mathbf{e}}} \prod O_i(\mathbf{x}_i; \hat{\mathbf{e}}_i) exp\{\frac{1}{g^2}Tr\sum_p(U_p + U_p^\dagger)\},$$

$$Z = \int \prod_{\mathbf{x};\hat{\mathbf{e}}} dU_{\mathbf{x};\hat{\mathbf{e}}} exp\{\frac{1}{g^2}Tr\sum_p(U_p + U_p^\dagger)\}.$$

$$\tag{B26}$$

This formulation is equivalent to the Hamiltonian formulation *for small g only.* Although the quantum mechanics based on the different formulations is expected to be the same in the continuum limit, various approximate results do depends on the formalism chosen. In many cases, it is useful to compare them. The operator characterizing the confinement properties of the theory is the Wilson loop integral, of which lattice analogue is

$$W[c] = Tr\prod_{(\mathbf{x},\hat{e})\in c} U_{\mathbf{x},\hat{\mathbf{e}}}. \tag{B27}$$

The product is ordered along the closed curve c.

The strong coupling expansion of the Euclidean theory may be obtained by expanding the exponent in Eq. (B26) as a power series in $1/g^2$. Using the orthogonality property for the representation matrices, we find that the lowest order nonvanishing contribution to $< W[c] >$ is of the order $(1/g^2)^{N(c)}$, where

$N(c)$ is the number of plaquettes in the planar area surrounded by the closed curve c. (Bander 1981) Relating the number of plaquettes to the area, i.e., $A = N(c)a^2$, we obtain

$$W[c] = exp(-ln\, g^2\, A/a^2),\qquad\qquad (B28)$$

which results in confinement with the energy of separation

$$E(R) = \frac{1}{a^2} ln\, g^2\, R.\qquad\qquad (B29)$$

Lattice theories, both in the Hamiltonian and in the Euclidean-Lagrangian formulation lead to quark confinement at strong couplings. A crucial question, which we do not address here, is how these results can be extrapolated smoothly to the weak coupling continuum region.

References

Weyl, H., *Ann. Physik* **59**, 101 (1919); Weyl, H., *Zeits. f. Physik* **56**, 330 (1929). The latter is the classic paper on gauge invariance (Eichinvariance).

Fock, V., *Zeits. f. Physik* **39**, 226 (1927).

London, F., *Zeits. f. Physik* **42**, 375 (1927).

Yang, C.N., and Mills, R.L., *Phys. Rev.* **96**, 191 (1954).

Faddeev, L.D. and Popov, V.N., Phys. Lett. **25B**, 29 (1967); on the Faddeev-Popov ghosts.

't Hooft, G., Nucl. Phys. **B33**, 173 (1971); on the Feynman rules for massless Yang-Mill fields.

Baker, M. and Johnson, K., Phys. Rev. **183**, 1292 (1969); on the β-function for QED.

Belavin, A.A. and Migdal, A.A., Pis'ma Zh. Eksp. Teor. Fiz **19**, 317 (1974) [JETP Lett. **19**, 181 (1974)]; Caswell, W.E., Phys. Rev. Lett. **33**, 244 (1974); Jones, D.R.T., Nucl. Phys. **B75**, 531 (1974); on the β-function for $SU(n)$ gauge field theories.

Mueller, A.H., *"Perturbative QCD at high energies"*, Phys. Rep. **73C**, 237 (1981); Reya, E., *"Perturbative quantum chromodynamics"*, Phys. Rep. **69C**, 195 (1981); on perturbative QCD.

Feynman, R.P., and Hibbs, A.R., *Quantum Mechanics and Path Integrals* (McGraw Hill, New York, 1965); Abers, E., and Lee, B.W., *Gauge Theories*, Phys. Rep. **9C**, 1 (1973); on the method of path integrals.

Wilson, K.G., Phys. Rev. **D10**, 2445 (1974); Kogut, J., and Susskind, L., Phys. Rev. **D11**, 395 (1975); Bander, M., Phys. Rep. **75**, 205 (1981); Kogut, J.B., Rev. Mod. Phys. **51**, 659 (1979); *ibid.* **55**, 775 (1983); in *Nuclear and Particle Physics on the Light Cone*, eds. Johnson, M.B., and Kisslinger, L.S. (World Scientific, Singapore, 1989), p. 239; Negele, J.W., in *Quarks, Mesons, and Nuclei I: Strong Interactions*, eds. Hwang, W-Y. P., and Henley, E.M. (World Scientific, Singapore, 1989), p. 1; on the treatment of lattice gauge theories.

Exercises: *Chapter 12*

12.1. Prove that the lagrangian, Eq. (8), is invariant under gauge transformations, Eqs. (2) and (5).

12.2. Derive the Feynman rules (13e) and (13f) by calculating $< 0 \mid S^{(1)} \mid i >$ with the suitable three-gluon and four-gluon initial states.

12.3. Use Eq. (17) in connection with Eq. (21). Discuss the asymptotically free behavior of the final result.

Chapter 13.
The Glashow-Salam-Weinberg Electroweak Theory

So far we have introduced two very important examples of gauge field theories, QED and QCD. The gauge fields in both cases are massless vector bosons. If both strong and electromagnetic interactions can be described successfully in terms of gauge field theories, it is natural to speculate that weak interactions, or any other fundamental interactions that may exist in nature, can also be phrased in a similar vein. It turns out that this might indeed be the case. Historically, Weinberg and Salam proposed independently in 1967 an $SU(2) \times U(1)$ gauge field theory that unifies the electromagnetic and weak interactions among leptons, before QCD was introduced. The structrue of the electroweak interactions in the Weinberg-Salam $SU(2) \times U(1)$ model agrees to lowest order with what Glashow obtained in 1961 by attempting to unify the electromagnetic and weak interactions, although Weinberg and Salam invoked the concept of the Higgs mechanism (Higgs 1964 and others) in their construction, a key ingredient that 't Hooft used later in 1971 to prove the renormalizability of the model.

The major obstacle for describing weak interactions in terms of a gauge field theory stems from the fact that the vector bosons mediating weak interactions are not massless. Thus, it was not clear whether gauge symmetry is relevant at all in the case of weak interactions. Invocation of Higgs mechanism, as in the case of the Weinberg-Salam $SU(2) \times U(1)$ model, allows for implementation of gauge symmetry at the lagrangian level. However, such symmetry must be eventually broken in the sense that the ground-state solution to the problem does not respect the symmetry. The specific way of symmetry breaking, in which the lagrangian (hamiltonian) respects the symmetry but the ground-state solution does not, is called "spontaneous symmetry breaking." The Glashow-Salam-Weinberg (GSW) electroweak theory is a spontaneously broken $SU(2) \times U(1)$ gauge field theory.

The GSW electroweak theory predicts the existence of neutral weak interactions, the existence of the charm quark, and the existence of $W^{\pm}$ and Z^0, all of which have been substantiated in quantitative terms by experiments. Although the physics related to the Higgs sector remains elusive, there is little doubt that any better theory proposed in the foreseeable future must reproduce successes of the GSW electroweak theory.

§.13.1. Higgs Mechanism in an $SU(2) \times U(1)$ Gauge Theory.

Consider an $SU(2) \times U(1)$ gauge theory. The gauge fields are to be denoted by $A_\lambda^i(x)$ and $B_\lambda(x)$, respectively. Following the procedure given earlier in Ch. 12 for QCD, we may write the lagrangian for gauge fields,

$$\mathcal{L}_{gauge} = -\frac{1}{4}F_{\mu\nu}^i F_{\mu\nu}^i - \frac{1}{4}B_{\mu\nu}B_{\mu\nu}, \qquad (1)$$

where

$$F_{\mu\nu}^i \equiv \partial_\mu A_\nu^i - \partial_\nu A_\mu^i + g\epsilon^{ijk}A_\mu^j A_\nu^k, \qquad (2a)$$

$$B_{\mu\nu} = \partial_\mu B_\nu - \partial_\nu B_\mu. \qquad (2b)$$

A general gauge transformation is specified by

$$\frac{\tau}{2} \cdot \mathbf{A}_\mu(x) \to \frac{\tau}{2} \cdot \mathbf{A}'_\mu(x) = U(\theta)\{\frac{\tau}{2} \cdot \mathbf{A}_\mu(x) \\ - \frac{i}{g}U^{-1}(\theta)\partial_\mu U(\theta)\}U^{-1}(\theta), \qquad (3a)$$

$$B_\mu(x) \to B'_\mu(x) = B_\mu(x) - \frac{1}{g'}\partial_\mu\varphi(x) \qquad (3b)$$

with

$$U(\theta) = exp[-i\theta(x) \cdot \frac{\tau}{2}]. \qquad (3c)$$

It is a routine exercise to demonstrate that the lagrangian $\mathcal{L}_{gauge}$ is gauge invariant, i.e. invariant under a local gauge transformation specified by Eqs. (3a)-(3c). Introduction of a mass term such as $-\frac{1}{2}m^2\mathbf{A}_\mu \cdot \mathbf{A}_\mu$ destroys gauge invariance so that weak boson masses can only be taken into account in a specific manner.

To incorporate weak boson masses, we introduce a pair of complex scalar fields,

$$\phi = \begin{pmatrix} \phi^+ \\ \phi^0 \end{pmatrix}, \qquad (4)$$

which transforms like a doublet under $SU(2)$ and possesses a weak hypercharge $Y_W = 1$ under $U(1)$. Thus, the gauge-invariant derivative for ϕ is given by

$$D_\mu = \partial_\mu - ig\frac{\tau}{2} \cdot \mathbf{A}_\mu(x) - i\frac{g'}{2}B_\mu(x). \qquad (5)$$

We choose

$$\mathcal{L}_{scalar} = -[D_\mu\phi]^\dagger[D_\mu\phi] - V(\phi), \qquad (6)$$

where

$$V(\phi) = \mu^2 \phi^\dagger \phi + \lambda (\phi^\dagger \phi)^2, \tag{7}$$

with $\mu^2 < 0$ and $\lambda > 0$. The lagrangian $\mathcal{L}$ is gauge invariant, i.e. invariant under the gauge transformation of Eqs. (3a)-(3c) and the equation given below:

$$\phi(x) \rightarrow \phi'(x) = U(\theta) \cdot exp(-i\frac{1}{2}Y_W \varphi(x))\phi(x). \tag{8}$$

However, the potential $V(\phi)$ has a minimum at

$$< \phi_0 > = \begin{pmatrix} 0 \\ v/\sqrt{2} \end{pmatrix}, \quad with \quad v = \sqrt{-\frac{\mu^2}{\lambda}}, \tag{9}$$

so that the physical ground state $< \phi_0 >$ differs from the trivial ground state, i.e. $< \phi >= 0$. Note that the explicit form for $< \phi_0 >$ varies with the gauge. In the unitary gauge (U-gauge), $< \phi_0 >$ assumes the form given by Eq. (9). In other gauges, Eq. (8) may be used to generate $< \phi_0' >$.

The situation is typical for the so-called "spontaneous symmetry breaking": The lagrangians $\mathcal{L}_{gauge}$ and $\mathcal{L}_{scalar}$ are invariant under an arbitrary gauge transformation but the ground-state solution to the problem varies with the gauge, i.e., breaks gauge symmetry. As we shall see shortly, such spontaneous breaking of gauge symmetry generates masses for weak bosons. The mechanism has been named "Higgs mechanism" (Higgs 1964; Anderson 1963; Englert et al. 1964; Kibble et al. 1964).

We rewrite Eq. (4) as follows,

$$\phi(x) = \begin{pmatrix} \phi^+(x) \\ \phi^0(x) \end{pmatrix} \equiv exp[+\frac{i}{2v}\xi(x) \cdot \tau] \begin{pmatrix} 0 \\ \frac{1}{\sqrt{2}}(v + \eta(x)) \end{pmatrix}, \tag{10}$$

with $\xi(x)$ and $\eta(x)$ four real functions. Accordingly, we make a gauge transformation,

$$U(\theta) = U(\xi/v) = exp[-\frac{i}{2v}\xi(x) \cdot \tau], \tag{11}$$
$$\varphi(x) = 0,$$

such that

$$\phi'(x) = U(\xi/v)\phi(x) = \begin{pmatrix} 0 \\ \frac{1}{\sqrt{2}}(v + \eta(x)) \end{pmatrix}. \tag{12}$$

For the sake of simplicity, we denote gauge fields as $\mathbf{A}_\mu(x)$ and $B_\mu(x)$ in this specific gauge (U-gauge). We introduce

$$W_\mu^\pm = \frac{1}{\sqrt{2}}(A_\mu^1 \pm iA_\mu^2), \tag{13a}$$

$$Z_\mu^0 = cos\theta_W A_\mu^3 - sin\theta_W B_\mu, \tag{13b}$$

$$A_\mu = sin\theta_W A_\mu^3 + cos\theta_W B_\mu, \tag{13c}$$

with

$$sin\theta_W = \frac{g'}{\sqrt{g^2 + g'^2}}, \tag{14a}$$

$$cos\theta_W = \frac{g}{\sqrt{g^2 + g'^2}}. \tag{14b}$$

Substituting Eq. (10) back into Eq. (6) and using Eqs. (13)-(14), we find, with $v^2 + \frac{\mu^2}{\lambda} = 0$,

$$
\begin{aligned}
\mathcal{L}_{scalar} =& \{-\frac{1}{2}\partial_\mu\eta\partial_\mu\eta - \frac{1}{2}(-2\mu^2)\eta^2 - \frac{\lambda}{4}(\eta^4 + 4v\eta^3)\} \\
&+ \frac{\mu^4}{4\lambda} \\
&- \frac{1}{8}\{v^2 + (2v\eta + \eta^2)\}\{2g^2 W_\mu^+ W_\mu^- + [g^2 + (g')^2]Z_\mu^0 Z_\mu^0\}.
\end{aligned}
\tag{15}
$$

Accordingly, three gague bosons become massive while the remaining one massless:

$$M_{W^\pm} = \frac{1}{2}gv, \tag{16a}$$

$$M_{Z^0} = \frac{1}{2}[g^2 + (g')^2]^{1/2}v = \frac{M_{W^\pm}}{cos\theta_W}, \tag{16b}$$

$$M_A = 0. \tag{16c}$$

The field $A_\mu(x)$ is indentified with the photon field while the massive gauge fields are indentified as weak bosons. What really happens is that, in the U-gauge, three degrees of freedom associated with $\phi(x)$ have been absored into $W_\mu^\pm$ and Z_μ^0 (as their longitudinal components) and the remaining one, namely $\eta(x)$, acquires a mass $-2\mu^2$.

The interaction terms in Eq. (15) read

$$
\begin{aligned}
\mathcal{L}_{\phi g}^{int} = & -\frac{\lambda}{4}(\eta^4 + 4v\eta^3) \\
& -\frac{1}{8}(2v\eta + \eta^2)\{2g^2 W_\mu^\pm W^- + [g^2 + (g')^2]Z_\mu^0 Z_\mu^0\}.
\end{aligned}
\tag{17}
$$

The Feynman rules in the U-gauge yield Green's functions which are unrenormalizable. A generalized renormalizable gauge formulation of spontaneously broken gauge theories leads to the so-called "R_ξ gauge" (Fujikawa, Lee, and Sanda 1972), which is to be described in the Appendix at the end of this chapter.

The coupling of weak bosons to the photon field is of some interest. Substituting Eqs. (13a)-(13c) back into Eq. (1), we find

$$
\begin{aligned}
\mathcal{L}_{gauge} = & -\frac{1}{4}\{F_{\mu\nu}F_{\mu\nu} + Z_{\mu\nu}^0 Z_{\mu\nu}^0 + 2W_{\mu\nu}^+ W_{\mu\nu}^-\} \\
& - ei\{F_{\mu\nu}W_\mu^+ W_\nu^- + (W_{\mu\nu}^+ W_\mu^- - W_{\mu\nu}^- W_\mu^+)A_\nu\} \\
& - ei\cot\theta_W\{Z_{\mu\nu}^0 W_\mu^+ W_\nu^- + (W_{\mu\nu}^+ W_\mu^- - W_{\mu\nu}^- W_\mu^+)Z_\nu^0\} \\
& - e^2\{W_\mu^+ W_\mu^-(A_\nu + \cot\theta_W Z_\nu^0)^2 \\
& \qquad - W_\mu^+ W_\nu^-(A_\mu + \cot\theta_W Z_\mu^0)(A_\nu + \cot\theta_W Z_\nu^0)\} \\
& - \frac{1}{2}(\frac{e}{\sin\theta_W})^2(W_\mu^+ W_\mu^- W_\nu^+ W_\nu^- - W_\mu^+ W_\mu^+ W_\nu^- W_\nu^-).
\end{aligned}
\tag{18}
$$

Here we have used

$$
e = g\sin\theta_W;
\tag{19}
$$

$$
\begin{aligned}
F_{\mu\nu} &\equiv \partial_\mu A_\nu - \partial_\nu A_\mu, \\
Z_{\mu\nu}^0 &\equiv \partial_\mu Z_\nu^0 - \partial_\nu Z_\mu^0, \\
W_{\mu\nu}^\pm &\equiv \partial_\mu W_\nu^\pm - \partial_\nu W_\mu^\pm.
\end{aligned}
\tag{20}
$$

Note that the weak bosons $W_\mu^\pm$ also couple to the photon field through a magnetic-moment coupling, namely $-eiF_{\mu\nu}W_\mu^+ W_\nu^-$. In addition, there are "seagull" terms [$\propto WW\ AA$] which may also be of some interest.

To sum up, we wish to mention that we have succeeded in constructing an $SU(2) \times U(1)$ gauge theory in which three gauge bosons [$W_\mu^\pm$ and Z_μ^0] are massive while the remaining one [A_μ or the photon field] is massless. The lagrangians $\mathcal{L}_{gauge}$ [Eq. (1)] and $\mathcal{L}_{scalar}$ [Eq. (6)] are gauge invariant but the ground-state solution to the problem varies with the gauge [i.e., breaks gauge symmetry spontaneously]. The mechanism yields two important relations: $M_{Z^0} = M_{W^\pm}/\cos\theta_W$ [Eq. (16b)] and $e = g\sin\theta_W$ [Eq. (19)].

§.13.2. The $SU(2) \times U(1)$ Electroweak Theory with Two Generations of Fermions

We wish to consider an application of the $SU(2) \times U(1)$ gauge theory introduced earlier by identifying the four gague bosons with the observed weak bosons $\{W_\mu^\pm, Z_\mu^0\}$ and the photon $\{A_\mu \text{ or } \gamma\}$. Specifically, we consider how quarks and leptons couple to these gauge bosons. For the sake of clarity, it is useful to study in some detail the case with two generations of fermions :

leptons:

$$\begin{pmatrix} \nu_e \\ e^- \end{pmatrix}, \qquad \begin{pmatrix} \nu_\mu \\ \mu^- \end{pmatrix};$$

Quarks:

$$\begin{pmatrix} u \\ d \end{pmatrix}, \qquad \begin{pmatrix} c \\ s \end{pmatrix}. \tag{21}$$

It is a straightforward task to generalize the scheme to incorporate the third generation of fermions.

First, we consider leptons. The $\mu - e$ university may be assumed so that the electroweak structure in the muon sector is identical with that in the electron sector. For a given Dirac field, we introduce the left-handed (L) and right-handed (R) components,

$$\psi_L \equiv \frac{1}{2}(1 + \gamma_5)\psi, \qquad \psi_R \equiv \frac{1}{2}(1 - \gamma_5)\psi. \tag{22}$$

In the electron sector, the physical fields relevant for weak interactions include e_L^-, e_R^-, and ν_{eL} only. A natural assignment is therefore given by

$$L = \begin{pmatrix} \nu_{eL} \\ e_L^- \end{pmatrix}, \quad SU(2) \quad doublet;$$
$$R = e_R^-, \quad SU(2) \quad singlet. \tag{23}$$

Note that, if ν_{eL} and e_L^- both were SU(2) singlets, then there would not be any coupling between e_L^- and $W^\pm$.

The fact that the photon remains massless upon spontaneous symmetry breaking implies that there remains an exact U(1) gauge symmetry with a quantum number which can be identified as the electric charge,[1]

---

[1] *Without loss of generality, it may be assumed that the conserved quantum number is $Q = aT_3^W + bY$. Application of this formula to the Higgs doublet, Eq.(4), yields a=1 and b=1/2.*

$$Q = T_3^W + \frac{Y}{2}. \tag{24}$$

Accordingly, we have

$$Y_L = -1, \quad and \quad Y_R = -2 \tag{25}$$

The fermion lagrangian in the electron sector is therefore given by

$$\begin{aligned}
\mathcal{L}_e = &- \bar{R}\gamma_\mu \{\partial_\mu + ig'B_\mu\}R \\
&- \bar{L}\gamma_\mu \{\partial_\mu - ig\frac{\tau}{2} \cdot \mathbf{A}_\mu + i\frac{g'}{2}B_\mu\}L.
\end{aligned} \tag{26}$$

A little algebra yields

$$\begin{aligned}
\mathcal{L}_e = &- \{\bar{e}\gamma_\mu\partial_\mu e + \bar{\nu}_L\gamma_\mu\partial_\mu\nu_L\} + \frac{g}{\sqrt{2}}\{i\bar{e}_L\gamma_\mu\nu_L W_\mu^- + h.c.\} \\
&+ \frac{e}{sin\theta_W cos\theta_W}Z_\mu^0\{i\bar{L}\frac{\tau_3}{2}\gamma_\mu L + sin^2\theta_W i\bar{e}\gamma_\mu e\} \\
&+ eA_\mu(-i)\bar{e}\gamma_\mu e.
\end{aligned} \tag{27}$$

Next, we consider quarks. The standard assignment is given by

$$\begin{aligned}
\begin{pmatrix} u_L \\ d_{cL} \end{pmatrix}, \begin{pmatrix} c_L \\ s_{cL} \end{pmatrix} &\;:\; SU(2) \;\; doublets, \;\; Y = \frac{1}{3}; \\
u_R, c_R &\;:\; SU(2) \;\; singlets, \;\; Y = \frac{4}{3}; \\
d_R, s_R &\;:\; SU(2) \;\; singlets, \;\; Y = -\frac{2}{3},
\end{aligned} \tag{28}$$

where

$$\begin{aligned}
d_c &= d\cos\theta_c + s\sin\theta_c, \\
s_c &= -d\sin\theta_c + s\cos\theta_c,
\end{aligned} \tag{29}$$

with θ_c the Cabibbo angle. This yields

$$\begin{aligned}
\mathcal{L}_Q = &- \{\bar{u}\gamma_\mu\partial_\mu u + \bar{d}\gamma_\mu\partial_\mu d + \bar{c}\gamma_\mu\partial_\mu c + \bar{s}\gamma_\mu\partial_\mu s\} \\
&+ \frac{g}{\sqrt{2}}\{i\bar{d}_{cL}\gamma_\mu u_L W_\mu^- + i\bar{s}_{cL}\gamma_\mu c_L W_\mu^- + h.c.\} \\
&+ \frac{e}{\sin\theta_W \cos\theta_W}Z_\mu^0\{\frac{i}{2}[\bar{u}_L\gamma_\mu u_L - \bar{d}_L\gamma_\mu d_L \\
&+ \bar{c}_L\gamma_\mu c_L - \bar{s}_L\gamma_\mu s_L] - \sin^2\theta_W J_\mu^{e.m.}\} \\
&+ eA_\mu J_\mu^{e.m.},
\end{aligned} \tag{30}$$

where the electromagnetic current for quarks is given by

$$J_\mu^{e.m.} = \frac{2}{3} i\bar{u}\gamma_\mu u - \frac{1}{3} i\bar{d}\gamma_\mu d + \frac{2}{3} i\bar{c}\gamma_\mu c - \frac{1}{3} i\bar{s}\gamma_\mu s. \tag{31}$$

Note that the neutral weak current, i.e., the current which couples to Z_μ^0, is flavor-conserving [i.e., $\triangle I = 0, \triangle I_3 = 0, \triangle S = 0$, and $\triangle C = 0$]. Introduction of the charm quark to make up another left-handed doublet allows one to avoid a sizable flavor-changing neutral weak current, so that consistency with experimental observations may be achieved. This is the so-called "GIM mechanism" (Glashow et al. 1970). It is worth mentioning that introduction of the charm quark in this context preceded the discovery of the ψ/J family (i.e., a family of quarkonium states consisting of a charm quark and an anticharm quark) in 1974.

The fermion-mass terms present some problem because the left-handed and right-handed components of a fermion belong to different representations of SU(2)×U(1). To preserve gauge symmetry, we may write, in the electron sector,

$$\begin{aligned}
\mathcal{L}_e^m &= -G_e(\bar{R}\phi^\dagger L + \bar{L}\phi R) \\
&= -\frac{G_e v}{\sqrt{2}}\bar{e}e + ..., \quad in \ the \ U-gauge,
\end{aligned} \tag{32}$$

This may explain to some extent why $m(\nu_e) = 0$ [and $m(\nu_\mu) = 0$], provided that right-handed neutrinos do not exist. However, there are many masses which are known to differ from zero: $m_e, m_\mu, m_u, m_d, m_c$, and m_s for the first two generations. If the mass generation mechanism similar to Eq. (32) is used, there are six parameters, one for each nonzero mass. It is clear that such mass generation mechanism is *not* natural, but ideas for a better picture remain to be both speculative and qualitative.

§.13.3. Weak Interactions at Low Energies

The GSW SU(2)×U(1) electroweak theory is renormalizable, allowing for calculations of higher order graphs for a given physical process. Elements for discussing renormalizability of the GSW theory are similar to those given earlier in Ch. 10 for QED. Here we choose not to discuss the subject any further because such discussion is necessarily highly technically involved. Instead, we wish to consider mainly weak interactions at low energies, i.e., $E \ll M_{W^\pm}, M_{Z^0}$, where existing experimental data are all about. To this end, we shall consider three generations of fermions (cf. §.0.1.), *leptons:*

$$\begin{pmatrix} \nu_e \\ e^- \end{pmatrix}, \quad \begin{pmatrix} \nu_\mu \\ \mu^- \end{pmatrix}, \quad \begin{pmatrix} \nu_\tau \\ \tau^- \end{pmatrix};$$

quarks:

$$\begin{pmatrix} u \\ d \end{pmatrix}, \quad \begin{pmatrix} c \\ s \end{pmatrix}, \quad \begin{pmatrix} t \\ b \end{pmatrix}. \tag{33}$$

We assume the $e - \mu - \tau$ universality for the sake of simplicity. In addition, the Cabibbo rotation in the case of two generations,

$$\begin{pmatrix} d_c \\ s_c \end{pmatrix} = \begin{pmatrix} cos\theta_c & sin\theta_c \\ -\sin\theta_c & \cos\theta_c \end{pmatrix} \begin{pmatrix} d \\ s \end{pmatrix}, \tag{34}$$

is to be replaced by a general Kobayashi-Maskawa rotation in the case of three generations (Kobayashi and Maskawa 1973),

$$\begin{pmatrix} d' \\ s' \\ b' \end{pmatrix} = \begin{pmatrix} c_1 & s_1 c_3 & s_1 s_3 \\ -s_1 c_2 & c_1 c_2 c_3 - s_2 s_3 e^{i\delta} & c_1 c_2 s_3 + s_2 c_3 e^{i\delta} \\ -s_1 s_2 & c_1 s_2 c_3 + c_2 s_3 e^{i\delta} & c_1 s_2 s_3 - c_2 c_3 e^{i\delta} \end{pmatrix} \begin{pmatrix} d \\ s \\ b \end{pmatrix}, \tag{35}$$

where $c_i \equiv cos\theta_i$, $s_i \equiv sin\theta_i$ $(i = 1, 2, 3)$, and δ is the CP-violating phase. There are many other parametrizations of the matrix, as briefly discussed in the 1988 publication of Particle Data Group. In the parametrization given above (and some others as well), the angle θ_1 may be identified with the Cabibbo angle θ_c.

Generalizing Eqs. (27) and (30) to the case of three generations, we obtain the weak-interaction lagrangian,

$$\mathcal{L}_W = \frac{1}{2\sqrt{2}} \frac{e}{\sin\theta_W} \{ \tilde{J}_\lambda^{(-)}(x) W_\lambda^{(-)}(x) + h.c. \} + \frac{1}{2} \frac{e}{\sin\theta_W \cos\theta_W} Z_\lambda^0(x) \tilde{N}_\lambda(x), \tag{36}$$

with $e = g \sin\theta_W = g' \cos\theta_W$ and

$$\tilde{J}_\lambda^{(-)} = \ell_\lambda^{(-)} + J_\lambda^{(-)}, \tag{37a}$$

$$\tilde{N}_\lambda = \ell_\lambda^{(0)} + N_\lambda. \tag{37b}$$

Here we have

$$\ell_\lambda^{(-)} = i\bar{e}\gamma_\lambda(1 + \gamma_5)\nu_e + i\bar{\mu}\gamma_\lambda(1 + \gamma_5)\nu_\mu + i\bar{\tau}\gamma_\lambda(1 + \gamma_5)\nu_\tau. \tag{38a}$$

$$J_\lambda^{(-)} = i\bar{d}'\gamma_\lambda(1+\gamma_5)u + i\bar{s}'\gamma_\lambda(1+\gamma_5)c$$
$$+ i\bar{b}'\gamma_\lambda(1+\gamma_5)t. \tag{38b}$$

$$\ell_\lambda^{(0)} = \frac{i}{2}\{\bar{\nu}_e\gamma_\lambda(1+\gamma_5)\nu_e - \bar{e}\gamma_\lambda(1+\gamma_5)e\} + 2sin^2\theta_W i\bar{e}\gamma_\lambda e$$
$$+ \{e \to \mu\} + \{e \to \tau\}. \tag{38c}$$

$$N_\lambda = \frac{i}{2}\{\bar{u}\gamma_\lambda(1+\gamma_5)u - \bar{d}\gamma_\lambda(1+\gamma_5)d$$
$$+ \bar{c}\gamma_\lambda(1+\gamma_5)c - \bar{s}\gamma_\lambda(1+\gamma_5)s$$
$$+ \bar{t}\gamma_\lambda(1+\gamma_5)t - \bar{b}\gamma_\lambda(1+\gamma_5)b\}$$
$$- 2sin^2\theta_W J_\lambda^{e.m.}. \tag{38d}$$

The hadronic electromagnetic current $J_\lambda^{e.m.}$ is given by

$$J^{e.m.} = \frac{2}{3}i\bar{u}\gamma_\lambda u - \frac{1}{3}i\bar{d}\gamma_\lambda d + \frac{2}{3}i\bar{c}\gamma_\lambda c$$
$$- \frac{1}{3}i\bar{s}\gamma_\lambda s + \frac{2}{3}i\bar{t}\gamma_\lambda t - \frac{1}{3}i\bar{b}\gamma_\lambda b. \tag{39}$$

In cases where $W^\pm$ and Z^0 are not directly observed, we write the second order S-matrix as follows:

$$S^{(2)} = -\frac{e^2}{8sin^2\theta_W}\int d^4x d^4y T(\tilde{J}_\lambda^{(-)}(x)\tilde{J}_\eta^{(+)}(y))$$
$$\cdot \frac{1}{(2\pi)^4}\int d^4k e^{-ik\cdot(x-y)}\frac{1}{i}\frac{\delta_{\lambda\eta}}{M_W^2 + k^2 - i\varepsilon}$$
$$- \frac{e^2}{8sin^2\theta_W cos^2\theta_W}\int d^4x d^4y T(\tilde{N}_\lambda(x)\tilde{N}_\eta(y))$$
$$\cdot \frac{1}{(2\pi)^4}\int d^4k e^{-ik\cdot(x-y)}\frac{1}{i}\frac{\delta_{\lambda\eta}}{M_Z^2 + k^2 - i\varepsilon}. \tag{40}$$

At energies well below M_{Z^0} and $M_{W^\pm}$, we may neglect k^2 as compared to M_{Z^0} or $M_{W^\pm}$ since the four-momentum k can be expressed as a simple linear combination of external momenta (by virtue of energy-momentum conservation at each vertex). Thus, the integration over d^4k yields $\delta^4(x-y)$. Define an effective lagrangian,

$$S^{(2)} \approx i\int d^4x \mathcal{L}_W^{eff}(x), \quad \text{for} \quad k \ll M_{W^\pm}, M_{Z^0}. \tag{41}$$

We find

$$\mathcal{L}_W^{eff}(x) = \frac{G_F}{\sqrt{2}}(\tilde{J}_\lambda^{(-)}(x)\tilde{J}_\lambda^{(+)}(x) + \tilde{N}_\lambda(x)\tilde{N}_\lambda(x)), \tag{42}$$

with the Fermi coupling constant G_F given by

$$\frac{G_F}{\sqrt{2}} = \frac{e^2}{8M_W^2 sin^2\theta_W} = \frac{e^2}{8M_Z^2 sin^2\theta_W cos^2\theta_W}. \tag{43}$$

A precise value for G_F can be extracted from studies of muon decay $\mu^- \to e^-\bar{\nu}_e\nu_\mu$ (Particle Data Group 1988; see Ch. 14):

$$G_F = (1.16637 \pm 0.00004) \times 10^{-5} GeV^{-2}c^4. \tag{44}$$

We may use Eqs. (37) and (38) to rewrite Eq. (42) as follows:

$$\begin{aligned}
\mathcal{L}_W^{eff}(x) = \frac{G_F}{\sqrt{2}}\{&[\ell_\lambda^{(-)}\ell_\lambda^{(+)} + \ell_\lambda^{(0)}\ell_\lambda^{(0)}] \\
&+ [\ell_\lambda^{(-)}J_\lambda^{(+)} + \ell_\lambda^{(+)}J_\lambda^{(-)} + 2\ell_\lambda^{(0)}N_\lambda] \\
&+ [J_\lambda^{(-)}J_\lambda^{(+)} + N_\lambda N_\lambda]\},
\end{aligned} \tag{45}$$

suggesting a classification of weak interactions into three distinct categories: (1) purely leptonic, (2) semileptonic, and (3) purely hadronic (or nonleptonic). Examples of these reactions are listed below. Formulae on decay rates, cross sections, and asymmetries are the subject of discussions in particle physics textbooks (e.g., Commins and Bucksbaum 1983); among them, most commonly used ones are listed in the 1988 publication of Particle Data Group and will be later reproduced in the next chapter. Note that we are already fully equipped to derive these formulae except that further discussion on such derivation will certainly divert our attention too much away from the focus of this book.

(a) Purely Leptonic Weak Interactions:

Examples of purely leptonic interactions which have been subject to experimental studies include:

$$muon\ decay: \ \mu^- \to e^-\bar{\nu}_e\nu_\mu, \tag{46a}$$

$$(\nu_\mu e)\ scattering: \ \nu_\mu + e^- \to \nu_\mu + e^-, \tag{46b}$$

$$(\nu_e e) \quad scattering: \quad \nu_e + e^- \rightarrow \nu_e + e^-, \tag{46c}$$

$$Weak \quad interaction \quad studies \quad in \quad e^+ e^- \rightarrow \mu^+ \mu^-. \tag{46d}$$

Note that only $W^\pm$ exchange is responsible for muon decay (46a) while only Z^0 exchange is allowed in the case of $(\nu_\mu e)$ scattering. However, both $W^\pm$ and Z^0 exchanges must be considered in the case of $(\nu_e e)$ scattering and both Z^0 and γ exchanges are involved in weak interaction studies associated with $e^+ e^- \rightarrow \mu^+ \mu^-$. Accordingly, it is not possible to further distinguish purely leptonic weak interactions into charged and neutral weak interactions.

(b) Semileptonic Weak Interactions:

Semileptonic weak interactions can be classified further into two distinct subcategories:

(b.1) semileptonic charged weak interactions

Examples of charged weak interactions include neutron β-decay $n \rightarrow p + e^- + \bar{\nu}_e$, β-decays of mesons such as $\pi^- \rightarrow \pi^0 + e^- + \bar{\nu}_e$ and $K^- \rightarrow \pi^0 + e + \bar{\nu}_e$, pion decay $\pi^+ \rightarrow \mu^+ \nu_\mu$, nuclear β-decays such as $^{12}B \rightarrow^{12} C + e^- + \nu_e$ and $^{12}N \rightarrow^{12} C + e^+ + \nu_e$, muon capture in hydrogen $\mu^- + p \rightarrow \nu_\mu + n$, nuclear muon capture, and so on. Comparison of the ^{14}O $\beta - decay$ rate with that for muon decay yields a value on the Cabibbo angle,

$$\theta_c = 0.210 \pm 0.025. \tag{47}$$

(b.2) semileptonic neutral weak interactions

Examples of semileptonic neutral weak interactions include

$$\begin{aligned}
&\nu(\bar{\nu}) + N \rightarrow \nu(\bar{\nu}) + X, \\
&\nu(\bar{\nu}) + p \rightarrow \nu(\bar{\nu}) + p, \\
&\nu(\bar{\nu}) + p \rightarrow \nu(\bar{\nu}) + N + \pi, \\
&\nu(\bar{\nu}) +^{12} C(g.s.) \rightarrow \nu(\bar{\nu}) +^{12} C^*(15.110), \\
&\vec{e} + p \rightarrow e + p, \\
&etc.
\end{aligned} \tag{48}$$

A "world average" value for the electroweak mixing parameter $sin^2\theta_W$ as obtained from neutral weak interaction experiments is given by

$$sin^2\theta_W = 0.230 \pm 0.005. \qquad (49)$$

It is useful to mention that classification of semileptonic weak inteactions into exclusive ones such as $\nu_\mu + p \rightarrow \nu_\mu + p$ and inclusive ones such as $\nu_\mu + p \rightarrow \nu_\mu + X$ (with X denoting unobserved hadrons) derives mostly from the difference in theoretical treatments (rather than in physics contents).

(c) Nonleptonic Weak Interactions:

Among all weak interactions, nonleptonic or purely hadronic weak interactions are least understood since the last term in Eq. (45) must be augmented by QCD corrections before a qualitative description can even be perceived. Examples of such interactions include hadronic decays of hyperons such as $\Lambda^0 \rightarrow p + \pi^-$ and $\sum^- \rightarrow n + \pi^-$, $K_{2\pi}$ decays such as $K^\pm \rightarrow \pi^\pm + \pi^0$ and $K_s^0 \rightarrow \pi^+ + \pi^-$, $K_{3\pi}$ decays such as $K_L^0 \rightarrow \pi\pi\pi$, hadronic decays of heavy-quark systems, and flavor-conserving nonleptonic weak interactions. Note that only $W^\pm$ exchanges contribute to flavor-changing nonleptonic weak interactions while both $W^\pm$ and Z^0 exchanges enter the problem of flavor-conserving nonleptonic weak interactions. In any event, complications arising from importance of QCD corrections prevent us from drawing a definitive support toward the GSW electroweak theory on the basis of nonleptonic weak interaction studies.

In summary, experimental data on weak interactions at low energies are described very well by the GSW $SU(2) \times U(1)$ electroweak theory. An additional boost came from recent direct observations of the $W^\pm$ and Z^0 weak bosons (at the predicted masses). Results of analyses of experimental data, which support quantitatively the validity of the GSW electroweak theory, are summarized in the 1988 publication of Particle Data Group and some portion of it will be reproduced and discussed in the subsequent chapter. With the coming generation of e^+e^- colliders with the center-of-mass energies $\sqrt{s}$ greater than M_{Z^0} such as SLC at Stanford and LEP at CERN, it is expected that the theory will be subject to much more severe scrutiny. Owing to the fact that weak interaction studies over the last century constantly produced one surprise after another, one might suspect that the GSW $SU(2) \times U(1)$ electroweak theory may mark the beginning of an unfolding mystery, rather than the completion of a zigsaw puzzle.

Appendix
Feynman Rules in the R_ξ Gauge

The Glashow-Salam-Weinberg (GSW) $SU(2) \times U(1)$ electroweak theory is a spontaneously broken gauge theory. For $\mu < 0$, three of the four real fields associated with the complex weak-isodoublet scalar field are absorbed as the longitudinal polarization degrees of freedom, one for each of three gauge fields, while the fourth one becomes a massive real scalar Higgs field. The idea of the "Higgs mechanism", as a special case of spontaneous symmetry breaking, is usually explained in the context of a classical field theory, where the ground state of the classical field is found by minimizing the potential. To compute quantum corrections or to treat the problem of renormalization in general, we need to work with the quantized version of the theory.

To quantize a gauge field theory with a general multicomponent field ϕ_i and corresponding set of sources J_i, we may work with the partition functional $Z[J] = exp\{iW[J]\}$ and define a set of quantities

$$\Phi_i(x) = \frac{\delta W[J]}{\delta J_i(x)}. \qquad (A1)$$

It may be shown that $\Phi_i(x)$ is the vacuum expectation value of ϕ_i in the presence of $J_i(x)$. Thus, for $J_i(x) = 0$, $\Phi_i(x) \mid_{J_i=0} = v_i$, which may differ from zero, is the vacuum expectation value of ϕ_i in the absence of sources. Following the standard exercise in classical mechanics, we may define a Legendre transformation $\Gamma[\Phi]$ by

$$\Gamma[\Phi] = W[\mathbf{J}] - \int d^4x J(x) \cdot \Phi(x), \qquad (A2)$$

and a "superpotential" $\mathcal{V}$ by

$$\Gamma[\Phi = 0] = -(2\pi)^4 \delta^4(0) \mathcal{V}(\phi). \qquad (A3)$$

It is then possible to demonstrate that $\mathcal{V}$ has properties strictly analogous to the classical potential $V(\phi_i)$. In other words, the simple classical analysis of $V(\phi_i)$ that describes the spontaneous symmetry breaking of Higgs mechanism (§.13.1) remains valid in the quantized version of the theory (Abers and Lee 1973).

Although the classical theory was presented in unitary gauge, or "U gauge", the quantum theory may be formulated in a variety of gauges. In addition to the U gauge, there are the R gauge (or the Landau gauge) and the 't Hooft-Feynman gauge, each of which is a special case of the so-called generalized renormalizable gauge, or the "R_ξ" gauge. The R_ξ gauge is characterized by continuous real parameters α, ξ, and η, which arise from choice of gauge-fixing terms to eliminate the mixing between gauge bosons and the unphysical scalar bosons (i.e., unoberved Higgs fields in the general gauge). The Faddeev-Popov ghost fields are then calculated accordingly (using the method introduced in Appendix A of Chapter 12). Although the Green's functions of the theory in general depend on the gauge-fixing parameters, the physical results, or the S-matrix elements, are gauge invariant (Fujikawa et al. 1972). In what follows, we wish to describe, without proof, Feynman rules in the R_ξ gauge.

In the R_ξ gauge, the various "particles" include the usual fermions, gauge bosons ($W^\pm$, Z^0, and γ), the Higgs scalar σ, three unphysical scalar bosons $s^\pm$ and χ, and an isotriplet of Faddeev-Popov ghosts. In the limit $\xi, \eta \to 0$, the unphysical scalar bosons disappear as a result of being absorbed as the longitudinal components of the gauge bosons. This is just the U gauge. On the other hand, for $\eta, \xi \to \infty$ (which gives rise to the R gauge), the unphysical scalar bosons become Goldstone bosons.

The propagators for vector mesons, Higgs, unphysical scalars, and ghost fields are given in the R_ξ gauge as follows:

(i) Vector mesons:

$$D^{W^\pm}_{\mu\nu}(k) = \frac{1}{i}[\delta_{\mu\nu} - \frac{k_\mu k_\nu}{k^2 + m_W^2/\xi}(1 - \frac{1}{\xi})]\frac{1}{k^2 + m_W^2 - i\epsilon}. \qquad (A4)$$

$$D^Z_{\mu\nu}(k) = \frac{1}{i}[\delta_{\mu\nu} - \frac{k_\mu k_\nu}{k^2 + m_Z^2/\eta}(1 - \frac{1}{\eta})]\frac{1}{k^2 + m_Z^2 - i\epsilon}. \qquad (A5)$$

$$D^\gamma_{\mu\nu}(k) = \frac{1}{i}[\delta_{\mu\nu} - \frac{k_\mu k_\nu}{k^2}(1 - \alpha)]\frac{1}{k^2 - i\epsilon}. \qquad (A6)$$

(ii) Higgs boson (mass m_σ):

$$\sigma: \qquad D^\sigma(k) = \frac{1}{i}\frac{1}{k^2 + m_\sigma^2 - i\epsilon}. \qquad (A7)$$

356

(iii) Unphysical scalar bosons:

$$s^{\pm} : \qquad D^{s^{\pm}}(k) = \frac{1}{i} \frac{1}{k^2 + m_W^2/\xi - i\epsilon}. \qquad (A8)$$

$$\chi : \qquad D^{\chi}(k) = \frac{1}{i} \frac{1}{k^2 + m_Z^2/\eta - i\epsilon}. \qquad (A9)$$

(iv) Fermion scalar ghosts:

$$D_{\pm}^{g}(k) = \frac{1}{i} \frac{1}{k^2 + m_W^2/\xi - i\epsilon}. \qquad (A10)$$

$$D_{0}^{g}(k) = \frac{1}{i} \frac{1}{k^2 + m_Z^2/\eta - i\epsilon}. \qquad (A11)$$

The gauge-fixing parameter α appearing in the photon propagator (A6) has no effect on S-matrix elements since $D_{\mu\nu}^{\gamma}(k)$ is always sandwiched between conserved currents. Or, the term $(k_{\mu}k_{\nu}/k^2)(1 - \alpha)$ always yields zero in S-matrix elements. For $\xi, \eta \neq 0$, all other propagators vary as k^{-2} for large k^2, suggesting renormalizability by "naive power counting".

Note that we may rewrite Eq. (A4) as follows:

$$D_{\mu\nu}^{W^{\pm}}(k) = \frac{1}{i} \frac{\delta_{\mu\nu} + (1/m_W^2)k_{\mu}k_{\nu}}{k^2 + m_W^2 - i\epsilon} - \frac{1}{i} \frac{k_{\mu}k_{\nu}}{m_W^2} \frac{1}{k^2 + m_W^2/\xi - i\epsilon}, \qquad (A12)$$

where the second term on the right-hand side has a pole that is to be canceled in the S-matrix element by the pole in the $s^{\pm}$ propagator (A8). Eq. (A12) indicates that the R gauge is recovered in the limit $\xi, \eta \to \infty$.

$$\lim_{\xi \to \infty} D_{\mu\nu}^{W^{\pm}}(k) = \frac{1}{i}(\delta_{\mu\nu} - \frac{k_{\mu}k_{\nu}}{k^2}) \frac{1}{k^2 + m_W^2 - i\epsilon}, \qquad (A13)$$

which is the propagator in the R gauge. Similar results hold, of course, also for the Z^0 boson. For $\xi = 1$ and $\eta = 1$, we obtain the 't Hooft-Feynman gauge

$$D_{\mu\nu}^{W^{\pm}}(k) = \frac{1}{i} \frac{\delta_{\mu\nu}}{k^2 + m_W^2 - i\epsilon}. \qquad (A14)$$

$$D^{s^{\pm}}(k) = \frac{1}{i} \frac{1}{k^2 + m_W^2 - i\epsilon}. \qquad (A15)$$

Finally, we consider the limit as $\xi, \eta \to 0$, which must be taken with care (i.e., after S-matrix elements are computed) to avoid ambiguities. Formally, we have, in the U gauge,

$$\lim_{\xi \to 0} D_{\mu\nu}^{W^\pm}(k) = \frac{1}{i} \frac{\delta_{\mu\nu} + k_\mu k_\nu / m_W^2}{k^2 + m_W^2 - i\epsilon}, \qquad (A16)$$

which is the propagator familiar for a massive vector meson. This gauge has the advantage that ghost and unphysical boson propagators vanish identically. However, for large k^2, $D_{\mu\nu}^{W^\pm}(k^2) = O(1)$ and renormalizability is not transparent by naive power counting. Since the S-matrix is independent of ξ and η, it has become possible to prove renormalization for $\xi, \eta \neq 0$ and then to take the limit $\xi, \eta \to 0$.

We proceed to consider the vertex factors associated with the GSW electroweak theory. Using the (e^-, ν_e) to illustrate the couplings of fermions to gauge bosons, we use Eq. (27) in the text to obtain the following Feynman rules.

Electron-Photon Coupling:

$$\Longleftrightarrow \quad e\gamma_\mu \qquad (A17a)$$

Electron-Z^0 Coupling:

$$\Leftrightarrow \quad i\frac{ie}{sin\theta_W \, cos\theta_W}\{-\tfrac{1}{4}\gamma_\mu(1 + \gamma_5) + sin^2\theta_W \gamma_\mu\} \qquad (A17b)$$

358

Neutrino-Z^0 Coupling:

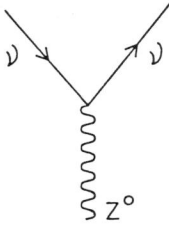

$$\leftrightarrow \quad i\frac{ie}{sin\theta_W\, cos\theta_W} \cdot \frac{1}{4}\gamma_\mu(1+\gamma_5) \qquad (A17c)$$

Fermion-$W^\pm$ Coupling:

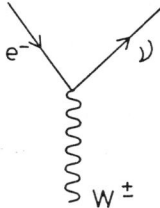

$$\Longleftrightarrow \quad i\frac{ig}{2\sqrt{2}}\gamma_\mu(1+\gamma_5) \qquad (A17d)$$

Couplings of quarks to gauge bosons may be obtained using Eq. (30) (instead of Eq. (27)) in the text.

In view of Eq. (18) in the text, there are trilinear and quartilinear self couplings among gauge bosons. These couplings are illustrated in Fig. 1 with momenta and internal indices shown explicitly. We introduce

$$T_{\alpha\beta\gamma} = (p-q)_\beta\delta_{\alpha\mu} + (q-r)_\alpha\delta_{\mu\beta} + (r-p)_\mu\delta_{\beta\alpha}. \qquad (A18a)$$

$$S_{\mu\nu,\lambda\rho} = 2\delta_{\mu\nu}\delta_{\lambda\rho} - \delta_{\mu\lambda}\delta_{\nu\rho} - \delta_{\mu\rho}\delta_{\nu\lambda}. \qquad (A18b)$$

Using Eq. (18) given in the text, we assign to the trilinear $W^+W^-\gamma$ coupling the factor $e\,T_{\alpha\beta\gamma}$ and the trilinear $W^+W^-Z^0$ coupling the factor $e\,cos\theta_W\,T_{\alpha\beta\gamma}$. On the other hand, Feynman rules for the $WWWW$, $WW\gamma\gamma$, and $WWZZ$ couplings shown in Fig. 1 are given, respectively, by $-g^2 S_{\mu\nu,\lambda\rho}$, $e^2 S_{\mu\nu,\lambda\rho}$, $e^2 cot^2\theta_W S_{\mu\nu,\lambda\rho}$.

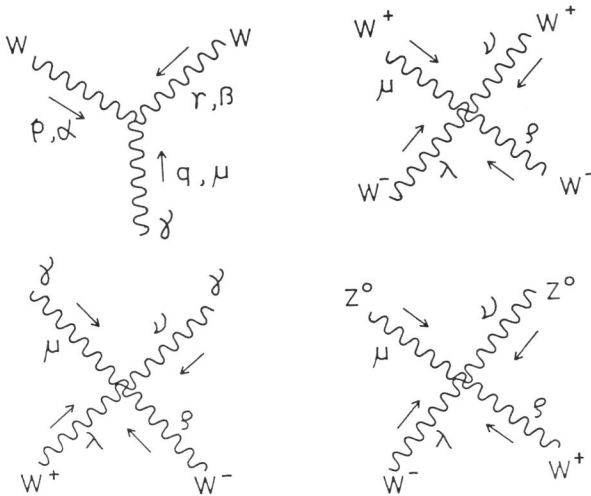

Fig. 1. Trilinear and quartilinear couplings in the GSW $SU(2) \times U(1)$ electroweak theory. p, q, and r are momentum variables while α, μ, etc. are Lorentz indices.

These rules should be contrasted with the result obtained when the minimal substitution $\partial_\mu W_\nu^\pm \rightarrow (\partial_\mu \mp ieA_\mu)W_\nu^\pm$ is made in the Lagrangian for the old-fashioned intermediate boson theory. It is found that in the new theory an additional term $-ieW_\mu^+ W_\nu^- F^{\mu\nu}$ arises, implying an additional contribution of $e\hbar/2m_W c$ to the magnetic moment of W^-, over and above the "normal" magnetic moment of $\mu_{W^0} = e\hbar/2m_W c$. The total W^- magnetic moment is thus expected to be

$$\mu_W = 2\mu_{W^0}. \qquad (A19)$$

It is important to know that there are many more trilinear and quartilinear couplings as we include the Higgs boson σ and the three unphysical bosons. Feynman rules for these couplings can be inferred from the original Higgs lagrangian, Eq. (6) in the text. The generic forms for the trilinear couplings are illustrated in Fig. 2.

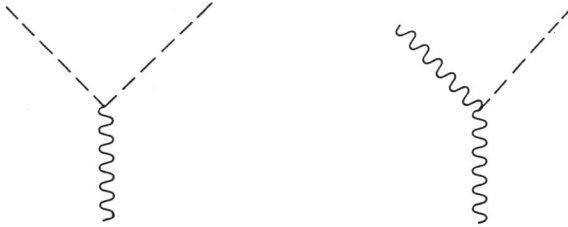

Fig. 2. Trilinear couplings which involves the Higgs boson σ and the three unphysical scalar bosons.

The couplings of the first type (LHS of Fig. 2) include $s^+s^-\gamma$, $s^+s^-Z^0$, $\sigma\chi Z^0$, σs^+W^-, χs^+W^-, σs^-W^+, and χs^-W^+ while those of the second type include γW^-s^+, $Z^0W^-s^+$, $W^+W^-\sigma$, and $Z^0Z^0\sigma$.

The generic form for the quartilinear couplings are illustrated by Fig. 3, where all possible combinations consistent with the interaction lagrangian must be taken into account.

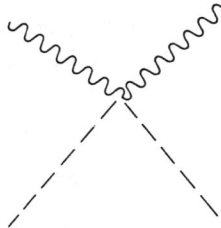

Fig. 3. quartilinear couplings which involves the Higgs boson σ and the three unphysical scalar bosons.

In addition, there are of course couplings of fermions to the Higgs boson σ and any of the three unphysical scalar bosons $s^\pm$ and χ. The diagrams in the electron sector are illustrated in Fig. 4. Feynman rules may be inferred from Eq. (32) in the text.

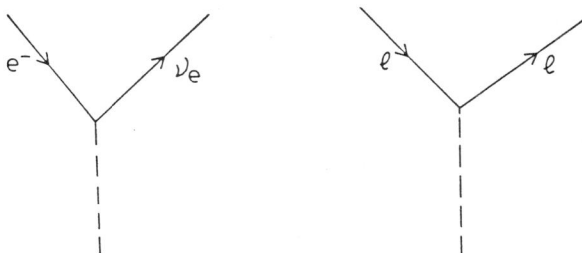

Fig. 4. Couplings of fermions to the Higgs boson σ or the unphysical scalar boson.

Finally, there are three Faddeev-Popov ghost fields which are yet to be incorperated to complete the set of Feynman rules for the GSW electroweak theory. In view of the large number of fields involved in the game, it is often convenient to focus on a specific higher-order loop calculation in order to decide how the various symmetries, such as gauge invariance, are maintained after all relevant diagrams are taken into account.[2]

---

[2] *As pointed out by Cheng and Tsai (1987), one must exercise care in obtaining Feynman rules in certain gauges such as the Coulomb gauge and the axial gauge. This is related to the fact that the method of path integrals suffers from ambiguities in these gauges.*

References

Weinberg, S., Phys. Rev. Lett. **19**, 1264 (1967); Salam, A., in *Elementary Particle Physics* (Nobel Symposium No. 8), ed. N. Svartholm (Almquist and Wiksell, Stockholm, 1968), p. 367; and Glashow, S. L., Nucl. Phys. **22**, 519 (1961); original references on the $SU(2) \times U(1)$ electroweak theory.

't Hooft, G., Nucl. Phys. **B33**, 173 (1971) and **B35**, 167 (1971) on the original proof of the renormalizability of the GSW electroweak theory.

Higgs, P.W., Phys. Lett. **12**, 132 (1964); Phys. Rev. Lett. **13**, 508 (1964); and Phys. Rev. **145**, 1156 (1966); Anderson, P.W., Phys. Rev. **130**, 439 (1963); Englert F. and Brout, R., Phys. Rev. Lett. **13**, 321 (1964); Englert, F., Brout, R., and Thiry, M.F., Nuovo Cimento **48**, 244 (1966); Guralnik, G.S., Hagen, C.R., and Kibble, T.W.B., Phys. Rev. Lett. **13**, 585 (1964); Kibble, T.W.B., Phys. Rev. **155**, 1554 (1967); on the Higgs mechanism.

Glashow, S.L., Iliopoulos, J., and Maiani, L., Phys. Rev. **D2**, 1285 (1970) on the GIM mechanism to suppress flavor-changing neutral weak currents.

Kobayashi, M., and Maskawa, T., Prog. Theor. Phys. Japan **49**, 652 (1973); Particle Data Group, *Review of Particle Properties*, Phys. Lett. **204B**, 1 (1988); on the Kobayashi-Maskawa matrix for three generations of fermions and other possible parametrizations.

Commins, E.D., and Bucksbaum, P.H., in *Weak Interactions of Leptons and Quarks* (Cambridge University Press, England, 1983) on an extensive review of weak interactions up to 1982.

Arnison, G., et al., Phys. Lett. **122B**, 103 (1983) and **126B**, 398 (1983); Banner, M., et al., Phys. Lett. **122B**, 476 (1983); Bagnaia, P., et al. Phys. Lett. **129B**, 130 (1983); on the experimental discovery of $W^\pm$ and Z^0 weak bosons.

Fujikawa, K., Lee, B.W., and Sanda, A.I., Phys. Rev. **D6**, 2923 (1972) on the R_ξ gauge.

H. Cheng and E.-C. Tsai, Chinese J. Phys. **25**, 95 (1987); on canonical quantization of non-abelian gauge field theories and Feynman rules.

Exercises: *Chapter 13*

13.1. Show that Eq. (18) follows from Eqs. (1) and (13a) - (13c). Note that $e = g\sin\theta_W$ arises naturally.

13.2. (a) Prove Eq. (30) from Eqs. (28) and (29).

(b) Suppose that the charm quark were absent. s_{cL} must then be considered as an SU(2) singlet. Derive $\mathcal{L}_Q$ in this case. Is there any way to suppress flavor-changing neutral weak currents (other than the GIM mechanism)?

13.3. Consider $e^+e^- \to \mu^+\mu^-$ for $m_e, m_\mu \ll \sqrt{s} \ll M_{Z^0}$. (There are two Feynman diagrams: $e^+e^- \to \gamma \to \mu^+\mu^-$ and $e^+e^- \to Z^0 \to \mu^+\mu^-$.)

(a) Use Feynman rules to write down the T-matrix element.

(b) Obtain $d\sigma/d\Omega$ with Ω the outgoing μ^+ (or μ^-) solid angle.

(c) Derive the charge asymmetry as defined by

$$\mathcal{A} \equiv \frac{\sigma(\cos\theta > 0) - \sigma(\cos\theta < 0)}{\sigma(\cos\theta > 0) + (\sigma(\cos\theta < 0)}.$$

13.4. Consider $e^+e^- \to W^+W^-$ at sufficiently high collider energies such as the proposed LEP II.

(a) Draw all Feynman diagrams relevant for the process.

(b) Use Feynman rules to write down the T-matrix element.

(c) Obtain $d\sigma/d\Omega$ with Ω the outgoing W^+ (or W^-) solid angle. Discuss your result.

Chapter 14.
Experimental Tests of the Standard Model

In this chapter, we wish to summarize briefly the status of the standard model as of 1989. To make our presentations up to date, we rely heavily on the 1988 publication of Particle Data Group on "Review of Particle Properties". Since many relevant materials in this useful reference manual cannot be included here because of the limitation in space and the level of our discussion, an interested reader should consult this publication which was updated from the 1986 version and will be updated further on a regular basis.

§.14.1. Quantum Chromodynamics

(1) QCD and the Parton Model

In §.12.2., we learned that quantum chromodynamics, or QCD, is asymptotically free. At sufficiently large Q^2 where the strong coupling $\alpha_s(Q^2)$ ($\equiv g^2/(4\pi)$) is sufficiently small compared to unity, a hadron must then look like a collection of non-interacting quarks, antiquarks, and gluons (or, in a collective term, "partons"), *at least to leading order in* $\alpha_s(Q^2)$. The parton picture of a hadron must be contrasted with its quark-model description at low Q^2, in which for example a baryon is considered as a system of three dressed, valence quarks confined to within a region specified by the baryon size and a meson is a quark-antiquark pair restricted to the meson volume. The two very different pictures of a single hadron, i.e. the parton-model description at large Q^2 and the valence quark-model language at small Q^2, may be made consistent with each other presumably by the asymptotically free nature of QCD. Nevertheless, reconciliation between the two very different pictures of a single hadron depends critically on how we specify the models and the subject is clearly beyond the scope of our present discussions.

In the parton-model language which is relevant at large Q^2, it is customary to work with the infinite-momentum frame in which the momentum of the hadron under study along a chosen $z-$direction (longitudinal direction) is infinitely large. Each parton will then carry a fraction x of the longitudinal momentum of the hadron. The information on parton's transverse momentum may be neglected at least in the first approximation. In addition, averaging over spins may also be carried out for the experiments which we shall mention in this chapter. Thus, a hadron, such as a proton, is characterized, at a given Q^2, by a set of distribution functions (or structure functions): $\{q_i(x), \bar{q}_i(x), G(x)\}$ with q_i, $\bar{q}_i$, and G the quark of flavor i, the antiquark of flavor i, and the gluon, respectively.

Suppose that we do an experiment on deep inelastic electron-proton scattering. In the infinite-momentum frame, the virtual photon strikes a quark or an antiquark of flavor i inside the target proton. Assuming that the electron-proton deep inelastic scattering (DIS) cross section is an incoherent sum of elementary electron-parton scattering cross sections, we find that the ep DIS cross section is proportional to a well-known structure function called $F_2(x, Q^2)$:

$$F_2(x, Q^2) = \sum_i e_i^2 \{q_i(x) + \bar{q}_i(x)\}, \tag{1}$$

where e_i is the electric charge of a quark of flavor i. Bjorken scaling says that $F_2(x, Q^2)$ depends only on x in the limit of large Q^2 and ν ($\equiv E_e - E_e'$) with x ($= Q^2/(2m_p\nu)$) held fixed.

As Q^2 is very large, we may choose to neglect quark-mass terms in QCD (at least for the light quarks). A massless quark can easily split into another quark of less energy and a gluon. Similarly, a massless gluon may also split into a quark-antiquark pair. Owing to energy-momentum conservation, these splitting processes occur collinearly (i.e., all particles involved line up along the same direction). In dimensional regularization, these collinear processes give rise to $\frac{1}{\epsilon}$ divergences. Thus, in massless QCD, we must absorb collinear divergences into a set of redefined structure functions, which depend explicitly on Q^2 causing violation of Bjorken scaling.

To describe the way in which scaling is broken in QCD, it is convenient to define the nonsinglet and singlet quark distributions:

$$\begin{aligned} F^{NS}(x, Q^2) &= q_i(x, Q^2) - q_j(x, Q^2), \\ F^S(x, Q^2) &= \sum_i \{q_i(x, Q^2) + \bar{q}_i(x, Q^2)\}. \end{aligned} \tag{2}$$

The nonsinglet structure functions have nonzero values of flavor quantum numbers such as isospin, strangeness, charm, or baryon number. The variation with

Q^2 of these structure functions is described by the so-called Altarelli-Parisi evolution equations (Altarelli and Parisi, 1977).

$$Q^2 \frac{\partial F^{NS}}{\partial Q^2} = \frac{\alpha_s(Q^2)}{2\pi} P^{qq} \circ F^{NS},$$

$$Q^2 \frac{\partial}{\partial Q^2} \begin{pmatrix} F^S \\ G \end{pmatrix} = \frac{\alpha_s(Q^2)}{2\pi} \begin{pmatrix} P^{qq} & 2N_f P^{qg} \\ P^{gq} & P^{gg} \end{pmatrix} \circ \begin{pmatrix} F^S \\ G \end{pmatrix}, \tag{3}$$

where $G(x, Q^2)$ is the gluon distribution function and "$\circ$" denotes a convolution integral:

$$f \circ g = \int_x^1 \frac{dy}{y} f(y) g(\frac{x}{y}). \tag{4}$$

The splitting functions to leading order in α_s may be obtained by considering leading-order QCD corrections to DIS. Collinear divergences, which appear as $\frac{1}{\epsilon}$ singularities in the dimensional regularization scheme, must be subtracted and absorbed into the redefined structure functions, yielding explicit expressions for the splitting functions. This was done by Altarelli and Parisi (1977), who obtained

$$\begin{aligned} P^{qq} &= \frac{4}{3}(\frac{1+x^2}{1-x})_+ + 2\delta(1-x), \\ P^{qg} &= \frac{1}{2}(x^2 + (1-x)^2), \\ P^{gq} &= \frac{4}{3}(\frac{1+(1-x)^2}{x}), \\ P^{gg} &= 6(\frac{1-x}{x} + x(1-x) + (\frac{x}{1-x})_+ + \frac{11}{12}\delta(1-x)) - \frac{N_f}{3}\delta(1-x). \end{aligned} \tag{5}$$

Here $\frac{1}{(1-x)_+}$ is defined by

$$\int_0^1 dx \frac{f(x)}{(1-x)_+} = \int_0^1 dx \frac{f(x) - f(1)}{(1-x)}.$$

The evolution equations can easily be integrated to yield the expressions which determines the distribution functions at Q^2 in terms of those at a given Q_0^2, the latter being obtained by a global fit to the existing experimental data. Thus, one may assume that the structure function $F_2(x, Q^2)$ is determined by distribution functions which depend explicitly on Q^2:

$$F_2(x, Q^2) = \sum_i e_i^2 \{q_i(x, Q^2) + \bar{q}_i(x, Q^2)\}. \tag{6}$$

Note that Eq. (6) determines a specific subtraction scheme, i.e., the scheme in which how collinear divergences are absorbed into a set of redefined structure functions. The other schemes are of course possible; among them, the "modified minimum subtraction" scheme is often adopted.

The above results are for massless quarks. Algorithms may be obtained for inclusion of nonzero quark masses (Glück et al., 1982). In addition, "higher-twist" contributions of the form

$$F_i(x, Q^2) = F_i^{(LT)}(x, Q^2) + \frac{F_i^{(HT)}(x, Q^2)}{Q^2} + ..., \tag{7}$$

are also of numerical importance at low Q^2 (e.g. $< 10 GeV^2$) or at x very close to unity.

We shall not discuss technical details concerning these corrections although the quality of the existing data already calls for suitable treatment of these higher-order corrections.

(2) Tests of QCD in Deep Inelastic Scattering

The original and still one of the most powerful quantitative tests of perturbative QCD comes from violation of Bjorken scaling observed in deep inelastic lepton-hadron scattering. The quality of contemporary experimental data already calls for proper inclusion of higher order corrections (Curci et al., 1980; Furmanski and Petronzio, 1980, 1982; Floratos, et al., 1981; Herrod and Wada, 1981). The current status of the experimental data has been reviewed by, for example, Voss (1987).

From the evolution equations (3), it is clear that a nonsinglet structure function offers in principle a better test of the theory since its Q^2 evolution is insensitive to the poorly known gluon distribution. However, such a measurement involves differences between cross sections, making a precision test less feasible. In practice, the most accurate measurements, involving singlet-dominated structure functions such as F_2, have resulted in strongly correlated measurements of the QCD renormalization scale $\Lambda_{\overline{MS}}$ in the modified minimum subtraction scheme[1] and the gluon distribution. The most accurate data

---

[1] It is well known that one must choose a renormalization scale, say Λ, in order to make sense out of massless QCD which is invariant under scale transformations.

currently available are from the BCDMS collaboration. The result obtained is,[2]

$$\Lambda = 230 \pm 20(stat.) \pm 60(sys.)MeV, \tag{8}$$

which is consistent with the world average value of $\Lambda = (238 \pm 43)MeV$, (statistical and systematic uncertainty added in quadrature) obtained from the existing deep-inelastic experiments. Here and in the rest of the present chapter Λ is to be understood as the QCD scale parameter in the modified minimum subtraction scheme with four effective massless quarks.[3]

The impact on the measurement of α_s due to the higher order corrections may be estimated as follows. One may use the leading-order evolution equations in the data analysis in order to extract α_s, which is then compared with the result obtained by using the next-to-leading-order equations. In this way, BCDMS obtained $\alpha_s(5GeV) = 0.240$ in the leading-order approximation, whereas their next-to-leading-order fit yields $\alpha_s(5GeV) = 0.191$.

Typically, Λ is extracted from the data by parametrizing the parton distributions in simple analytic forms at some Q_0^2, evolving to higher Q^2 using the next-to-leading-order evolution equations, and fitting globally to the measured structure functions to obtain Λ. Thus an important task of such studies is the extraction of parton distributions at a fixed reference value of Q_0^2. These can then be evolved in Q^2 and used as input for phenomenological studies in hadron-hadron collisions. It is useful to have a simple analytic approximation to the parton distributions valid over a range of x and Q^2 values. Indeed, such parametrizations are available in the literature (Glück et al., 1982; Duke and Owens 1984; Eichten et al., 1984; Diemoz et al., 1987; Martin et al. 1988).

---

[2] *BCDMS collaboration: Benvenuti, A.C., et al., Phys. Lett. **B195**, 97 (1987).*

[3] *We adopt the standard approximation in which quarks of mass greater than μ are neglected completely while quarks of mass less than μ are treated as massless. This means, for example, that the QCD parameter in the range of having five effective massless quarks, i.e. above the b-quark production threshold, is different from that with four effective massless quarks, although the two is related in a simple manner.*

(3) Tests of QCD in High Energy Hadron Collisions

There are many ways in which perturbative QCD can be tested in high energy hadron colliders. The production of single large-transverse-momentum photons offers an interesting possibility, since the leading-order QCD subprocesses are $q\bar{q} \to \gamma g$, $qg \to \gamma q$, and $\bar{q}g \to \gamma\bar{q}$. Explicit expressions for the corresponding scattering amplitudes can be found, for example, in a review article of Owen (1987). If the parton distributions are taken from other processes and a value of Λ is assumed, then an absolute prediction is obtained. Conversely, the data can be used to extract information on parton distributions and a value for Λ. This is also one of the few hard scattering processes for which the next-to-leading-order corrections are known (Aurenche et al., 1984, 1988), so that a precision test is possible in principle. In practice, however, the uncertainties on the most accurate experimental data available to date are in the order of $20 - 30\%$, which is too large to limit the accuracy of the extracted value on α_s (or, equivalently, Λ). At this point, a value for Λ in the range $100 - 300 MeV$ accounts for a wide range of data.

The production of hadrons with large transverse momentum in hadron-hadron collisions provides a direct probe of the scattering of quarks and gluons: $qq \to qq$, $qg \to qg$, $gg \to gg$, etc. The QCD prediction combines the parton distributions with the leading-order $2 \to 2$ parton hard scattering amplitudes. The present generation of $p\bar{p}$ colliders provide center-of-mass energies which are sufficiently high that these processes can be unambiguously identified in two-jet production at large transverse momentum. Corrected inclusive jet cross sections have been directly compared to the calculated parton cross sections, and the agreement is impressive. Data are also available on the angular distribution of jets; they are also in agreement with QCD predictions.[4]

Many authors (Altarelli et al., 1978, 1979, 1984, 1985; etc.) have considered QCD corrections to Drell-Yan type cross sections, i.e., the lepton-pair production in hadron collisions by quark-antiquark annihilation into virtual photons, or W or Z bosons. The $O(\alpha_s)$ QCD corrections are sizable and approximately constant over the lepton-pair mass range probed by experiments, leading to a simple parametrization:

---

[4] *UA1 collaboration: Arnison, G., et at., Phys. Lett.* **B177**, *244 (1986).*

$$\sigma_{DY} = \sigma_{DY}^{(0)} [1 + \frac{\alpha_s(Q^2)}{2\pi} C + ...]. \tag{9}$$

It is interesting to note that the corresponding correction to W and Z production, as measured at $p\bar{p}$ colliders, has essentially the same theoretical form and is of order 30%. Total W and Z production cross sections to be measured soon will be accurate enough to be sensitive to such 30% QCD correction effects and can in principle offer another test of the theory. However, the calculation is still lacking on the complete $O(\alpha_s^2)$ QCD correction which is of potential importance in view of the large $O(\alpha_s)$ term.

Finally, QCD effects are also observable in the production of W and Z bosons with large transverse momentum (Halzen and Scott, 1978; Altarelli et al., 1984). There is good qualitative agreement, although the statistics is rather poor at present (UA1 collaboration, 1984; Bawa and Stirling, 1988).

(4) Tests of QCD in Heavy Quarkonium Decay

Potential model dependences often cancel out in the ratios of decay widths of heavy quarkonium states. Important examples of such ratios are

$$\frac{\Gamma(1^{--} \to ggg)}{\Gamma(1^{--} \to \mu^+\mu^-)}, \quad \frac{\Gamma(1^{--} \to \gamma gg)}{\Gamma(1^{--} \to ggg)}. \tag{10}$$

Thus, the ratios of partial decay widths allow for a determination of α_s at the heavy quark mass scale. The most precise data come from the decay widths of the $J^{PC} = 1^{--}$ J/ψ and Υ resonances. A summary of quarkonium decay rates can be found in a review article by, e.g., Kwong et al. (1987a).

However, the perturbative corrections to these ratios are rather large (Lepage and Mackenzie, 1981). They change the predictions by a factor of 1.64 and 0.77 respectively in the case of Υ decay. The corrections in the J/ψ case are even larger. In addition, relativistic corrections are unknown and could be substantial in the J/ψ case.

A recent analysis (Kwong et al., 1987b) of bottomonium decay-width ratios from CUSB, CLEO, and ARGUS[5] yields, upon neglecting the theoretical uncertainties,

---

[5] *Csorna, S.E., et al., Phys. Rev. Lett.* **56**, *1222 (1986); Schamberger, R.D., et al., Phys. Lett.* **138B**, *225 (1984); Albrecht, H., et al., Phys. Lett.* **B199**, *291 (1987).*

$$\alpha_s(m_b) = 0.179 \pm 0.009. \tag{11}$$

(5) Tests of QCD in e^+e^- Collisions

As a well-known result (Ch. 9), the total cross section for $e^+e^- \rightarrow hadrons$ may be obtained by multiplying the muon-pair cross section by the famous R-factor $R = 3\sum_q e_q^2$. The higher order QCD corrections to this quantity have been calculated, and the result may be expressed as follows:

$$R = R^{(0)}[1 + \frac{\alpha_s}{\pi} + C_2(\frac{\alpha_s}{\pi})^2 + C_3(\frac{\alpha_s}{\pi})^3 + ...].$$
$$C_2 = (\frac{2}{3}\varsigma(3) - \frac{11}{12})N_f + \frac{365}{24} - 11\varsigma(3). \tag{12}$$

Numerically, $C_2 = 1.41$ in the modified minimum subtraction scheme. Recently C_3 has also been computed (Gorishny et al., 1988); numerically (for $N_f = 5$) $C_3 = 64.7$ in the modified minimum subtraction scheme. Quark mass effects are neglected in these calculations.

At the highest energies currently accessible, the corrections from QCD and Z exchange are comparable. A comparison of the theoretical prediction of Eq. (12) (corrected for the b-quark mass effect) with all the available data (including those from TRISTAN at $\sqrt{s} = 50\,GeV$) has been performed by the CELLO Collaboration.[6] The result is a correlated measurement of α_s and $sin^2\theta_W$. Fixing $sin^2\theta_W$ at the world-average value of 0.23, one obtains

$$\alpha_s(34GeV) = 0.132 \pm 0.016. \tag{13}$$

Two remarks are useful here. First, the principal advantage of determining α_s from R in e^+e^- annihilation is that there is no dependence on fragmentation models, jet algorithms, etc. Second, the order α_s^3 term in Eq. (12) is numerically twice as large as the order α_s^2 term, throwing the accuracy of the QCD prediction into serious doubt.

The traditional method of determining α_s in e^+e^- annihilation is from measuring quantities which are sensitive to the relative rate of two- and three-jet events. There are many possible choices of such "shape variables"; thrust

---

[6] *CELLO Collaboration: Behrend, H.J., et al., Phys. Lett. **B183**, 400 (1987); and de Boer, W., SLAC-PUB-4428 (1987).*

(Farhi 1977), energy-energy correlations (Basham et al., 1978), planar triple-energy correlations (Csikor et al., 1985), average jet mass. etc. Calculations of these observables are free of infrared divergences, so that they can be reliably calculated in perturbation QCD. The starting point for calculating all these quantities is the simple "three-jet" cross section for $e^+e^- \to q\bar{q}g$ (which may be verified using the procedure introduced in Ch. 9 together with Feynman rules for QED and QCD):

$$\frac{1}{\sigma}\frac{d^2\sigma}{dx_1 dx_2} = \frac{2\alpha_s}{3\pi}\frac{x_1^2 + x_2^2}{(1-x_1)(1-x_2)},\tag{14}$$

where

$$x_i = \frac{2E_i}{\sqrt{s}}$$

are the center-of-mass energy fractions of the final-state (massless) quarks. A distribution in a "three-jet" variable, such as those mentioned above, is then obtained by integrating this differential cross section over an appropriate phase space region for a fixed value of the variable.

A compilation of all the available data and a complete list of references can be found in, e.g., review papers by Stirling and Whalley (1987) and by Wu (1987). A "world average" is

$$\alpha_s(34 GeV) = 0.14 \pm 0.02,\tag{15}$$

with the error being the spread between the different experiments including the fragmentation uncertainty, but not that due to the size of the higher order corrections, which might be somewhat larger than this error. Notice that this value of α_s is in agreement with the value obtained from the measurement of R described above.

(6) Conclusions

There are many other ways in which QCD can be tested in lepton-hadron deep inelastic scattering, hadron-hadron collisions, and electron-positron collisions. We may mention in particular the study of exclusive processes (form factors, elastic scattering, ...), the behavior of quarks and gluons in nuclei, the spin properties of the theory and the importance of polarized scattering data, QCD effects in hadron spectroscopy, etc.

The global consistency among the existing data in the framework of perturbative QCD suggest that QCD *describes* strong interactions among quarks. Nevertheless, there are still many important tests to be made before the role of QCD is firmly and uniquely established.

§.14.2. The Glashow-Salam-Weinberg $SU(2) \times U(1)$ Electroweak Theory

The standard Glashow-Salam-Weinberg (GSW) $SU(2) \times U(1)$ electroweak model has been introduced and elucidated in some detail in Ch. 13. Apart from the Higgs mass M_H, fermion masses, and generation mixing parameters, the theory has three parameters. A particularly useful set is:

(a) *the fine structure constant* $\alpha = 1/137.036$,[7] determined from electron magnetic moment anomaly $(g - 2)$.

(b) *the Fermi coupling constant,* $G_F = 1.16637 \times 10^{-5} \, GeV^{-2}$, determined from the muon lifetime formula (including lepton mass and $O(\alpha)$ radiative corrections):[8]

$$
\begin{aligned}
\tau_\mu^{-1} = \frac{G_F^2 m_\mu^5}{192\pi^3} [1 &+ \frac{\alpha}{2\pi}(\frac{25}{4} - \pi^2)(1 + \frac{2\alpha}{3\pi} \ln \frac{m_\mu}{m_e})] \\
&\times [1 - \frac{8m_e^2}{m_\mu^2}] \times [1 + \frac{3}{5}\frac{m_\mu^2}{M_W^2}].
\end{aligned}
\tag{16}
$$

and

(c) *the elecroweak mixing parameter* $sin^2\theta_W$, determined from neutral-current processes and the W and Z^0 masses. Note that the value of $sin^2\theta_W$ depends on the renormalization prescription. A very useful scheme (Sirlin 1980 and 1984) is to take the tree-level formula $sin^2\theta_W = 1 - M_W^2/M_Z^2$ as the definition of

---

[7] α *depends upon the energy scale under investigation. This standard value is good only at very low energies. For example, the value* $1/128$ *is appropriate at energies of order* M_W.

[8] *See Eq. (44), Ch. 13.*

the renormalized $sin^2\theta_W$ to all orders in perturbation theory.[9] Alternatively, one may take M_Z rather than $sin^2\theta_W$ as the third fundamental parameter. This may become useful when a very precise value of M_Z is determined from experiments at SLC and LEP in the immediate future.

The quality of the existing data already demands that complete $O(\alpha)$ radiative corrections must be suitably taken into account. Accordingly, the tree-level expressions for M_W and M_Z, as given previously by Eqs. (16) and (43) in Ch. 13, are modified:

$$M_W = \frac{A_0}{sin\theta_W(1 - \triangle r)^{1/2}},$$
$$M_Z = \frac{M_W}{cos\theta_W}, \tag{17}$$

where $A_0 = (\pi\alpha/\sqrt{2}G_F)^{1/2} = 37.281\,GeV$. The radiative correction parameter $\triangle r$ is predicted to be 0.0713 ± 0.0013 for $m_t = 45\,GeV$ and $M_H = 100\,GeV$, while $\triangle r \to 0$ for $m_t \sim 245\,GeV$. If M_Z is regarded as a fundamental parameter, then

$$sin^2\theta_W = \frac{1}{2}\{1 - [1 - \frac{4A_0^2}{M_Z^2(1 - \triangle r)}]^{1/2}\} \tag{18}$$

is a derived parameter, and $M_W = M_Z\,cos\theta_W$.

(1) Formulae on cross sections and asymmetries

Following the 1988 publication of Particle Data Group, we shall first introduce the notations adopted by researchers working on different facets of the electroweak theory and then move on to list the basic formulae on cross sections and asymmetries which may easily be understood in terms of what we have learned in previous chapters.

It is convenient to write the four-fermion interactions relevant to ν-hadron (νH), νe, and e-hadron (eH) processes in a form that is valid in an arbitrary

---

[9] *An alternative is to use the modified minimal subtraction quantity $sin^2\theta_W(\mu)$, where μ is conveniently chosen to be M_W for electroweak processes. The two definitions are related by $sin^2\theta_W(M_W) = C(m_t, M_H)sin^2\theta_W$, where $C = 0.9907$ for $m_t = 45GeV$, $M_H = 100GeV$.*

gauge theory (assuming massless left-handed neutrinos). One has

$$\mathcal{L}^{\nu H} = \frac{G_F}{\sqrt{2}} i\bar{\nu}\gamma_\mu(1+\gamma_5)\nu$$
$$\times \left\{ \sum_i [\epsilon_L(i)\, i\bar{q}_i\gamma_\mu(1+\gamma_5)q_i + \epsilon_R(i)\, i\bar{q}_i\gamma_\mu(1-\gamma_5)q_i] \right\}, \tag{19}$$

$$\mathcal{L}^{\nu e} = \frac{G_F}{\sqrt{2}}\, i\bar{\nu}_\mu\gamma_\mu(1+\gamma_5)\nu_\mu\, i\bar{e}\gamma_\mu(g_V^e + g_A^e\gamma_5)e, \tag{20}$$

$$\mathcal{L}^{eH} = \frac{G_F}{\sqrt{2}} \sum_i [C_{1i}\, i\bar{e}\gamma_\mu\gamma_5 e\, i\bar{q}_i\gamma_\mu q_i + C_{2i}\, i\bar{e}\gamma_\mu e\, i\bar{q}_i\gamma_\mu\gamma_5 q_i]. \tag{21}$$

Note that, for $\nu_e\bar{e}$ or $\bar{\nu}_e e$, the charged-current contribution must be included.

The standard model expressions for $\epsilon_{L,R}(i), g_{V,A}^e$ and C_{ij} are summarized in the following table.

Table 1. Standard model expressions for the neutral-current parameters for ν-hadron, νe, and e-hadron processes. Entries are from Particle Data Group (1988).

Quantity	Standard Model Expression
$\epsilon_L(u)$	$\rho_{\nu N}^{NC}[\frac{1}{2} - \frac{2}{3}\kappa_{\nu N}\sin^2\theta_W + \lambda_{uL}]$
$\epsilon_L(d)$	$\rho_{\nu N}^{NC}[-\frac{1}{2} + \frac{1}{3}\kappa_{\nu N}\sin^2\theta_W + \lambda_{dL}]$
$\epsilon_R(u)$	$\rho_{\nu N}^{NC}[-\frac{2}{3}\kappa_{\nu N}\sin^2\theta_W + \lambda_{uR}]$
$\epsilon_R(d)$	$\rho_{\nu N}^{NC}[\frac{1}{3}\kappa_{\nu N}\sin^2\theta_W + \lambda_{dR}]$
g_V^e	$\rho_{\nu e}[-\frac{1}{2} + 2\kappa_{\nu e}\sin^2\theta_W]$
g_A^e	$\rho_{\nu e}[-\frac{1}{2}]$
C_{1u}	$\rho'_{eq}[-\frac{1}{2} + \frac{4}{3}\kappa'_{eq}\sin^2\theta_W]$
C_{1d}	$\rho'_{eq}[\frac{1}{2} - \frac{2}{3}\kappa'_{eq}\sin^2\theta_W]$
C_{2u}	$\rho_{eq}[-\frac{1}{2} + 2\kappa_{eq}\sin^2\theta_W]$
C_{2d}	$-C_{2u}$

Note that, if radiative corrections are ignored, $\rho = \kappa = 1, \lambda = 0$, as from Eqs. (45), (38a)-(38d), and (39) in Ch. 13. At $O(\alpha)$, $\rho_{\nu N}^{NC} = 1.00074$, $\kappa_{\nu N} = 0.9902$, $\lambda_{u_L} = -0.0031$, $\lambda_{d_L} = -0.0026$, and $\lambda_{u_R} = (1/2)\lambda_{d_R} = 3.5 \times 10^{-5}$ for $m_t = 45\,GeV$, $M_H = 100\,GeV$, $\sin^2\theta_W = 0.23$, and $<Q^2> = 20\,GeV^2$. For

νe scattering, $\kappa_{\nu e} = 0.9897$ and $\rho_{\nu e} = 1.0054$ (at $< Q^2 >= 0$). For atomic parity violation, $\rho'_{eq} = 0.9793$ and $\kappa'_{eq} = 0.9948$. For the SLAC polarized electron experiment, $\rho'_{eq} = 0.970$, $\kappa'_{eq} = 0.993$, $\rho_{eq} = 0.993$, and $\kappa_{eq} = 1.03$ after incorporating additional QED corrections.

At present, the most precise determinations of $sin^2\theta_W$ are from deep inelastic neutrino scattering from isoscalar nuclear targets such as deuteron, carbon, or iron (the last being an approximate isoscalar target). The ratio $R_\nu = \sigma_{\nu N}^{NC}/\sigma_{\nu N}^{CC}$ of neutral- to charged-current cross sections has been measured to 1% accuracy by the CDHS and CHARM Collaborations,[10] so that it is imperative to obtain theoretical expressions for R_ν and $R_{\bar\nu}$ ($\equiv \sigma_{\bar\nu N}^{NC}/\sigma_{\bar\nu N}^{CC}$) as functions of $sin^2\theta_W$ to similar accuracy. Fortunately, most of the uncertainties arise from the strong interactions and neutrino spectra; they cancel in the ratio.

Using the quark parton model, one finds, as a simple zeroth-order approximation,

$$R_\nu \equiv \sigma_{\nu N}^{NC}/\sigma_{\nu N}^{CC} = g_L^2 + g_R^2 r$$

$$R_{\bar\nu} \equiv \sigma_{\bar\nu N}^{NC}/\sigma_{\bar\nu N}^{CC} = g_L^2 + \frac{g_R^2}{r}, \tag{22}$$

where

$$g_L^2 \equiv \epsilon_L(u)^2 + \epsilon_L(d)^2 \simeq \frac{1}{2} - sin^2\theta_W + \frac{5}{9}sin^4\theta_W,$$
$$g_R^2 \equiv \epsilon_R(u)^2 + \epsilon_R(d)^2 \simeq \frac{5}{9} - sin^4\theta_W, \tag{23}$$

and $r \equiv \sigma_{\bar\nu N}^{CC}/\sigma_{\nu N}^{CC}$ is the ratio of $\bar\nu$ and ν charged-current cross sections, which can be measured directly. Note that, in practice, Eq. (22) must be corrected for quark mixing, the s and c seas, c-quark threshold effects (which mainly affect σ^{CC} - these turn out to be the largest theoretical uncertainty), nonisoscalar target effects, $W - Z$ propagator differences, and radiative corrections (which lower the extracted value of $sin^2\theta_W$ by ~ 0.009). Details of the neutrino spectra, experimental cuts, x and Q^2 dependence of structure functions, and longitudinal

---

[10] *CDHS: Abramowicz, H., et al., Phys. Rev. Lett.* **57**, *298 (1986); CHARM: Allaby, J.V., et al., Phys. Lett.* **B177**, *446 (1986).*

structure functions enter only at the level of these corrections and therefore lead to very small uncertainties (Amaldi *et al.* 1987). Altogether, the theoretical uncertainty is $\triangle sin^2\theta_W \sim \pm 0.005$, which would be very hard to improve in the forseeable future.

The laboratory cross section for $\nu_\mu e \to \nu_\mu e$ or $\bar{\nu}_\mu e \to \bar{\nu}_\mu e$ elastic scattering is given by

$$\frac{d\sigma_{\nu_\mu, \bar{\nu}_\mu}}{dy} = \frac{G_F^2 m_e E_\nu}{2\pi} \times [(g_V^e \pm g_A^e)^2 + (g_V^e \mp g_A^e)^2 (1-y)^2 - (g_V^{e2} - g_A^{e2}) \frac{y m_e}{E_\nu}], \tag{24}$$

where the upper (lower) sign refers to $\nu_\mu (\bar{\nu}_\mu)$, and $y \equiv E_e/E_\nu$ [which runs from 0 to $(1 + m_e/2E_\nu)^{-1}$] is the ratio of the kinetic energy of the recoil electron to the incident ν or $\bar{\nu}$ energy. For $E_\nu \gg m_e$ this yields a total cross section

$$\sigma = \frac{G_F^2 m_e E_\nu}{2\pi} [(g_V^e \pm g_A^e)^2 + \frac{1}{3}(g_V^e \mp g_A^e)^2]. \tag{25}$$

The most accurate leptonic measurements[11] of $sin^2\theta_W$ come from the ratio $R \equiv \sigma_{\nu_\mu e}/\sigma_{\bar{\nu}_\mu e}$ in which many of the systematic uncertainties cancel. Radiative corrections, which are small compared to the precision of present experiments, increase the extracted $sin^2\theta_W$ by $\simeq 0.002$. The cross section for $\nu_e e$ and $\bar{\nu}_e e$ may be obtained from Eq. (24) by replacing $g_{V,A}^e$ by $g_{V,A}^e + 1$, according to the Fierz reordering theorem (which enables one to take into account the charged-current contribution in a simple manner).

The SLAC polarized-electron experiment (Prescott et al., 1979) measured the parity-violating asymmetry

$$\mathcal{A} = \frac{\sigma_R - \sigma_L}{\sigma_R + \sigma_L}, \tag{26}$$

where $\sigma_{R,L}$ is the cross section for the deep-inelastic scattering of a right- or left-handed electron: $e_{R,L} N \to eX$. In the quark parton model, we have

$$\frac{\mathcal{A}}{Q^2} = a_1 + a_2 \frac{1 - (1-y)^2}{1 + (1-y)^2}, \tag{27}$$

---

[11] *CHARM; Bergsma, F., et al., Phys. Lett. **147B**, 481 (1984); and BNL E734: Ahrens, L.A., et al., Phys. Rev. Lett. **54**, 18 (1985).*

where $Q^2 > 0$ is the momentum transfer and y is the fractional energy transfer from the electron to the hadron. For the deuteron or other isoscalar target, one has, neglecting the s quark and antiquarks,

$$
\begin{aligned}
a_1 &= \frac{3G_F}{5\sqrt{2}\pi\alpha}(C_{1u} - \frac{1}{2}C_{1d}) \simeq \frac{3G_F}{5\sqrt{2}\pi\alpha}(-\frac{3}{4} + \frac{5}{3}sin^2\theta_W), \\
a_2 &= \frac{3G_F}{5\sqrt{2}\pi\alpha}(C_{2u} - \frac{1}{2}C_{2d}) \simeq \frac{9G_F}{5\sqrt{2}\pi\alpha}(sin^2\theta_W - \frac{1}{4}).
\end{aligned}
\tag{28}
$$

Radiative corrections lower the extracted value of $sin^2\theta_W$ by ~ 0.005.

Experiments which detect atomic parity violation (Bouchiat and Pottier, 1986; Piketty 1986) are now quite precise, and the uncertainties associated with atomic wave functions are relatively small (especially for cesium). For heavy atoms one determines the "weak charge"

$$
\begin{aligned}
Q_W &= -2[C_{1u}(2Z + N) + C_{1d}(Z + 2N)] \\
&\simeq Z(1 - 4sin^2\theta_W) - N.
\end{aligned}
\tag{29}
$$

Radiative corrections increase the extracted $sin^2\theta_W$ by ~ 0.008.

The forward-backward asymmetry for $e^+e^- \rightarrow \ell\bar{\ell}$, $\ell = \mu$ or τ is defined as

$$
\mathcal{A}_{FB} \equiv \frac{\sigma_F - \sigma_B}{\sigma_F + \sigma_B},
\tag{30}
$$

where $\sigma_F(\sigma_B)$ is the cross section for ℓ^- to travel forward (backward) with respect to the e^- direction. The charge asymmetry $\mathcal{A}_{FB}$ and the ratio of the total cross section relative to pure QED, R, are given, from Problem 13.3. in Ch. 13, by

$$
\begin{aligned}
R &= F_1, \\
\mathcal{A}_{FB} &= \frac{3}{4}\frac{F_2}{F_1},
\end{aligned}
\tag{31}
$$

with

$$
\begin{aligned}
F_1 &= 1 - 2\chi_0 V^e V^\ell \cos\delta_R + \chi_0^2(V^{e2} + A^{e2})(V^{\ell2} + A^{\ell2}), \\
F_2 &= -2\chi_0 A^e A^\ell \cos\delta_R + 4\chi_0^2 A^e A^\ell V^e V^\ell,
\end{aligned}
\tag{32}
$$

where

$$
\begin{aligned}
tan\delta_R &= \frac{M_Z\Gamma_Z}{M_Z^2 - s} \\
\chi_0 &= \frac{G_F}{2\sqrt{2}\pi\alpha}\frac{sM_Z^2}{[(M_Z^2 - s)^2 + M_Z^2\Gamma_Z^2]^{1/2}}
\end{aligned}
\tag{33}
$$

and $\sqrt{s}$ is the CM energy. Eq. (32) is valid in the tree approximation. If QED radiative corrections are taken into account for the data, then the remaining electroweak corrections can be incorporated (Lynn and Stuart 1985; in an approximation adequate for existing PEP and PETRA data) by replacing χ_0 by $\chi(s) \equiv \chi_0(s)\alpha/\hat{\alpha}(s)$, where $\hat{\alpha}(s)$ is the running QED coupling. Numerically, $\alpha/\hat{\alpha}(s) \sim 1 - \Delta_r$ if Δ_r is evaluated for $m_t < 100\,GeV$.

At SLC and LEP, A_{FB} for $e^+e^- \to \ell^+\ell^-$ near the Z pole will be measured to high precision. Similarly, it is expected that the left-right asymmetry

$$A_{LR} \equiv \frac{\sigma_L - \sigma_R}{\sigma_L + \sigma_R}, \tag{34}$$

where $\sigma_L(\sigma_R)$ is the cross section for a left(right)-handed incident electron, is to be measured very precisely at SLC and possibly at LEP. Neglecting terms of order $(\Gamma_Z/M_Z)^2$, one has, at tree level,

$$A_{FB} \simeq 3\eta_e \frac{\eta_e + \frac{1}{2}P_e}{1 + 2P_e\eta_e},$$

$$A_{LR} \simeq 2\eta_e, \tag{35}$$

where P_e is the initial e^- polarization and

$$\eta_i \equiv \frac{V^i A^i}{V^{i2} + A^{i2}}, \quad i = e, \mu, \tau. \tag{36}$$

In Eq. (36), we have used the notations:

$$V^i \equiv t_{3L}(i) - 2e_i sin^2\theta_W,$$

$$A^i \equiv t_{3L}(i), \tag{37}$$

with $t_{3L}(i)$ the 3rd component of the weak isospin for the fermion of flavor i and e_i its electric charge. The high-precision measurements will require careful application of both QED and electroweak radiative corrections to Eq. (35).

To complete our list on cross sections and asymmetries, we should also mention the width for gauge bosons to decay into massless fermions $f_1\bar{f}_2$, which is given by

$$\Gamma(W^+ \to e^+\nu_e) = \frac{G_F M_W^3}{6\sqrt{2}\pi} \simeq 230 MeV$$

$$\Gamma(W^+ \to u_i \bar{d}_i) = \frac{C G_F M_W^3}{6\sqrt{2}\pi} \mid V_{ij} \mid^2 \simeq 717 \mid V_{ij} \mid^2 MeV$$

$$\Gamma(Z \to \psi_i \bar{\psi}_i) = \frac{C G_F M_Z^3}{6\sqrt{2}\pi} [V^{i2} + A^{i2}]$$

$$\simeq \begin{cases} 170\,MeV\,(\nu\bar{\nu}), & 85.4\,MeV\,(e^+e^-) \\ 305\,MeV\,(u\bar{u}), & 394\,MeV\,(d\bar{d}). \end{cases}$$

(38)

For leptons $C = 1$, while for quarks $C = 3[1 + \frac{\alpha_s(M_W)}{\pi}]$, where the factor 3 is due to color and the factor in parentheses is a QCD correction (Albert et al., 1980; Consoli et al., 1983; Güsken et al., 1985; Jegerlehner, 1986). The remaining corrections are negligible.

Assuming three fermion families, we obtain the total widths as follows:

$$\Gamma_Z \sim (2.58 - 2.55)GeV,$$
$$\Gamma_W \sim (2.52 - 2.12)GeV,$$

(39)

where the range arises from a variation of the top quark mass m_t, and the other fermion masses have been neglected. The data up to the year of 1988 are severely limited by the small number of the observed events:

$$\Gamma_{Z\,exp} < 6.5 GeV,$$
$$\Gamma_{W\,exp} < 5.6 GeV.$$

(40)

Nevertheless, results from Fermilab Tevatron ($p\bar{p}$ collider), Stanford SLC and CERN LEP (e^+e^- colliders) are piling up rapidly beginning at the summer of 1989, providing stringent experimental tests of Eq. (39).[12]

(2) Standard-model analyses of experimental results

The electroweak mixing parameter $sin^2\theta_W$ and equivalently, the Z^0 mass, M_Z, have been determined from the W and Z masses and from a variety of neutral-current processes covering a very wide Q^2 range. The results (Amaldi et al., 1987), as taken from Particle Data Group (1988) and shown in Table 2,

---

[12] *See Eq. (41) below.*

are in impressive agreement with each other, indicating the quantitative success of the GSW electroweak theory. The best fit to all data yields $sin^2\theta_W = 0.230 \pm 0.0048$ which corresponds to $M_Z = 92.0 \pm 0.7\,GeV$, where the errors (as well as those given below for other neutral-current parameters) include full statistical, systematic, and theoretical uncertainties.

Table 2. Determination of $sin^2\theta_W$ and M_Z (in GeV) from various reactions. Entries are from from Particle Data Group (1988).

Reaction	$sin^2\theta_W$	M_Z
Deep inelastic (isoscalar)	$0.233 \pm 0.003 \pm [0.005]$	$91.6 \pm 0.4 \pm [0.8]$
$\nu_\mu p \to \nu_\mu p$	0.210 ± 0.033	95.0 ± 5.2
$\bar{\nu}_\mu p \to \bar{\nu}_\mu p$	0.210 ± 0.033	95.0 ± 5.2
$\nu_\mu e \to \nu_\mu e$	$0.223 \pm 0.018 \pm [0.002]$	93.0 ± 2.7
$\bar{\nu}_\mu e \to \bar{\nu}_\mu e$	$0.223 \pm 0.018 \pm [0.002]$	93.0 ± 2.7
W, Z	$0.228 \pm 0.007 \pm [0.002]$	92.3 ± 1.1
Atomic parity violation	$0.209 \pm 0.018 \pm [0.014]$	95.1 ± 3.9
$SLAC\ eD$	$0.221 \pm 0.015 \pm [0.013]$	93.3 ± 2.7
μC	0.25 ± 0.08	89.6 ± 9.7
All data	0.230 ± 0.0048	92.0 ± 0.7

Note that the central values of all fits shown in Table 2 are obtained by assuming, in the determination of radiative corrections, that the top quark mass $m_t = 45\,GeV$ and the Higgs boson mass $M_H = 100\,GeV$. Whereas the first of the two errors shown is experimental, the second (in square brackets) is theoretical, as computed by assuming that (1) there are three fermion families and that (2) the top-quark mass and the Higgs boson mass are allowed to vary but subject to the arbitrarily chosen constraint $m_t < 100\,GeV$, and $M_H < 1\,TeV$. In the other cases the theoretical and experimental uncertainties are combined. When m_t is allowed to be totally arbitrary, the fits to all data yield $sin^2\theta_W = 0.229 \pm 0.007$ and $M_Z = 91.8 \pm 0.9\,GeV$. The existing e^+e^- data do not yield a useful determination of $sin^2\theta_W$: At PEP and PETRA energies, all values of $sin^2\theta_W$ from 0.1 to 0.4 give a good description of the existing data on the observed asymmetries.

The radiative corrections are sensitive to the isospin breaking associated with a large m_t. Consistency of the $sin^2\theta_W$ values derived from the various reactions requires $m_t < 180\,GeV$ at $90\,\%\,C.L.$ for $M_H \leq 100\,GeV$, with a slightly weaker limit for larger M_H. Similar limits hold for the mass splittings between fourth-generation quarks or leptons.

The measured values of M_W and M_Z are given in Table 3. They are in excellent agreement with the predictions of the standard model when full radiative corrections (to both the W and Z mass formulas and to deep inelastic scattering) are included, but in severe disagreement when the corrections are excluded.

Table 3. The W and Z masses (in GeV). The first uncertainties are mainly statistical and the second are energy calibration uncertainties that are 100% correlated between M_W and M_Z for each group. The last two rows are predictions of the GSW eletroweak theory, using $sin^2\theta_W$ determined from deep inelastic scattering, with and without radiative corrections, respectively. The entries are from Particle Data Group (1988).

	Group	M_W	M_Z
	UA2	$80.2 \pm 0.8 \pm 1.3$	$91.5 \pm 1.2 \pm 1.7$
	UA1	$83.5^{+1.1}_{-1.0} \pm 2.7$	$93.0 \pm 1.4 \pm 3.0$
UA1+	UA2 combined	80.9 ± 1.4	91.9 ± 1.8
Prediction with radiative		80.2 ± 1.1	91.6 ± 0.9
corrections			
Prediction without radiative		75.9 ± 1.0	87.1 ± 0.7
corrections			

At the moment of this writing, the latest published data[13] on the mass and width of the Z^0 boson come from Fermilab Tevatron and Stanford SLC (August

[13]*Abrams, G.S., et al., Phys. Rev. Lett. **63**, 2173 (1989); Abe, F., et al., Phys. Rev. Lett. **63**, 720 (1989).*

1989):

$$M_Z = 91.14 \pm 0.12 \, GeV; \quad \Gamma_Z = 2.42^{+0.45}_{-0.35} \, GeV,$$

$$(Stanford \; SLC);$$

$$M_Z = 90.9 \pm 0.3(stat + sys) \pm 0.2(scale) \, GeV;$$

$$\Gamma_Z = 3.8 \pm 0.8 \pm 1.0 \, GeV; \quad (Fermilab \; Tevatron).$$

(41)

Note that the newly observed Z^0 mass values are in excellent agreement with the GSW model prediction with radiative corrections (the entry in Table 3). The observed widths also compare well with the prediction given previously in Eq. (39). The need to reduce the error of the theoretical prediction has become obvious.

In view of the fact that both Stanford SLC and CERN LEP, $e^+ e^-$ colliders capable of producing Z^0 copiously, already started producing extensive experimental results, it is clear that we may soon witness significant advances, or even another breakthrough, in the area of electroweak physics.

(3) Model-independent analyses of experimental results

The W and Z masses and neutral-current data can be used to set limits on possible deviations from the standard model. For example, the relation in Eq. (17) between M_W is modified if there are Higgs multiplets with weak isospin $> 1/2$ with significant vacuum expectation values. In order to calculate to higher orders in such theories one must select a set of four fundamental renormalized parameters. It is convenient to take these as α, G_F, M_Z, and M_W, especially because M_W and M_Z are directly measurable. Then $sin^2\theta_W$ and ρ can be considered dependent parameters defined by

$$sin^2\theta_W \equiv \frac{A_0^2}{M_W^2}(1 - \triangle r),$$

(42)

and

$$\rho = M_W^2 / (M_Z^2 \cos^2\theta_W).$$

(43)

As long as the new physics which yields $\rho \neq 1$ is a small perturbation which does not significantly affect the radiative corrections, ρ can be regarded as a phenomenological parameter which multiplies G_F in Eqs. (19)-(21) and (33).[14] The allowed regions in the $\rho - sin^2\theta_W$ plane are obtained and a global fit to all data (Amaldi et al. 1987) yields

$$sin^2\theta_W = 0.229 \pm 0.0064,$$
$$\rho = 0.998 \pm 0.0086,$$
(44)

which is remarkably close to what we have obtained earlier in the context of the standard model. (It justifies to some extent the neglect of $\rho - 1$ due to radiative corrections).

Table 4. Values of the model-independent neutral-current parameters, compared with the GSW electroweak theory with $sin^2\theta_W = 0.230$. $\theta_i, i = L$ or R, is defined as $tan^{-1}[\epsilon_i(u)/\epsilon_i(d)]$. Entries are from Particle Data Group (1988).

Quantity	Experimental Value	Standard Model Prediction
$\epsilon_L(u)$	0.339 ± 0.017	0.345
$\epsilon_L(d)$	-0.429 ± 0.014	-0.427
$\epsilon_R(u)$	-0.172 ± 0.014	-0.152
$\epsilon_R(d)$	$-0.011^{+0.081}_{-0.057}$	0.076
g_L^2	0.2996 ± 0.0044	0.301
g_R^2	0.0298 ± 0.0038	0.029
θ_L	2.47 ± 0.004	2.46
θ_R	$4.65^{+0.48}_{-0.32}$	5.18
g_A^e	-0.498 ± 0.027	-0.503
g_V^e	-0.044 ± 0.036	-0.045
C_{1u}	-0.249 ± 0.071	-0.191
C_{1d}	0.381 ± 0.064	0.340
$C_{2u} - \frac{1}{2}C_{2d}$	0.19 ± 0.37	-0.039

Most of the parameters relevant to ν-hadron, νe, e-hadron, and e^+e^- processes are now determined uniquely and precisely from the data in "model independent" fits (i.e., fits which allow for an arbitrary electroweak gauge theory).

---

[14] *In addition, the expression for M_Z in Eq. (17) is divided by $\sqrt{\rho}$; the M_W formula is unchanged.*

The values for the parameters defined in Eq. (19)-(21) are given in Table 4 along with the predictions of the GSW electroweak theory with $sin^2\theta_W = 0.230$. Note that there is a second $g^e_{V,A}$ solution, given approximately by $g^e_V \leftrightarrow g^e_A$, which is eliminated by e^+e^- data under the assumption that the neutral current is dominated by the exchange of a single Z.

The agreement is excellent between the results from the model-independent fits and the standard model predictions with $sin^2\theta_W = 0.230$, indicating the amazing success of the GSW electroweak theory. Note that it is difficult to present the e^+e^- results in a model-independent way as Z-propagator effects are non-negligible at PETRA and PEP energies. However, assuming $e - \mu - \tau$ universality, the lepton asymmetries imply $A^e = -0.511 \pm 0.013$, in good agreement with the standard model prediction $-1/2$. The vector coupling is not well determined by existing e^+e^- data: values of V^e from -0.3 to 0.3 are allowed.

(4) Conclusions

In summary, it is clear that the existing experimental data on weak interactions are described remarkably well by the GSW $SU(2) \times U(1)$ electroweak theory. A major boost came from direct observations of the $W^\pm$ and Z^0 weak bosons at the predicted masses. With the coming generation of e^+e^- colliders with the center-of-mass energies $\sqrt{s}$ greater than M_{Z^0} (SLC and LEP), it is expected that the theory will be subject to much more severe scrutiny. Owing to the fact that weak interaction studies over the last century constantly produced one surprise after another, one might suspect that the GSW $SU(2) \times U(1)$ electroweak theory may mark the beginning of an unfolding mystery, rather than the completion of a zigsaw puzzle.

386

§.14.3. Concluding Remarks

Our presentations in §.14.1 and §.14.2.[15] should have made it clear why quantum chromodynamics (QCD) and the Glashow-Salam-Weinberg (GSW) electroweak theory altogether have assumed the status of the "Standard Model". Both QCD and the GSW electroweak theory are phrased in terms of renormalizable gauge field theories. Thus, we might wonder upon a number of basic questions, a few of which we wish to mention here from our own perspectives.

First of all, we do not understand the reason why gauge principle must be invoked in constructing a theory of elementary particles. This is against the so-called "operational point of view", since there is a gauge degree of freedom which is not directly observable.

Nor we understand the reason why the principle of renormalizability play such an essential role here. In fact, the statement posed by Dirac on that physics must be based on strict mathematics may be a serious challenge here, in view of all the divergences appearing in the present theory such as ultraviolet divergences, infrared divergences, and, even more in massless QCD, collinear divergences. It is indeed difficult to imagine why, in such mathematically ill-behaved theory, some finite parts make physical sense while a good chunk of infinite parts are only renormalizations of those entities which we do not understand, such as mass and electric charge. To some extent, all these divergences are tied to the fact that we are trying to use a local field theory to describe a "pointlike" particle, but we must pardon ourselves that the concept of a "pointlike" object does not make any physical sense at all. In this regard, we might hope that adoption of a better definition for "renormalization" and the possible close tie with the so-called "renormalization group equation" may turn the present mathematically ill-defined theory into the one that would please Dirac and many of us. But the quest for mathematical rigour is yet to be satisfied.

---

[15] *The presentation in §.14.1. has much in common with "Review of Particle Properties", Particle Data Group, pp. 96 - 101 (Phys. Lett. **204**, 1988), prepared April 1988 by Barnett, R.M., Hinchliffe, I., and Stirling, W.J. On the other hand, much of the material for the section on the GSW electroweak theory comes from pp. 102 - 106 of the same publication, prepared April 1988 by Langacker, P. The interested reader may consult the original references for further details which have been left out in our discussion.*

Of course, we may quit being too sophisticated and shall have no more digging into the bottom of it. However, we may still wonder whether strong interactions and electroweak interactions may be unified into a grand unified theory (GUT) with a gauge group having better looking than $SU(3) \times SU(2) \times U(1)$, since after all the running $\alpha_s(Q^2)$ must meet the slowly-changing electroweak $\alpha(Q^2)$ at some Q^2, the so called "grand unification scale". Georgi and Glashow (1974) led the efforts on this front by proposing the simplest $SU(5)$ grand unification scheme which predicts, among others, that a proton would decay into a system of mesons and leptons with a net zero baryon number. Experimental searches for proton decay or baryon number nonconservation set a limit of proton lifetime greater than $(10^{31} - 10^{32})$ years, contradicting the prediction of the simplest $SU(5)$ grand unified model. Of course, there is plenty of room to get life going again by mixing into the GUT scheme ideas of "supersymmetry" (fermion-boson symmetry), or cooking the pot more thoroughly by trying to put everything altogether including this time the feeble gravitational force in the framework of supersymmetric string models ("superstrings"). Of course, we might succeed along one of these routes for searching for a better theory, but the deeper meaning for why a gauge field theory has been so successful remains to be unraveled.

Closely tied to the the questions listed above, there are unsatisfactory features, or even failures, related to the Standard Model. For example, we have thus far failed to identify the existence of the Higgs particle predicted by the GSW electroweak theory. Searches for the top (t) quark have yet to turn up a positive signature. There are coupling constants, masses, etc., so many of them which we do not have any good idea about why they are there and why they take on specific values as they have had. To move forward, we need yet better ideas: ideas that allow us to go beyond the GSW electroweak theory, ideas that will help us to beat the highly nonlinear QCD thoroughly in order to get to the bottom of it, etc. We are stalled if, for example, we need to rely on a computer that is so many orders of magnitude more powerful than the existing ones in order to understand QCD in quantitative terms; but, fortunately, we all know, or believe, that we are still far from hitting the limit of our capability to understand.

Thus, it might be worth repeating the statement which we already used earlier in a slightly different context: If the history could be a guide, the Stan-

388

dard Model might just mark the beginning of another unfolding mystery, rather than completion of a seemingly-never-ending zigsaw puzzle that the Nature has given all of us to play.

References

Particle Data Group, *Review of Particle Properties*, Physics Letters **B204**, 1 (1988).

Altarelli, G., and Parisi, G., Nucl. Phys. **B126**, 298 (1977); Parisi, G., *Proc. 11th Rencontre de Moriond*, 1976, ed. J. Tran Thanh Van; on the QCD evolution equations.

Glück, M., Hoffmann, E., and Reya, E., Z. Phys. **C13**. 119 (1981); on the nonzero quark mass effect on the QCD evolution equations.

Curci, G., Furmanski, W., and Petronzio, R., Nucl. Phys. **B175, 27** (1980); Furmanski, W. and Petronzio, R., Phys. Lett. **97B**, 437 (1980), and Z. Phys. **C11**, 293 (1982); Floratos, E.G., Kounnas, C., and Lacaze, R., Phys. Lett. **98B**. 89 (1981), Phys. Lett. **98B,** 285 (1981), and Nucl. Phys. **B192, 417** (1981); and Herrod, R.T. and Wada, S., Phys. Lett. **96B,** 195 (1981), and Z. Phys. **C9**, 351 (1981); on higher order corrections to QCD.

Voss, R., talk at *the 1987 International Symposium on Lepton and Photon Interactions at High Energies*, Hamburg 1987; on the current status of the experimental data on deep inelastic scattering.

Glück, M., Hoffmann, E., and Reya, E., Z. Phys. **C13**, 119 (1982); Duke, D.W. and Owens, J.F., Phys. Rev. **D30**, 49 (1984); Eichten, E., Hinchliffe, I., Lane, K., and Quigg, C., Rev. Mod. Phys. **56**, 579 (1984); Erratum: **58**, 1065 (1986); Diemoz, M., Ferroni, F., Longo, E., and Martinelli, G., CERN preprint CERN-TH-4751-87 (1987); Martin, A.D., Roberts, R.G., and Stirling, W.J., Phys. Rev. **D37**, 1161 (1988); on parametrizations of parton distributions.

Owens, J.F., Rev. Mod. Phys. **59**, 465 (1987); on the production of single large-transverse-momentum photons in hadron colliders.

Aurenche, P., Baier, R., Douiri, A., Fontannaz, M., and Schiff, D., Phys. Lett. **140B**, 87 (1984); Nucl. Phys. **B297**, 661 (1988); on next-to-leading-order corrections in the production of single large-transverse-momentum photons in hadron colliders.

Altarelli, G., Ellis, R.K., and Martinelli, G., Nucl. Phys. **B143**, 521 (1978); Nucl. Phys. **B157**, 461 (1979); Altarelli, G., Ellis, R.K., Greco, M., and

Martinelli, G., Nucl. Phys. **B246**, 12 (1984); Altarelli, G., Ellis, R.K., and Martinelli, G., Phys. Lett. **151B**, 457 (1985); Kubar-André, J. and Paige, F.E., Phys. Rev. D19, 221 (1979); Kubar-André, J., Le. Bellac, M., Meunier, J.L., and Plaut, G., Nucl. Phys. **B157**, 251 (1980); Dokshitzer, Yu. L., Dyakonov, D.I., and Troyan, S.I., Phys. Lett. **78B**, 290 (1978); Phys. Report **58**, 269 (1980); Parisi, G. and Petronzio, R., Nucl. Phys. **154**, 427 (1979); Curci, G., Greco, M., and Srivastava, Y., Phys. Rev. Lett. **43**, 434 (1979); Nucl. Phys. **B159**, 451 (1979); Kodaira, J., and Trentadue, L., Phys. Lett. **112B**, 66 (1982); **123B**, 335 (1983); Davies, C.T.H. and Sterling, W.J., Nucl. Phys. **B244**, 337 (1984); Davies, C.T.H., Webber, B.R., and Sterling, W.J., CERN preprint TH-3987/84; Collins, J., and Soper, D.E., Nucl. Phys. **B193**, 381 (1981); **B194**, 445 (1982); **B197**, 446 (1982); Collins, J., Soper, D.E., and Sterman, G., CERN preprint TH-3923/84; Collins, J., Invited talk at LAMPF Conference on Quarks and Gluons in Nuclei, February 1986 (unpublished); Bodwin, G.T., Brodsky, S.J., and Lepage, G.P., Phys. Rev. Lett. **47**, 1799 (1981); Bodwin, G.T., Phys. Rev. D31, 2616 (1985); W-Y. P. Hwang and L.S. Kisslinger, Phys. Rev. D38, 788 (1988); W-Y. P. Hwang, talk at the 1988 LAMPF Workshop on "Nuclear and Particle Physics on the Light Cone" (World Scientific, Singapore, 1989); on QCD corrections to Drell-Yan lepton-pair production in hadron-hadron collisions.

Halzen, F., and Scott, W., Phys. Rev. D18. 3378 (1978); Altarelli, G., et al., Nucl. Phys. **B246**, 12 (1984); on QCD effects in the production of W and Z bosons with large transverse momentum.

UA1 collaboration: Arnison, G., et al., Phys. Lett. **139B**, 115 (1984); Bawa, A.C., and Stirling, W.J., Phys. Lett. **B203**, 172 (1988); on the data in the production of W and Z bosons with large transverse momentum.

Kwong, W., Quigg, C., and Rosner, J.L., Ann. Rev. Nucl. and Part. Sci. **37**, 325 (1987); on a summary of quarkonium decay rates. (1987a)

Lepage, P. and Mackenzie, P., Phys. Rev. Lett. **47**, 1244 (1981); on perturbative QCD corrections to quarkonium decay ratios.

Kwong, W., Mackenzie, P.B., Rosenfeld, R., and Rosner, J.L., University of Chicago preprint EFI 87-31 (1987); on a recent analysis of bottomonium decay-width ratios from CUSB, CLEO, and ARGUS. (1987b)

Gorishny, S.G., Kataev, A., and Larin, S.A., INR preprint, Moscow, 1988; for the calculation of $C_3 = 64.7$ in the modified minimum subtraction scheme. (See Eq. (12).)

Farhi, E. Phys. Rev. Lett. **39**, 1587 (1977); on the thrust variable in the production of jets in e^+e^- collisions.

Basham, C.L., Brown, L.S., Ellis, S.D., and Love, S.T., Phys. Rev. **D17**, 2298 (1978); on energy-energy correlations in the produiction of jets in e^+e^- collisions.

Csikor, F., et al., Phys. Rev. **D31**, 1025 (1985); on planar triple-energy correlations in the produiction of jets in e^+e^- collisions.

Stirling, W.J. and Whalley, M. R., Durham-RAL Database Publication RAL-87 -107 (1987); and Wu, S.L., in *"Proceedings of the 1987 International Symposium on Lepton and Photon Interactions at High Energies"* (Hamburg 1987); on a comprehensive review of the available data on jet productions in e^+e^- collisions.

Sirlin, A., Phys. Rev. **D22**, 971 (1980); **D29**, 89 (1984). Extensive references to other papers are given in *"Radiative Corrections in $SU(2)_L \times U(1)$"*, eds. Lynn, B.W. and Wheater, J.F. (World Scientific, Singapore. 1984); on the renormalization prescription for $sin^2\theta_W$ in the GSW electroweak theory.

Prescott, C.Y., et al., Phys. Lett. **84B**, 524 (1979); on the SLAC polarized-electron experiment to measure the parity-violating asymmetry.

Bouchiat, M.A., and Pottier, L., *Science* **234**, 1203 (1986); and Piketty, C.A., *Weak and Electromagnetic Interactions in Nuclei,* ed. H. V. Klapdor (Springer-Verlag. Berlin, 1986), p. 603; for the reviews on the experiments measuring atomic parity violation.

Lynn, B.W. and Stuart, R.G., Nucl. Phys. **B253**, 216 (1985); and *Physics at LEP*, eds. Ellis, J. and Peccei, R., CERN 86-02, Vol. I; on electroweak corrections to the forward-backward asymmetry in $e^+e^- \to \ell\bar{\ell}$.

Amaldi, U., Böhm, A., Durkin, L.S., Langacker, P., Mann, A.K., Marciano, W.J., Sirlin, A., and Williams, H.H., Phys. Rev. **D36**, 1385 (1987); on analyses of the data related to the Glashow-Salam-Weinberg Electroweak Theory. Very similar conclusions are reached in an analysis by Costa, G., Ellis, J., Fogli,

G.L., Nanopoulos, D.V., and Zwirner, F., CERN preprint CERN-TH, 4675/87 (1987).

UA1: Arnison, G., et al., Phys. Lett. **166B**, 484 (1986); UA2: Ansari, R., et al., Phys. Lett. **B186**, 440 (1987); on the W and Z boson masses shown in Table 3.

Albert, D., Marciano, W.J., Wyler, D., and Parsa, Z., Nucl. Phys. **B166**, 460 (1980); Consoli, M., Lo Presti, S., and Maiani, L. Nucl. Phys. **B223**, 474 (1983);Güsken, S., Kühn, J.H., and Zerwas, P.M., Phys. Lett. **155B**, 185 (1985); and Jegerlehner, F., Z. Phys. **C32**, 425 (1986); on QCD corrections to $Z \to \psi\bar{\psi}$ decays.

Georgi, H., and Glashow, S. L., Phys. Rev. Lett. **33**, 438 (1974); on the simplest $SU(5)$ grand unification scheme.

Exercises: *Chapter 14*

14.1. Assume that the electron-proton deep inelastic scattering (DIS) cross section is an incoherent sum of elementary electron-parton scattering cross sections,

$$\sigma(e + P \to e' + X) = \int_0^1 dx\, f_{i/P}(x)\, \sigma(e + i \to e' + i'),$$

where $f_{i/P}(x)$ is the probability of finding a parton of flavor i and longitudinal momentum fraction x inside the proton. Show that the ep DIS cross section is proportional to a well-known structure function called $F_2(x, Q^2)$:

$$F_2(x, Q^2) = \sum_i e_i^2 \{q_i(x) + \bar{q}_i(x)\},$$

where e_i is the electric charge of a quark of flavor i.

14.2. Draw the Feynman diagrams which describe $e^+ e^- \to q\bar{q}g$. Use Feynman rules for QED (in Ch. 9) and for QCD (in Ch. 12) to obtain the T-matrix elements. As another major step, try to prove Eq. (14) quoted in the text.

14.3. As a simple exercise, show that entries in Table 1 follow from what we have obtained in §.13.3, in the absence of radiative corrections. Try to draw those Feynman diagrams which allow you to understand the need to introduce the factors in ρ, κ, and λ.

14.4. Prove Eq. (25) by starting from the appropriate T-matrix elements.

14.5. Prove Eq. (27), with Eq. (28), by starting from the appropriate T-matrix elements at the quark parton level. Consult Problem 13.3. as a reference.

14.6. Prove Eqs. (38) by starting from the appropriate T-matrix elements. Consult the Appendix in Chapter 13 on the Feynman rules for the GSW electroweak theory.

SUBJECT INDEX

NAME INDEX